Lecture Notes in Bioinformatics   10562

Subseries of Lecture Notes in Computer Science

More information about this series at http://www.springer.com/series/5381

Joao Meidanis · Luay Nakhleh (Eds.)

Comparative Genomics

15th International Workshop, RECOMB CG 2017
Barcelona, Spain, October 4–6, 2017
Proceedings

 Springer

Editors
Joao Meidanis
University of Campinas
Campinas, São Paulo
Brazil

Luay Nakhleh
Rice University
Houston, TX
USA

ISSN 0302-9743
Lecture Notes in Bioinformatics
ISBN 978-3-319-67978-5
DOI 10.1007/978-3-319-67979-2

ISSN 1611-3349 (electronic)

ISBN 978-3-319-67979-2 (eBook)

Library of Congress Control Number: 2017953424

LNCS Sublibrary: SL8 – Bioinformatics

Printed on acid-free paper

This Springer imprint is published by Springer Nature
The registered company is Springer International Publishing AG
The registered company address is: Gewerbestrasse 11, 6330 Cham, Switzerland

Preface

This volume contains the papers presented at RECOMBCG-17: the 15th RECOMB Comparative Genomics Satellite Workshop, held during October 4–6, 2017, in Barcelona, Spain.

There were 32 submissions. Each submission was reviewed by at least three reviewers, and some received up to five reviews. The reviews were conducted by members of the Program Committee as well as by additional reviewers who were sought based on their expertise for specific papers. The committee decided to accept 16 papers. The program also included four invited talks and a poster session.

We thank the members of the Program Committee as well as the additional reviewers for their diligent work reviewing the manuscripts and conducting thorough discussions on the manuscripts and their reviews that informed the decision process. We also thank the members of the Steering Committee, Jens Lagergren, Aoife McLysaght, and David Sankoff, for their guidance. Special thanks go to the local organizers, Laura Prat Busquets, Tomàs Marquès-Bonet, and Roderic Guigó Serra. Last but not least, we would like to thank the four keynote speakers who agreed to speak at the conference despite their very busy schedules: Iria Fernández Silva, Toni Gabaldón, Iñaki Ruiz-Trillo, and Wes Warren.

The workshop would not have been possible without the generous contribution of our sponsors, whose support is greatly appreciated:

- FEDER - Fondo Europeo de Desarollo Regional
- Goberno de España - Ministerio de Economía, Industria y Competitividad
- Instituto de Salud Carlos III
- PRB2 - Plataforma de Recursos Biomoleculares y Bioinformáticos
- ELIXIR Spain
- INB - Instituto Nacional de Bioinformática
- Illumina
- The Company of Biologists

We used the EasyChair system for submissions, reviews, and proceedings formatting.

August 2017

Joao Meidanis
Luay Nakhleh

Organization

Program Committee

Max Alekseyev	George Washington University, USA
Lars Arvestad	Stockholm University, Sweden
Anne Bergeron	Université du Quebec a Montreal, Canada
Marília Braga	Bielefeld University, Germany
Cedric Chauve	Simon Fraser University, Canada
Leonid Chindelevitch	Simon Fraser University, Canada
Miklos Csuros	Université de Montreal, Canada
Ingo Ebersberger	Goethe University, Germany
Nadia El-Mabrouk	University of Montreal, Canada
Guillaume Fertin	University of Nantes, France
Katharina Jahn	ETH Zurich, Switzerland
Asif Javed	GIS A-STAR, Singapore
Kevin Liu	Michigan State University, USA
Ketil Malde	Institute of Marine Research
Joao Meidanis	University of Campinas/Scylla Bioinformatics, Brazil
István Miklós	Rnyi Institute, Budapest, Hungary
Siavash Mirarab	University of Texas at Austin, USA
Luay Nakhleh	Rice University, USA
Aida Ouangraoua	Université de Sherbrooke, Canada
Marie-France Sagot	Inria, France
Michael Sammeth	Centre d'Anàlisí Genòmica (CNAG)
Jens Stoye	Bielefeld University, Germany
Krister Swenson	CNRS, Université de Montpellier, France
Eric Tannier	Inria, France
Glenn Tesler	University of California, San Diego, USA
Tamir Tuller	Tel Aviv University, Israel
Lusheng Wang	City University of Hong Kong, SAR China
Tandy Warnow	The University of Illinois at Urbana-Champaign, USA
Louxin Zhang	National University of Singapore, Singapore
Jie Zheng	Nanyang Technological University, Singapore

Additional Reviewers

Aganezov, Sergey	Durand, Dannie	Sinaimeri, Blerina
Alexeev, Nikita	Fransiskus Xavier, Ivan	Viduani Martinez,
Avdeyev, Pavel	Labarre, Anthony	Fábio Henrique
Bansal, Mukul S.	Moret, Bernard	Willing, Eyla
Bulteau, Laurent	Peterlongo, Pierre	Wittler, Roland
Coelho, Vitor	Sankoff, David	Zhou, Xinrui
Doerr, Daniel	Schulz, Tizian	

Contents

The Similarity Distribution of Paralogous Gene Pairs Created
by Recurrent Alternation of Polyploidization and Fractionation 1
   Yue Zhang and David Sankoff

A Tractable Variant of the Single Cut or Join Distance
with Duplicated Genes 14
   Pedro Feijão, Aniket Mane, and Cedric Chauve

Generation and Comparative Genomics of Synthetic Dengue Viruses 31
   Eli Goz, Yael Tsalenchuck, Rony Oren Benaroya, Shimshi Atar,
   Tahel Altman, Justin Julander, and Tamir Tuller

ASTRAL-III: Increased Scalability and Impacts of Contracting
Low Support Branches 53
   Chao Zhang, Erfan Sayyari, and Siavash Mirarab

Algorithms for Computing the Family-Free Genomic Similarity
Under DCJ 76
   Diego P. Rubert, Gabriel L. Medeiros, Edna A. Hoshino,
   Marília D.V. Braga, Jens Stoye, and Fábio V. Martinez

New Algorithms for the Genomic Duplication Problem 101
   Jarosław Paszek and Paweł Górecki

TreeShrink: Efficient Detection of Outlier Tree Leaves 116
   Uyen Mai and Siavash Mirarab

Rearrangement Scenarios Guided by Chromatin Structure 141
   Sylvain Pulicani, Pijus Simonaitis, Eric Rivals, and Krister M. Swenson

A Unified ILP Framework for Genome Median, Halving, and Aliquoting
Problems Under DCJ 156
   Pavel Avdeyev, Nikita Alexeev, Yongwu Rong, and Max A. Alekseyev

Orientation of Ordered Scaffolds 179
   Sergey Aganezov and Max A. Alekseyev

Finding Teams in Graphs and Its Application to Spatial Gene
Cluster Discovery 197
   Tizian Schulz, Jens Stoye, and Daniel Doerr

Inferring Local Genealogies on Closely Related Genomes 213
   Ryan A. Leo Elworth and Luay Nakhleh

Enhancing Searches for Optimal Trees Using SIESTA.   232
  Pranjal Vachaspati and Tandy Warnow

On the Rank-Distance Median of 3 Permutations   256
  Leonid Chindelevitch and João Meidanis

Statistical Consistency of Coalescent-Based Species Tree Methods
Under Models of Missing Data.   277
  Michael Nute and Jed Chou

Fast Heuristics for Resolving Weakly Supported Branches Using
Duplication, Transfers, and Losses   298
  Han Lai, Maureen Stolzer, and Dannie Durand

Author Index   321

The Similarity Distribution of Paralogous Gene Pairs Created by Recurrent Alternation of Polyploidization and Fractionation

Yue Zhang and David Sankoff[⊠]

University of Ottawa, Ottawa, Canada
{yzhan481,sankoff}@uottawa.ca

Abstract. We study modeling and inference problems around the process of fractionation, or the genome-wide process of losing one gene per duplicate pair following whole genome doubling (WGD), motivated by the evolution of plants over many tens of millions of years, with their repeated cycles of genome doubling and fractionation. We focus on the frequency distribution of similarities between the two genes, over all the duplicate pairs in the genome. Our model is fully general, accounting for repeated duplication, triplication or other k-tupling events (all subsumed under the term WGD), as well as a general fractionation rate in any time period among multiple progeny of a single gene. It also has a biologically and combinatorially well-motivated way of handling the tendency for at least one sibling to survive fractionation. We show how the method reduces to previously proposed models for special cases, and settles unresolved questions about the expected number of gene pairs tracing their ancestry back to each WGD event.

Keywords: Whole genome duplication · Gene loss · Birth and death process · Multinomial model · Paralog gene tree · Sequence divergence

1 Introduction

The basic data of comparative genomics and phylogenetics is the degree of identity between homologous genes or proteins, as assessed by any one of a number of quantitative measures. For example, identity among orthologous genes, singly or in combination with other genes, feeds directly into distance-based methods for phylogenetic inference.

An important way of visualizing and analyzing the relationship among and within genomes is based on the distribution of similarities (percent identity, K_S, 4dTv, etc.) across all pairs of detectible homologs. Similarities between homologs decay over time in a roughly synchronous manner due to sequence-level mutations - largely base replacement, insertion and deletions - so that similarities between *orthologs* in two genomes tend to be relatively highly concentrated around a value indicative of the time of speciation of the two organisms, while a concentration of similarities between *paralogs* in a single genome points to the

© Springer International Publishing AG 2017
J. Meidanis and L. Nakhleh (Eds.): RECOMB CG 2017, LNBI 10562, pp. 1–13, 2017.
DOI: 10.1007/978-3-319-67979-2_1

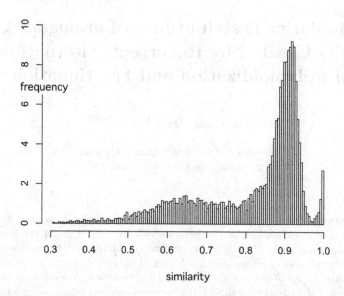

Fig. 1. Distribution of similarities in poplar, showing two peaks, one reflecting an early core eudicot triplication, the other a more recent duplication in the family Salicaceae. Additional peak near 100% similarity captures heterozygous maternal-paternal pairs, not duplicate genes.

time of a whole genome duplication (WGD) event. The concentration of these homolog similarities are reflected in local peaks, or local modes, in the frequency distributions of all similarities. There may be several peaks if there has been a history of recurrent whole genome doubling, tripling, quadrupling, etc. events (collectively referred to as WGD), as illustrated in Fig. 1. Except for very recent events, however, the evolutionary signal in the distribution may be largely or completely obscured by:

- the progressive flattening of the peaks as the genome-wide synchrony of duplicate gene similarity decays; that is, the variance of the duplicate similarities increases as a function of time, and the tails of neighbouring peaks increasingly overlap,
- the accumulation of "tandem" duplicate pairs that arise over time, not linked to a genome-wide event like speciation or polyploidization,
- the propagation of transposable elements,
- bioinformatic limitations in gene identification and gene comparison and, most important,
- the process of fractionation whereby either member of a duplicate pair disappears due to DNA excision [1,2] or pseudogenization.

All of these processes increase the level of background noise more severely for early events than for more recent ones.

The most directly applicable statistical method in wide use treats the distribution of similarities of duplicate pairs as a mixture of distributions, such as that implemented in the EMMIX software [3]. While this may produce optimum results for the mixture problem, it cannot take into account biological constraints and will often produce biologically uninterpretable results, such as lower variance components of the mixture for early events or inconsistently smaller numbers of pairs for recent events.

In this paper, we propose a general model for the repeated cycle of WGD events followed by fractionation. This allows an arbitrary number of events and arbitrary rates of fractionation of the progeny of any gene holding across the entire genome after each event. We calculate expected numbers of duplicate gene pairs at the time of observation originating at each of the WGD events, and from this predict the entire distribution of similarities, with the help of standard models of mutational processes. This distribution becomes the basis for principled inference of the parameters of the WGD and fractionation model, although the details of the statistical procedures are only touched upon here. The main mathematical advance here is the calculation of the expected number of duplicated pairs more efficiently, without explicitly counting pairs in each evolutionary history.

The present paper studies only WGD events in full detail rather than the extension to speciation events, although these can be handled by similar models and inference procedures [4,5].

2 The Model

We propose a model for the generation, maintenance and loss of duplicate genes through polyploidization (leading to WGD) and fractionation that is basically a continuous-time birth-and-death process with the entire population synchronized with respect to birth times and progeny number, but with deaths of individuals coordinated among siblings. Thus, the process starts with $m_1 \geq 1$ genes at time t_1; at times $t_1 < \cdots < t_{n-1}$ for some $n \geq 1$, each existing gene is replaced by $r_1, \ldots, r_{n-1}$ progeny, respectively, where each $r_i \geq 2$. For each gene's progeny, at least one and at most r_i genes survives until time t_{i+1}, as governed by a probability distribution $u^{(i)}(1), \ldots, u^{(i)}(r_i)$.

Although the data available for inference consists of similarity measures on *pairs* of paralogous genes, we must first develop an account of the fate of *individual* genes and their descendants in the environment of WGD and fractionation.

2.1 The Gene Tree of WGD Paralogs

Let $\mathbf{r} = (r_1, r_2, \ldots, r_{n-1})$ where each $r_i \geq 2$. Starting with a single diploid genome, where each of the m_1 genes is single copy, suppose we have a series of polyploidizations with ploidies $2r_1, 2r_2, \ldots, 2r_{n-1}$ at times $t_1 < t_2 < \cdots < t_{n-1}$. The results are observed at time t_n, namely a measure of residual identity between all pairs of descendants of each gene. Although there may be as many as $m_1 \prod_{i=1}^{n-1} r_i$ such descendants in the haploid genome observed at time t_n, there are likely far less, because of the process of fractionation, which potentially deletes single genes and all their descendants from the gene tree.

In the model, illustrated in Fig. 2, at each polyploidization time t_i, each existing gene, say gene g, that has survived fractionation up to t_i, is replaced by r_i copies, at least one of which survives until time t_{i+1}, where r_i is called the *ploidy*.

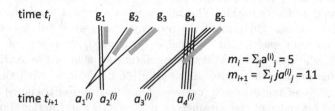

Fig. 2. Event with ploidy $r_i = 4$, showing population of $m_i = 5$ genes at time t_i, each giving rise to 4 progeny, of which $1 \leq j \leq 4$ survive until time t_{i+1}. $a_j^{(i)}$ is the number of times j progeny survive. In the diagram, thin solid lines represent individual progeny that survive, and thick grey lines represent the total progeny of a gene that do not survive. In this example the only gene, all of whose progeny survive, is g_4. Here $a_1^{(i)} = 2, a_2^{(i)} = a_3^{(i)} = a_4^{(i)} = 1, m_{i+1} = 2 \times 1 + 1 \times 2 + 1 \times 3 + 1 \times 4 = 11$.

If there are m_i genes at time t_i, let $a_1^{(i)}, \ldots, a_{r_i}^{(i)}$ be the number of cases where $1, \ldots, r_i$ copies survive fractionation until time t_{i+1}, so that $\sum_{j=1}^{r_i} a_j^{(i)} = m_i$. Note that there is no provision for g to have zero surviving descendants (i.e., $a_0^{(i)} \equiv 0$); these genes would be considered as leaving no evidence for their existence and are not counted in m_i.

It follows that $m_{i+1} = \sum_{j=1}^{r_i} j a_j^{(i)}$. The initial value is $m_1 \geq 1$.

We have

$$N(\mathbf{r}) = \prod_{i=1}^{n-1} \binom{m_i}{a_1^{(i)}, \ldots, a_{r_i}^{(i)}}, \tag{1}$$

where $0 \leq a_j^{(i)} \leq m_i$, $j = 1, \ldots, r_i$, and, for $i \geq 1$, $m_i = \sum_{j=1}^{r_i} a_j^{(i)} = \sum_{j=1}^{r_{i-1}} j a_j^{(i-1)}$, and $N(\mathbf{r})$ represents the number of different gene copy-and-fractionation histories.

We use $u_j^{(i)}$ to represent the probability that j of the r_i potential copies survive to time t_{i+1}, for $j = 1, \ldots, r_i$. To avoid excessive parametrization, we may set

$$u_j^{(i)} = \frac{\binom{r_i}{j} v^{(i)j}(1 - v^{(i)})^{r_i-j}}{1 - (1 - v^{(i)})^{r_i}}. \tag{2}$$

The parameter $v^{(i)}$ is the survival probability for a gene copy created at time t_i until t_{i+1}. It should be monotone decreasing in $t_{i+1} - t_i$ and, for empirical purposes, we will assume

$$v^{(i)} = \alpha + (1 - \alpha) \exp[-\beta(t_{i+1} - t_i)] \tag{3}$$

for some rate constant $\beta > 0$ and "background" level $0 < \alpha < 1$.

Thus the probability distribution of the evolutionary histories represented by $\mathbf{r}$ and the variable $\mathbf{a} = \{a_j^{(i)}\}_{j=1\ldots r_i}^{i=1\ldots n-1}$ is

$$P(\mathbf{r};\mathbf{a}) = \prod_{i=1}^{n-1} \binom{m_i}{a_1^{(i)},\ldots,a_{r_i}^{(i)}} \prod_{j=1}^{r_i} u_j^{(i)a_j^{(i)}}. \tag{4}$$

The expected number of genes at time t_n is then

$$\mathbf{E}(m_n) = \sum_{\mathbf{a}} P(\mathbf{r};\mathbf{a})\, m_n \tag{5}$$

Based on Eq. (1), for $1 \leq j \leq k$ we write

$$N^{(j,k)}(\mathbf{r},\mathbf{a}) = \prod_{i=j}^{k-1} \binom{m_i}{a_1^{(i)},\ldots,a_{r_i}^{(i)}} \tag{6}$$

to represent the number of different partial gene copy-and-fractionation histories starting with m_j genes and undergoing polyploidization with ploidies $r_j,\ldots,r_{k-1}$ at times $t_j,\ldots,t_{k-1}$. Similarly, based on Eq. (4), we write

$$P^{(j,k)}(\mathbf{r};\mathbf{a}) = \prod_{i=j}^{k-1} \binom{m_i}{a_1^{(i)},\ldots,a_{r_i}^{(i)}} \prod_{h=1}^{r_i} u_h^{(i)a_h^{(i)}} \tag{7}$$

for the probability measure over all events starting at t_j with m_j genes, and preceding t_k. In this case the expected number of genes at time t_k is

$$\mathbf{E}^{(j,k)}(m_k) = \sum_{\mathbf{a}} P^{(j,k)}(\mathbf{r};\mathbf{a})\, m_k. \tag{8}$$

Having characterized the origin and survival of individual genes and their descendants in the environment of WGD and fractionation, we may now focus on the main objects of study, the pairs of genes observed at time t_n. In previous work [4,5] we simply generated all evolutionary histories, and counted the pairs of various types in each history. This approach was limited computationally as the number of histories quickly becomes prohibitively large for even moderate size r_i and n. In the next section, we develop a formula for the number of pairs, which will allow for the automated counting of pairs in larger examples.

2.2 The Synchronous Gene Pairs

For each of the $a_j^{(i)}$ genes with j surviving copies, $j \geq 2$, there are $\binom{j}{2}$ surviving pairs of genes. If $j = 1$ there are no pairs. The total number of pairs created at t_i and surviving to t_{i+1} is thus

$$d^{(i,i+1)} = \sum_{j=2}^{r_i} \binom{j}{2} a_j^{(i)}. \tag{9}$$

These are called the t_i-pairs at time t_{i+1}. The expected number of such pairs is

$$\mathbf{E}(d^{(i,i+1)}) = \sum_{\mathbf{a}} P^{(1,i+1)}(\mathbf{r};\mathbf{a}) \sum_{j=2}^{r_i} \binom{j}{2} a_j^{(i)}. \tag{10}$$

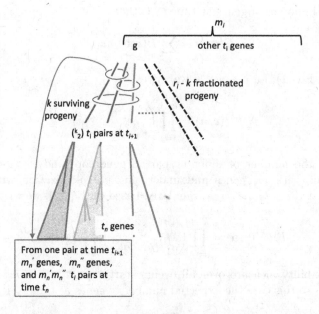

Fig. 3. Counting t_i-pairs. The three unfractionated progeny of gene g define three t_i-pairs, as indicated by three ovals. We follow the pair contained in the uppermost oval, as the two members at time t_{i+1} independently (shaded triangles) evolve into m_n' and m_n'' genes, respectively, at time t_n, defining $m_n' m_n''$ t_i-pairs at time t_n.

As illustrated in Fig. 3, at time $t_h, i+1 \le h \le n$, any two descendants of the two genes making up a t_i-pair *with no more recent common ancestor* is also called a t_i-pair (at time t_h). In other words, for any two genes at t_h, they form a t_i-pair if their most recent common ancestor underwent WGD at t_i.

For a given t_i-pair g' and g'' at time t_{i+1}, where $i < n-1$ the expected number of pairs of descendants $d^{(i,n)}$ having no more recent common ancestor than g' and g'', will be

$$\mathbf{E}(d^{(i,n)}) = \mathbf{E}(d^{(i,i+1)}) \mathbf{E}^{(i+1,n)}(m_n') \mathbf{E}^{(i+1,n)}(m_n'') \tag{11}$$

where $m_{i+1}' = m_{i+1}'' = 1$. This follows from the independence of the fractionation process between t_i and t_{i+1} and both parts of the process starting with g' and g''. Note that the two terms containing m_n' and m_n'' are identical. They could be collapsed into one quadratic term, but we write them separately just to emphasize the independence of the processes starting from g' and g''.

The terms $\mathbf{E}(d^{(i,i+1)})$ and $\mathbf{E}^{(i+1,n)}(m_n)$ in Eq. (11) both involve calculating probabilites with Eq. (7). As n and the r_i increase, so do the m_i, and this becomes computationally very expensive, due to the product of multinomial coefficients in Eq. (6) and the sum of many probabilities with the coefficient in Eqs. (4) and (7). Nevertheless, making use of the recursive nature of these calculations allows for more efficiency than the explicit generation of evolutionary histories and the counting of pairs within each one.

2.3 The Constraints Against Zero Progeny and on Fractionation Among Siblings only

There is strong biological motivation for retaining at least one offspring of most genes after WGD, as in our model. Our mathematical solution for this goes back to early (pre-genomic) work comparing human and mouse gene families [6]. The loss of all progeny of a gene with important functions would generally be lethal (although the cell sometimes has ways of buffering the loss of genes with essential functions). In our model, if zero progeny of a gene g were to survive, no trace of that g would remain in the "population" of genes, which in turn is equivalent to reducing the number of progeny of the parent of g by one. In the probabilities, this phenomenon is handled by truncating the binomial, as in Eq. (2).

The idea that a duplication event ends the fractionation among existing genes and replaces it with fractionation only among newly generated "siblings" is somewhat arbitrary, biologically speaking, but is not completely unrealistic. Certainly fractionation will be more intense among siblings than among "cousins", and though our model exaggerates this to some extent, inference of total fractionation rates should not be substantially affected. Modeling fractionation among successive generations of cousins would increase the complexity of the model and the number of parameters beyond any hope of inferential meaningfulness.

These two constraints on our model are the key to its analytical tractability.

2.4 The Distribution

Knowing the expected number of pairs of genes originating at each WGD in the past is the first step in predicting the full distribution F of similarities. The second step is to derive the actual distribution of gene pair similarities, or an appropriate approximation to it, for t_i-pairs.

Gene pair divergence may be measured in terms of a probability p reflecting *similarity* – the proportion of nucleotide positions that are occupied by the same base in the two orthologs (or paralogs). Alternative measures are *synonymous distance K_s* – the proportion of synonymous changes (not affecting translation to an amino acid) over all eligible positions, and *fourfold degenerate synonymous distance* 4dTv – the transversion rate at fourfold degenerate third codon positions [7], which require somewhat different analyses.

Besides p, the other important parameter is G, gene length in terms of the number of nucleotides in the genes' coding region, setting aside for the moment that this varies greatly from gene to gene. In the simplest case, p is the sum of

G binomial distributions, divided by G, and is related to the time $t \in [0, \infty)$ elapsed since the event that gave rise to the pair:

$$\text{mean} : \mathrm{E}[p] = \frac{1}{4} + \frac{3}{4}e^{-\lambda t} \in [0, 1] \tag{12}$$

$$\text{variance} : \mathrm{E}(p - \mathrm{E}[p])^2 = \frac{3}{16}\frac{(1 + 3e^{-\lambda t})(1 - e^{-\lambda t})}{G},$$

where $\lambda > 0$ is a divergence rate parameter. The binomial distribution may be approximated by a normal distribution or, if p is close to 1.0, by a Poisson.

Despite the similarity in form with Eq. (3), the processes of fractionation and divergence are orthogonal, neither influencing the other in our model. Formula (12) holds no matter how many duplications have occurred, and the rates u and v, which are proxies for time, are defined without reference to time, although they have a natural conversion to time rates, if desired, once t and λ are determined.

The densities of similarities of t_i-pairs can be approximated by a normal distribution $\mathbf{N}(\mathrm{E}[p], \mathrm{E}(p - \mathrm{E}[p])^2)$, and the expected frequency by

$$F_i = \mathrm{E}(d^{(i,n)})\mathbf{N}(\mathrm{E}[p], \mathrm{E}(p - \mathrm{E}[p])^2). \tag{13}$$

Finally, we can predict the entire frequency distribution over all t_i as:

$$F(s) = \sum_{i=1}^{n-1} \mathbf{E}(d^{(i,n)})F_i(s). \tag{14}$$

Although in its simplest interpretation G is gene length in terms of the number of nucleotides in the genes' coding region, in practice it is a parameter to be estimated.

3 Inference

Duplicate gene frequency distributions may be produced using the genome data base and analysis tools on the CoGe database [8,9]. Genomes for which coding sequences are known may be compared, giving complete lists of duplicate pairs and their similarities. K_s scores are also available. Self-comparison gives the type of data pertinent to the present paper.

In practice, p for duplicate gene pairs is generally much greater than 0.25, so we base our analysis on those pairs with similarity greater than, say, 0.5 or 0.6. We also endeavour to remove all tandem gene pairs. In addition, distribution of similarities within heterozygous genomes generally have a peak very close to 100%, as in Fig. 1, but these pairs consist mostly of maternal-paternal variants of a single heterozygous gene, and not pairs created by WGD.

The parameters in the model may be estimated using maximum likelihood or other methods. This includes

- $n - 1$: number of events affecting the genome,
- the r_i: ploidy of each event,
- the t_i: times of each event,

- the $u^{(i)}$ or $v^{(i)}$: fractionation rates,
- λ: the divergence rate, and
- G: the gene length parameter.

Of course, estimation may be computationally difficult if too many parameters are unknown. Moreover the form of Eqs. (3) and (12) suggest that only $U = ut$ and $\Lambda = \lambda t$ and not all of u, λ and t can be estimated.

4   Examples

4.1   Two Successive WGD

The simplest example involves two whole genome doublings. Set $u = u_1^{(1)}$ and $v = u_1^{(2)}$. For t_1-pairs

$$d^{(1,2)} = 0 \times a_1^{(1)} + 1 \times a_2^{(1)}$$
$$\mathrm{E}(d^{(1,2)}) = 0 \times (1-u) + 1 \times u$$
$$= u \tag{15}$$

If $m_2 = 1$, there are no t_1-pairs. Otherwise consider the single pair g', g''. Now, $m_3' = 1$ or $m_3' = 2$ with probabilities $1 - v$ and v, respectively, so that $\mathrm{E}(m_3') = 1 + v$. Similarly $\mathrm{E}(m_3'') = 1 + v$. So

$$\mathbf{E}(d^{(1,3)}) = \mathbf{E}(d^{(1,2)})\mathbf{E}^{(2,3)}(m_3')\mathbf{E}^{(2,3)}(m_3'')$$
$$= u(1+v)^2 \tag{16}$$

For t_2-pairs,

$$d^{(2,3)} = 0 \times a_1^{(2)} + 1 \times a_2^{(2)}$$
$$\mathrm{E}(d^{(2,3)}) = 0 \times (1-u)(1-v) + 2uv^2 + 2uv(1-v)$$
$$= v(1+u) \tag{17}$$

For this small example, a breakdown of the set of evolutionary histories, the t_1- and t_2-pairs and their probabilities is given in Fig. 4.

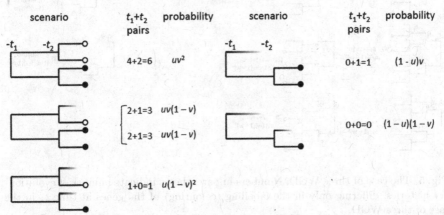

Fig. 4. The case of two WGD.

4.2   Three Successive WGD

A mathematically more interesting case adds a third doubling event to the first two. Set $u = u_1^{(1)}$ and $v = u_1^{(2)}$ as before, and $w = u_1^{(3)}$. Then the expected number of t_1-, t_2- and t_3-pairs works out to be

$$E(d^{(1,4)}) = 8uv^2w^3 + 6uv^2w^2 + 8uvw^3 + 4uv^2w + 2uv - 6uvw^2$$
$$+ 4uvw + 4uw^2 + 2uv^2$$
$$E(d^{(2,4)}) = v + 2uv^2 + 2vw + vw^2 - v^2 + 2v^2w - v^2w^2 - 4uv^2w^3$$
$$+ 2uv^2w^2 + 4uvw^3$$
$$E(d^{(3,4)}) = vw + uw^2 + w - 2uvw^2 - 3uv^2w^3 + uv^2w^2 + 3uvw^3$$
$$+ 2uv^2w \tag{18}$$

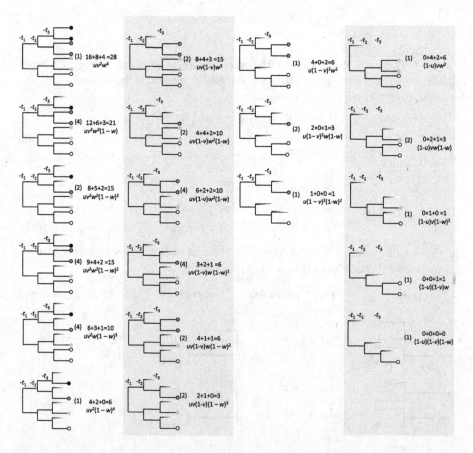

Fig. 5. The case of three WGD. Numbers in parentheses indicate number of evolutionary histories, differing only in the labelling (colouring) of the genes at time t_4 in the case of three WGD.

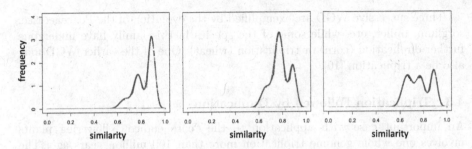

Fig. 6. Three models differing only in divergence rates. Left: $u = 0.1, v = 0.1, w = 0.2$. Middle: $u = 0.1, v = 0.3, w = 0.1$. Right: $u = 0.5, v = 0.1, w = 0.2$

Figure 5 displays a breakdown of the pairs in each evolutionary history. Beyond this case, $n = 4$, the number of evolutionary histories for successive WGD becomes too cumbersome to display.

The contrived examples in Fig. 6 demonstrate the importance of fractionation rates and divergence time t_i to the structure of the distribution of duplicate pair similarities with three successive WGD.

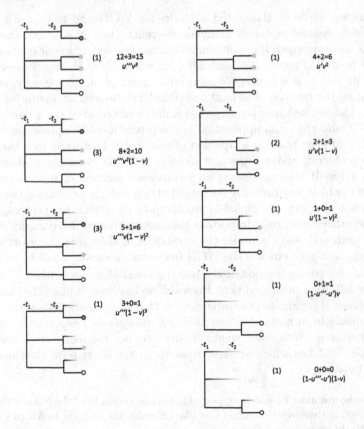

Fig. 7. A triplication followed by a duplication.

Three successive WGD are exemplified by the evolution of the Poaceae: rice, sorghum, millet, etc. while some of the species in this family have undergone further duplication (corn) or triplication (wheat). One of the earlier WGD may also be a triplication [10].

4.3 Triplication Followed by Duplication

An important case with application to the "core eudicot" flowering plants involves one whole genome triplication more than 100 million years ago. The descendants have been extraordinarily successful in diversifying and coloniz-ing in virtually* all terrestrial regions, and many of them have independently undergone a further duplication, like the Salicaceae lineage including the poplar genome depicted in Fig. 1, and other WGD events (Fig. 7).

This event, along with other hexaploidizations (e.g., tomato [11]), octo-ploidizations and decaploidizations (e.g., strawberry [12]), motivate the general-ity of our model with respect to the r_i.

5 Conclusions

In this paper, we have elaborated a model for WGD and fractionation that is completely general as far as numbers of events, their ploidy, and fractiona-tion rates are concerned. Its main simplification, the restriction of fractionation pressures to sibling genes, not something for which there is empirical evidence in either direction, is a minor, common sense assumption, biologically, and the constraint on the survival of at least one sibling can be seen as a notational con-venience. The key result is the expected number of observed pairs dating from each WGD, and the main application is the prediction of frequency curves for the number of gene pairs as a function of similarity. This leads to a variety of inference problems, about times and ploidies of WGD events, and about frac-tionation rates. It thus represents an advance on mixture of models methods like EMMIX, which may optimize statistical fitting criteria in locating the t_i, but which do not offer any principled approach to the fractionation process, nor to any of the other biologically important parameters that we have modeled.

As a birth-and-death process, the constraint against zero survivors among the progeny of a gene ensures the WGD-fractionation model tends to be super-critical for non-trivial fractionation rates (i.e., excluding cases like $a_1^{(i)} \equiv m_i$), under the definition presented here. However, we have not studied the inevitable disappearance of t_i pairs as time progresses so that their similarities are no longer distinguishable from noise. And we have not incorporated constraints on total gene complement which must eventually operate, nor the many other processes of gene gain and loss which are less dramatic in the short term than doubling and fractionation.

Acknowledgements. Research supported in part by grants from the Natural Sciences and Engineering Research Council of Canada. DS holds the Canada Research Chair in Mathematical Genomics.

References

1. Eckardt, N.: A sense of self: the role of DNA sequence elimination in allopolyploidization. Plant Cell **13**, 1699–1704 (2001)
2. Freeling, M., Woodhouse, M.R., Subramaniam, S., Turco, G., Lisch, D., Schnable, J.C.: Fractionation mutagenesis and similar consequences of mechanisms removing dispensable or less-expressed DNA in plants. Curr. Opin. Plant Biol. **15**(2), 131–139 (2012)
3. McLachlan, G.J., Peel, D., Basford, K.E., Adams, P.: The EMMIX software for the fitting of mixtures of normal and t-components. J. Stat. Softw. **4**(2), 1–14 (1999)
4. Zhang, Y., Zheng, C., Sankoff, D.: Evolutionary model for the statistical divergence of paralogous and orthologous gene pairs generated by whole genome duplication and speciation. IEEE/ACM Trans. Comput. Biol. Bioinform. (2017). doi:10.1109/TCBB.2017.2712695
5. Sankoff, D., Zheng, C., Zhang, Y., Meidanis, J., Lyons, E., Tang, H.: Models for similarity distributions of syntenic homologs and applications to phylogenomics. IEEE/ACM Trans. Comput. Biol. Bioinform. (2017, in press)
6. Nadeau, J.H., Sankoff, D.: Comparable rates of gene loss and functional divergence after genome duplications early in vertebrate evolution. Genetics **147**(3), 1259–1266 (1997)
7. Kumar, S., Subramanian, S.: Mutation rates in mammalian genomes. Proc. Nat. Acad. Sci. **99**(2), 803–808 (2002)
8. Lyons, E., Freeling, M.: How to usefully compare homologous plant genes and chromosomes as DNA sequences. Plant J. **53**(4), 661–673 (2008). doi:10.1111/j.1365-313X.2007.03326.x
9. Lyons, E., Pedersen, B., Kane, J., Alam, M., Ming, R., Tang, H., Wang, X., Bowers, J., Paterson, A., Lisch, D., Freeling, M.: Finding and comparing syntenic regions among Arabidopsis and the outgroups papaya, poplar and grape: CoGe with rosids. Plant Physiol. **148**, 1772–1781 (2008)
10. Murat, F., Armero, A., Pont, C., Klopp, C., Salse, J.: Reconstructing the genome of the most recent common ancestor of flowering plants. Nat. Genet. **49**(4), 490–496 (2017)
11. Tomato Genome Consortium: The tomato genome sequence provides insights into fleshy fruit evolution. Nature **485**, 635–641 (2012)
12. Hirakawa, H., Shirasawa, K., Kosugi, S., Tashiro, K., Nakayama, S., Yamada, M., Kohara, M., Watanabe, A., Kishida, Y., Fujishiro, T., et al.: Dissection of the octoploid strawberry genome by deep sequencing of the genomes of Fragaria species. DNA Res. **21**(2), 169–181 (2014)

A Tractable Variant of the Single Cut or Join Distance with Duplicated Genes

Pedro Feijão[1], Aniket Mane[2], and Cedric Chauve[2(✉)]

[1] School of Computing Science, Simon Fraser University,
8888 University Drive, Burnaby, BC V5A 1S6, Canada
pfeijao@sfu.ca
[2] Department of Mathematics, Simon Fraser University,
8888 University Drive, Burnaby, BC V5A 1S6, Canada
{amane,cedric.chauve}@sfu.ca

Abstract. In this work, we introduce a variant of the Single Cut or Join distance that accounts for duplicated genes, in the context of directed evolution from an ancestral genome to a descendant genome where orthology relations between ancestral genes and their descendant are known. Our model includes two duplication mechanisms: single-gene tandem duplication and creation of single-gene circular chromosomes. We prove that in this model, computing the distance and a parsimonious evolutionary scenario in terms of SCJ and single-gene duplication events can be done in linear time. Simulations show that the inferred number of cuts and joins scales linearly with the true number of such events even at high rates of genome rearrangements and segmental duplications. We also show that the median problem is tractable for this distance.

1 Introduction

The analysis of genome evolution by genome rearrangements has been subject to much research in the field of computational biology. Various rearrangement mechanisms explaining these variations have been proposed and studied, leading to a large corpus of algorithmic results [1]. The *pairwise genome rearrangement distance* problem aims at finding a most parsimonious or most likely sequence of genome rearrangements, within a given evolutionary model, that transforms one given genome into another given genome, thus giving a possible evolutionary scenario between the two given genomes. This problem has numerous applications toward unraveling important evolutionary mechanisms; recent examples include the nature of evolutionary breakpoints in bacteria [2] or the differentiated evolutionary mode of sex chromosomes and autosomes in *Anopheles* mosquitoes [3].

A number of rearrangement models have been studied from an algorithmic point of view, among them the reversal model was one of the first for which it was shown that the distance problem is tractable [4]; we refer to [1] for a review of the rich literature in this field up to 2009. For most evolutionary models that do not consider gene duplication, computing a rearrangement distance is tractable; additionally, some models can account for unequal gene content due to

© Springer International Publishing AG 2017
J. Meidanis and L. Nakhleh (Eds.): RECOMB CG 2017, LNBI 10562, pp. 14–30, 2017.
DOI: 10.1007/978-3-319-67979-2_2

gene gain or loss, for example [6,7]. However, when gene duplication is allowed as an evolutionary event, most rearrangement distance problems become NP-hard. For instance, whereas the distance between two genomes can be computed in linear time for genomes without duplicate genes under the Double-Cut and Join (DCJ) model, it becomes NP-hard to compute when duplicate genes are considered [9], even when the gene content in both genomes is the same [11]. So far limited tractability results for computing distances in the presence of duplicated genes exist only for simpler genome rearrangement models, such as the breakpoint (BP) distance and the Single Cut or Join (SCJ) distance [5]. Even in these simpler models, the general problem of computing a distance with duplicated genes is difficult [12–15], although exponential algorithms have been developed for some specific problems such as the Exemplar BP distance for example [16–19], as well as polynomial time algorithms for two extensions of the SCJ model that include large-scale duplications: the double distance [5], where duplicated genes occur through a whole genome duplication, and the SCJ and whole chromosome duplication problem [20].

In the present work, we consider a problem of rearrangement distance with duplicated genes that we prove to be tractable. Our evolutionary model is also an extension of the SCJ model, that includes single-gene duplications. In the problem we consider that one genome, say A, is duplication-free, while the other one, denoted by D, can contain duplicated genes. This setting is inspired by the recent development of algorithms that reconstruct ancestral gene orders along a given species phylogeny using reconciled gene trees that provide, for each gene family within the set of considered genomes, one-to-one or one-to-many orthology relations between each ancestral gene and its descendant gene(s), if any. This general framework was introduced by Sankoff and El-Mabrouk in [21] (see also [23]) and motivated the introduction of the EBD problem [22]; it was later implemented in the DeCo* family of algorithms [24] to reconstruct ancestral gene orders in a duplication-aware evolutionary model from data including extant gene orders and reconciled gene trees. In this context, the genome A represents an ancestral genome, reconstructed for example with DeCo*, the genome D represents a descendant of A, either extant or reconstructed too, and we are interested in computing a directed distance, from an ancestor to its descendant, where all members of a same gene family present in genome A are considered as distinguishable thanks to the information provided by the reconciled gene tree of this family. In the evolutionary model we consider, rearrangements are either Single Cuts or Single Joins, while duplications can only be single gene duplications, but of two different types, Tandem Duplications (TD) or Floating Duplications (FD) in which a new copy is introduced as a circular chromosome. We show that in this model the distance problem can be simply reduced to deciding, for each gene family with duplicates in D, the length of a tandem array of duplicates to introduce in A and we provide a polynomial time algorithm for this problem.

The remaining paper is organized as follows: in Sect. 2, we recall some basic definitions, describe our evolutionary model and the variant of the SCJ distance and its relation with existing problems. We also introduce the median problem for this distance. In Sect. 3 we present our theoretical results, a closed equation for the SCJ distance with duplications, a linear time algorithm to find an optimal scenario and an algorithm for the median problem. Finally, we provide preliminary experimental results in Sect. 4.

2   Preliminaries

2.1   Definitions and Problem Statements

Genes and Genomes. A genome consists of a set of chromosomes, each being a linear or circular ordered set of oriented genes[1]. In our examples, a circular chromosome is represented using round brackets (*e.g.* $(a, \bar{b}, c)$) while a linear chromosome is represented using square brackets (*e.g.* $[a, \bar{b}, c]$), where a gene b in reverse orientation is denoted by $\bar{b}$. Alternatively, a genome can be represented by a set of *gene extremity adjacencies.* In this representation, gene x is represented using a pair of gene extremities (x_t, x_h), x_t denotes the tail of the gene x and x_h denotes its head, and an *adjacency* is a pair of gene extremities that are adjacent in a genome. For example $(a, \bar{b}, c)$ is encoded by the set of adjacencies $\{a_h b_h, b_t c_t, c_h a_t\}$ and $[a, \bar{b}, c]$ by $\{a_h b_h, b_t c_t\}$.

We assume that a given gene a can have multiple copies in a genome, with its number of occurrences being called its *copy number.* A genome in which every gene has copy number 1 is a *trivial genome* [18]. In this context, a non-trivial genome sometimes cannot be represented unambiguously by a set of adjacencies unless we distinguish the copies of each gene, for example by denoting the copies of a gene a with copy number k by $a^1, \ldots, a^k$. For example, the genome $(a^1, \bar{b}, c^1, a^2), [\bar{a}^3, d, c^2]$, with two duplicated genes of respective copy numbers 3 and 2, is represented by $\{a_h^1 b_h, b_t c_t^1, c_h^1 a_t^2, a_h^2 a_t^1, a_t^3 d_t, d_h c_t^2\}$. We call a *gene family* the set of all copies of a gene that is present in both considered genomes. A gene family is trivial if it has exactly one copy in both genomes. From now, we identify a genome with its multi-set of adjacencies.

Let A and D be the respective adjacency sets for the genomes $[a, b, \bar{c}, d]$ and $[a, b, \bar{c}, \bar{d}][b, \bar{c}]$. Then the *multiset difference* between the two sets is denoted by $A - D$ (similarly $D - A$). Thus, if $A = \{a_h b_t, b_h c_h, c_t d_t\}$ and $D = \{a_h b_t, b_h c_h, c_t d_h, b_h c_h\}$ then $A - D = \{c_t d_t\}$ whereas $D - A = \{c_t d_h, b_h c_h\}$.

Evolutionary Model. We consider two types of evolutionary events: genome rearrangements and duplications. Genome rearrangements are modeled by *Single Cut or Join* (SCJ) operations, that either delete an adjacency from a genome (a cut) or join a pair of gene extremities that are not adjacent to any other gene extremity (a join), thus forming an adjacency. For duplication events, we consider two types of duplications, both creating an extra copy of a single gene: *Tandem Duplications* (TD) and *Floating Duplications* (FD). A tandem duplication of an existing gene g introduces an extra copy of g, say g' by adding an

---

[1] We use the generic term "gene" here to identify a genomic locus.

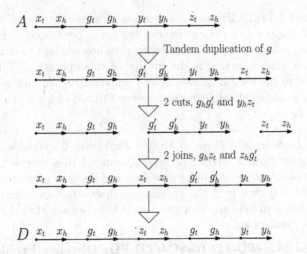

Fig. 1. In this example, a tandem duplicate of gene g is introduced. The adjacency $g_h y_t$ has been replaced by $g'_h y_t$ and an adjacency $g_h g'_t$ has been introduced. In this case the total number of operations to obtain D from A is 5. Note that the number of cut and join operations is dependent on the adjacencies of the gene g in A and D.

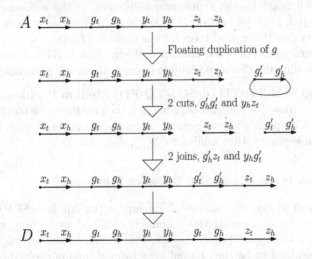

Fig. 2. In this example, a floating duplicate of gene g is introduced. An adjacency $g'_h g'_t$ has been added. In this case the total number of operations to obtain D from A is 5.

adjacency $g_h g'_t$, and, if there was an adjacency $g_h x$ by replacing it by the adjacency $g'_h x$, as shown in Fig. 1. A floating duplication introduces an extra copy g' of a gene g as a single-gene circular chromosome by adding the adjacency $g'_h g'_t$, as in Fig. 2. The motivation for this type of duplication is that gene insertions and gene deletions have been modeled with artificial circular chromosomes before, greatly simplifying how to deal with such type of operations. For instance, in the

Double-Cut and Join (DCJ) model, a deletion of a gene can be seen as a DCJ operation that applies two cuts to remove the given gene from a chromosome, followed by two joins to "repair" the broken chromosome and to circularize the deleted gene. A gene insertion is the inverse of this operation. This idea was effectively used in the DCJ indel model by Compeau [10]. We discuss the possibility of using a single-gene linear chromosome instead of a circular one at the end of Sect. 3.2.

The Pairwise Distance and Directed Median Problems. We consider the case of directed evolution from a trivial ancestral genome A to a descendant genome D. The evolutionary model excludes gene loss and de-novo gene creation, so we assume that every gene a in A has at least one descendant in D and conversely every gene D has a unique ancestor gene in A. If so, we say that A and D have the same *gene families set*.

The Directed SCJ-TD-FD (d-SCJ-TD-FD) Distance Problem. Let A be a trivial genome and D be a non-trivial genome, such that no gene family is absent from either A or D. Compute the minimum number of SCJ, TD and FD operations needed to transform A into D, denoted by $d_{\mathrm{DSCJ}}(A, D)$.

Note that if D is a trivial genome, the usual SCJ distance, denoted by $d_{\mathrm{SCJ}}(A, D)$ is defined by the symmetric differences of the adjacencies sets of A and D: $d_{\mathrm{SCJ}}(A, D) = |A - D| + |D - A|$ where the first term accounts for the number of cuts and the second term for the number of joins.

We now turn to the directed median problem, that is the natural extension of the pairwise directed distance problem towards the small parsimony problem.

The Directed SCJ-TD-FD (d-SCJ-TD-FD) Median Problem. Let $k \geq 2$ and $D_1, \ldots, D_k$ (possibly) non-trivial genomes, such that no gene family is absent from any D_i. Compute a trivial genome A on the same set of gene families as the non-trivial genomes, that minimizes $\sum_{i=1}^{k} d_{\mathrm{DSCJ}}(A, D_i)$.

2.2 Relation to the Exemplar Distance Framework

Sankoff [22] introduced the notion of Exemplar Breakpoint (EBP) distance, where an *exemplar* of a non-trivial genome is obtained by keeping exactly one gene copy from each gene family. In the directed evolution setting, an exemplar can be assumed to be the original gene from A having evolved into a gene now present in D, all other copies having been created by duplications. So the EBP distance problem aims to find an exemplar for each group of duplicates in D such that the trivial genome that results from deleting all non-exemplar copies minimizes the breakpoint distance to A. The notion of exemplar distance can naturally be used in conjunction with the SCJ distance instead of the BP distance, a problem we denote the ESCJ distance. The EBP distance problem has been shown to be NP-hard even in the directed evolution case where every duplicated gene has exactly two copies in D [12], and it is immediate to extend this hardness result to the directed ESCJ distance problem.

Intuitively, the directed ESCJ distance and the d-SCJ-FD-TD distance problems seem very similar. For example in the case of duplicated genes having exactly two copies in D, the later aims at deciding which copy in D is exemplar (*i.e.* evolved from the original copy in A) and then, for the second copy, if it originates from a TD or a FD, thus resulting in a matching between two genomes with two copies of each duplicate, opposed to the ESCJ setting where the matching is between genomes with one copy of each gene.

It is interesting to notice that both problems, although similar, have opposed properties in terms of tractability, and that the d-SCJ-TD-FD distance problem is tractable despite considering a larger solution space. Moreover, one can ask if there is a strong correlation between the distance obtained in both settings. It is not difficult to find examples that show that both distances can be quite different: the ESCJ distance between $A = [a, b, c, d]$ and $D = [a, c, b, d, c, a, d, b]$ is 0, whereas the SCJ-TD-FD distance between the same two genomes is 18 (4 duplications, 7 cuts, 7 joins). However, although the difference between both distances can be arbitrarily large, tight bounds can be derived.

Lemma 1. *Let A be a trivial genome, D be an arbitrary genome on the same set of gene families than A, $d_{DSCJ}(A, D)$ and $d_{ESCJ}(A, D)$ denote the d-SCJ-FD-TD and the ESCJ distances, respectively. Let k be the difference between the number of genes in D and the number of genes in A. The following bounds*

$$k \leq d_{DSCJ}(A, D) - d_{ESCJ}(A, D) \leq 5k$$

are tight.

Proof. First, we obtain the genome A' by applying $d_{ESCJ}(A, D)$ SCJ operations on A in a way that the duplicated genes in D are in the same order as the corresponding matched genes in A', as given by an optimal exemplar matching. In the SCJ-FD-TD model, we need to apply at least k duplications on A' to obtain D, so $d_{DSCJ}(A, D) \geq d_{ESCJ}(A, D) + k$ (otherwise $d_{ESCJ}(A, D)$ would not be optimal). To show that this bound is tight, we can see that the trivial case of no duplications holds. But, whenever A and D differ by only k tandem duplications, the bound is tight, since in this case $d_{ESCJ}(A, D) = 0$ and $d_{DSCJ}(A, D) = k$.

Now, from A', we can apply k free duplications, followed by k cuts on these duplications. Also, perform at most k cuts, between any two genes on A' if both have more than one copy on D. Since A' was ordered in relation to its corresponding copies on D, it is possible to join the "fragments" of A' that were created with the previous $2k$ cuts with $2k$ joins in a way to transform A' in D, and therefore we built a d-SCJ-FD-TD scenario from A to D with $d_{ESCJ}(A, D) + 5k$ operations. Any pair of circular genomes $A = (1, 2, \ldots, n)$ and $D = (1, n, 2, 1, \ldots, i, i - 1, \ldots, n, n - 1)$ satisfies the tight bound. □

3 Algorithmic Results

In this section, we show that, after a preprocessing step of removing obvious TD and FD in D, the d-SCJ-TD-FD distance can be calculated with the symmetric difference between the adjacency (multi)sets of the input genomes, with an extra factor to account for the gene duplications. We first focus on the preprocessing. Next we describe a linear time algorithm to compute a parsimonious scenario and a polynomial time algorithm for the directed median problem.

3.1   The Directed SCJ-TD-FD Distance

An *observed duplication* in D is defined as an adjacency of the form $g_h g_t$, that defines either a single-gene circular chromosome or a *tandem array* of two (or more) copies of a gene g that occur consecutively and with the same orientation. We denote by t the number of such adjacencies in D and by D' the genome obtained from D by removing first all genes but one from each tandem arrays, and then all single-circular chromosomes for genes from non-trivial families but one if all genes of the family are in such circular chromosomes. D can obviously be obtained from D' by t duplications and the following lemma is immediate:

Lemma 2. $d_{DSCJ}(A, D) = d_{DSCJ}(A, D') + t$.

As a consequence, we assume from now on that D has been preprocessed as described above and does not contain any tandem array or any extra copy of a non-trivial family that is in a single-gene circular chromosome. We say that D is *reduced*. Note that single-gene linear chromosomes are not impacted by this preprocessing as, in our setting, if the considered gene is from a non-trivial family, the linear chromosome it forms required at least a cut to be created.

Theorem 1. *Given a trivial genome A and a reduced non-trivial genome D such that no gene family is absent from either A or D and where D has n_d more genes than A, the d-SCJ-TD-FD distance between A and D is given by*

$$d_{DSCJ}(A, D) = |A - D| + |D - A| + 2n_d.$$

Proof. First, we show that $d_{DSCJ}(A, D) \geq |A - D| + |D - A| + 2n_d$. To obtain D from A, we need exactly n_d gene duplications. Each duplication of a gene g will create the adjacency $g_h g_t$, regardless of the type of the duplication or the timing of the duplication event. Therefore, n_d adjacencies of the type $g_h g_t$ will have to be cut, as D is reduced and has no adjacency of this type. In addition, any adjacency in $A - D$ and $D - A$ defines an unavoidable cut or join respectively. Therefore, we can not transform A into D with less than $|A - D| + |D - A| + 2n_d$ operations.

Now, we show that $d_{DSCJ}(A, D) \leq |A - D| + |D - A| + 2n_d$, by induction on n_d. For the base case $n_d = 0$, the result follows immediately as both genomes are trivial and $d_{DSCJ}(A, D) = d_{SCJ}(A, D)$.

We now assume that $n_d > 0$, and pick a gene g with one copy in A and more than one copy in D. Depending on how the adjacencies of g are conserved or not in D, we have a few different subcases to consider. However, in each subcase the general strategy remains the same, as follows. We build a genome A_2 from A by applying one duplication (FD or TD) and also relabeling the original copy

g as g', creating an adjacency $g_h g_t$ in the case of an FD or $g'_h g_t$ in the case of a TD. Then we build a genome D_2 from D by also relabeling one copy of g to g', thus creating a new trivial gene family and an instance of the d-SCJ-TD-FD problem with exactly $n_d - 1$ duplicated gene copies. We can apply the induction hypothesis, leading to the inequality

$$d_{\mathrm{DSCJ}}(A_2, D_2) \leq |A_2 - D_2| + |D_2 - A_2| + 2(n_d - 1).$$

Also, as D and D_2 are identical but for the relabeling of g, there is a scenario from A to D, going from A to A_2 and then to D, resulting in the upper bound

$$d_{\mathrm{DSCJ}}(A, D) \leq d_{\mathrm{DSCJ}}(A, A_2) + d_{\mathrm{DSCJ}}(A_2, D_2) = 1 + d_{\mathrm{DSCJ}}(A_2, D_2).$$

We will then show that we can build A_2 and D_2 in a way that they satisfy

$$|A - D| + |D - A| = |A_2 - D_2| + |D_2 - A_2| - 1,$$

where the -1 term is due to the extra $g_h g_t$ adjacency on A_2 created with the duplication. Together with the above inequalities this will lead to

$$d_{\mathrm{DSCJ}}(A, D) \leq 1 + d_{\mathrm{DSCJ}}(A_2, D_2) \leq |A - D| + |D - A| + 2n_d$$

and the result follows. To show that we can build A_2 and D_2 that satisfy the above conditions, we will consider three subcases.

Case (i): Assume that g is not a telomere (and so there are two adjacencies involving g in A, say xg_t and $g_h y$) and there is a copy of g in D whose extremities for also adjacencies xg_t and $g_h y$. We say that the context of g is *strongly conserved* between A and D. Note that x and y do not need to belong to trivial gene families and there might be several copies of x, y, g in D that conserve the context of g in A.

In this case, we build A_2 by applying an FD to create an extra copy of g and relabel the original copy of g in A as g'; we also relabel g' an arbitrary copy of g in D that has the same context than g in A, to obtain D_2 (see Fig. 3). Comparing the adjacency sets of A and D with A_2 and D_2, we can see that from A to A_2 two adjacencies where renamed from xg_t and $g_h y$ to xg'_t and $g'_h y$, and exactly the same change happened from D to D_2. Also, the adjacency $g_h g_t$ was added in A_2. As a result, $A_2 = A - \{xg_t, g_h y\} + \{xg'_t, g'_h y, g_h g_t\}$. Similarly, $D_2 = D - \{xg_t, g_h y\} + \{xg'_t, g'_h y\}$. Therefore, we have that $|A - D| + |D - A| = |A_2 - D_2| + |D_2 - A_2| - 1$. Note that this relabeling only works if we introduce a an extra copy of g in A with an FD here; if instead we introduce it with a TD, it would not be possible to get adjacencies xg'_t and $g'_h y$ in D_2, as the copy of g involved in both adjacencies would be different.

Case (ii): Assume that g is not a telomere in A, its context is not strongly conserved between A and D, but both adjacencies involving g, xg_t and $g_h y$, are present in D *on different copies* of g. We say that the context of g is *weakly conserved* between A and D. Again x and y need not to be trivial gene families and there might be several occurrences of adjacencies xg_t and $g_h y$ in D.

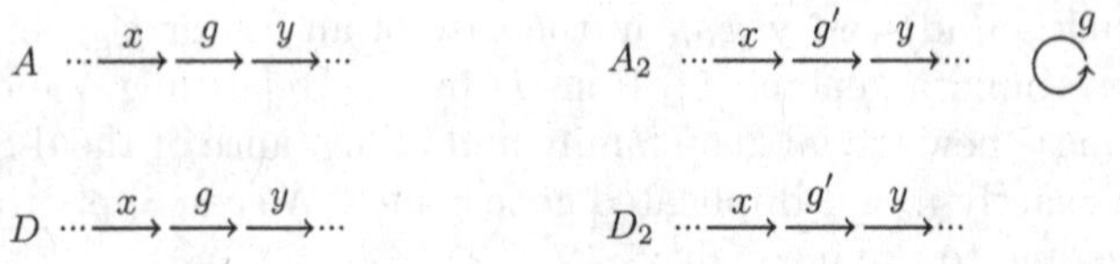

$$A \ \cdots \xrightarrow{x} \xrightarrow{g} \xrightarrow{y} \cdots \qquad\qquad A_2 \ \cdots \xrightarrow{x} \xrightarrow{g'} \xrightarrow{y} \cdots \quad \overset{g}{\circlearrowright}$$

$$D \ \cdots \xrightarrow{x} \xrightarrow{g} \xrightarrow{y} \cdots \qquad\qquad D_2 \ \cdots \xrightarrow{x} \xrightarrow{g'} \xrightarrow{y} \cdots$$

Fig. 3. The context of g is strongly conserved between A and D (Case (i)).

In this case, we build A_2 by applying a TD on g, relabeling the gene g that has the adjacency xg_t as a new gene g' in both A_2 and D_2, as shown on Fig. 4. Comparing the adjacency sets of A and A_2, we notice that the adjacency xg_t changes to xg_t', and $g_h g_t$ is added. Thus, $A_2 = A - \{xg_t\} + \{xg_t', g_h g_t\}$. From D to D_2 we also have the same change, and possibly one more, depending if g_h' is a telomere in D (no change) or if g_h' has an adjacency $g_h' w$. In the former case, $D_2 = D - \{xg_t\} + \{xg_t'\}$. Otherwise, $D_2 = D - \{xg_t, g_h w\} + \{xg_t', g_h' w\}$. In either case, the possible adjacency $g_h' w$ does not exist in A or A_2. Consequently, the equality $|A - D| + |D - A| = |A_2 - D_2| + |D_2 - A_2| - 1$ holds.

Note also that in this case an FD would not be optimal, because it would force the labeling of the adjacency $g_h y$ to $g_h' y$, and since the adjacency $g_h y$ on D cannot have the label $g_h' y$, this would force an extra pair of SCJ operations.

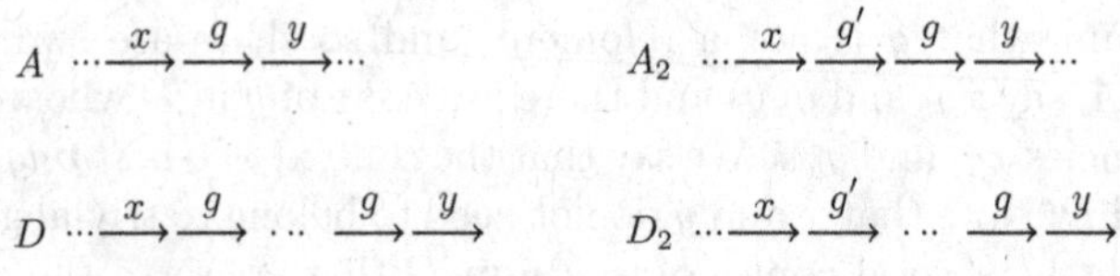

$$A \ \cdots \xrightarrow{x} \xrightarrow{g} \xrightarrow{y} \cdots \qquad\qquad A_2 \ \cdots \xrightarrow{x} \xrightarrow{g'} \xrightarrow{g} \xrightarrow{y} \cdots$$

$$D \ \cdots \xrightarrow{x} \xrightarrow{g} \ \cdots \ \xrightarrow{g} \xrightarrow{y} \qquad\qquad D_2 \ \cdots \xrightarrow{x} \xrightarrow{g'} \ \cdots \ \xrightarrow{g} \xrightarrow{y}$$

Fig. 4. The context of g is weakly conserved between A and D (Case (ii)).

Case (iii): We assume now that the context of g in A is neither strongly nor weakly conserved, and so at most one adjacency of g in A is also present in D.

This case is similar to case (i), if we assume that either xg_t or $g_h y$, are present in D, or neither. In the same way, we apply an FD on g, labeling the original copy as g', as shown in Fig. 5. On D, we pick a gene g that has an adjacency xg_t or $g_h y$ if any or, if no adjacency involving g is conserved in D, we pick an arbitrary g, and relabel it as g'.

Now, any adjacencies that were conserved between A and D will remain conserved between A_2 and D_2, and no new conserved adjacencies have been created. Since, as before, A_2 has a new $g_h g_t$ adjacency, the equality $|A - D| + |D - A| = |A_2 - D_2| + |D_2 - A_2| - 1$ holds.

These three cases cover all possible configurations for g, so the theorem is proved. $\qquad\qquad\qquad\qquad\qquad\qquad\qquad\qquad\qquad\qquad\qquad\qquad\qquad\qquad$ $\square$

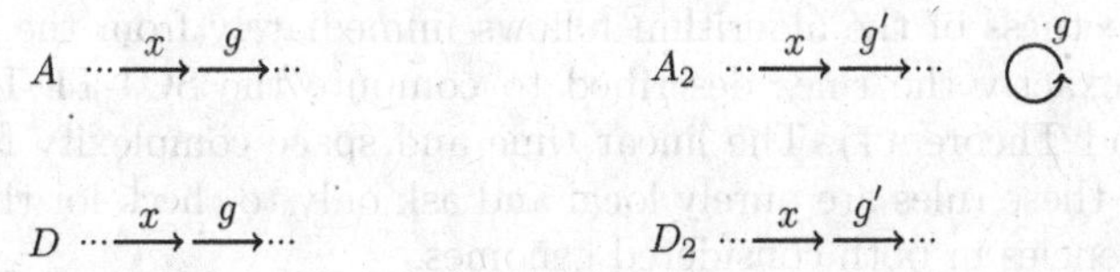

$$A \;\cdots\xrightarrow{x}\xrightarrow{g}\cdots \qquad\qquad A_2 \;\cdots\xrightarrow{x}\xrightarrow{g'}\cdots \quad \circlearrowright^{g}$$

$$D \;\cdots\xrightarrow{x}\xrightarrow{g}\cdots \qquad\qquad D_2 \;\cdots\xrightarrow{x}\xrightarrow{g'}\cdots$$

Fig. 5. At most one adjacency of g is conserved (Case (iii)).

3.2   Computing a Parsimonious Scenario

It follows from Lemma 2 and Theorem 1 that computing the d-SCJ-TD-FD distance can be done in linear time in the size of the considered genomes A and D. Moreover, they define a simple algorithm that computes a parsimonious scenario in terms of duplications, cuts and joins from A to D, described in Algorithm 1 below.

Algorithm 1. Compute an SCJ-TD-FD parsimonious scenario between a trivial genome A and a genome D

Reduce D into a reduced genome D'
Let $A' = A$ and $i = 1$
while (A', D') has a non trivial gene family **do**
    Let g be an arbitrary gene from a non trivial family in A'; relabel g by g^i.
    if the context of g is strongly conserved **then**
        relabel the corresponding copy of g in D' by g^i
        add to A' a single-gene circular chromosome g.
    else if the context of g is weakly conserved **then**
        create an extra copy of g^i with a TD
        relabel a copy of g involved in adjacency xg_t in D' by g^i.
    else if one adjacency of g is conserved in D' **then**
        relabel the corresponding copy of g in D' by g^i
        add to A' a single-gene circular chromosome g^i.
    else
        relabel an arbitrary copy of g in D' by g^i
        add to A' a single-gene circular chromosome g^i.
    end if
    $i = i + 1$
end while
Compute an SCJ scenario from A' to D'.
Recreate in D', the tandem arrays and single-gene circular chromosomes removed when reducing D into D'.

Theorem 2. *Given a trivial genome A with n_A genes and a possibly non-trivial genome D on the same set of gene families and with n_D genes, Algorithm 1 computes a parsimonious SCJ-TD-FD scenario that transforms A into D and can be implemented to run in time and space $O(n_D)$.*

The correctness of the algorithm follows immediately from the fact that it implements exactly the rules described to compute the SCJ-TD-FD distance (Lemma 2 and Theorem 1). The linear time and space complexity follows from the fact that these rules are purely local and ask only to check for the conservation of adjacencies in both considered genomes.

Every iteration of the while loop in Algorithm 1 takes place only if there is a non-trivial gene family left in D'. The maximum number of iterations is the number of duplicates genes, $n_d = n_D - n_A$ which is $O(n_D)$ when $n_D \geq n_A$. In each iteration, we check if the context of the chosen gene g is strongly conserved, weakly conserved or not conserved. This involves trying to match the adjacencies of g in A with those in the adjacency set of D' that involve a copy of g. This can be done in constant time, with a linear time preprocessing of the data. Hence, the worst-case time complexity is $O(n_D)$.

Remark 1. We have discussed in Sect. 2.1 the rationale to create duplicate genes with a FD creating a circular single-gene chromosome. However if the evolutionary model of the FD event created a linear single-gene chromosome, this would introduce a dissymetry between TD and FD (namely no adjacency is created with an FD), while in our model each created copy induces a cost of two due to the necessary break of the created adjacency required in the process of obtaining the reduced genome D'. We conjecture that the use of linear chromosomes would affect the choice of duplication event (FD or TD) only when the context is not conserved, which would result in a more complicated distance formula.

3.3   The Directed Median Problem

Let us remind that under the SCJ-TD-FD evolutionary model, the *directed median problem* asks, given k non-trivial genomes $D_1, \ldots, D_k, k \geq 2$, with the same gene families, to find a trivial common ancestor A, such that $\sum_{i=1}^{k} d_{DSCJ}(A, D_i)$ is minimized.

We first assume that the genomes $D_1, \ldots, D_k$ are reduced. We define the *score* $s(A)$ of a genome A as

$$s(A) = \sum_{i=1}^{k} d_{DSCJ}(A, D_i) = \sum_{i=1}^{k} \left(|A - D_i| + |D_i - A| + 2n_{d_i} \right)$$

where n_{d_i} is the number of extra gene copies in D_i compared to A, for $i = 1, \ldots, k$. Using the fact that $|A - D| + |D - A| = |A| + |D| - 2|A \cap D|$ we derive

$$s(A) = N_d - \left(2 \sum_{i=1}^{k} |A \cap D_i| - k|A| \right)$$

where $N_d = \sum_{i=1}^{k} (2n_{d_i} + |D_i|)$, and does not depend from A. Therefore, minimizing $s(A)$ is equivalent to maximizing $2 \sum_{i=1}^{k} |A \cap D_i| - k|A|$.

For a given adjacency a, let $\delta_i(a)$ be 1 if $a \in D_i$, and 0 otherwise. The score of a genome with a single adjacency a is $s(\{a\}) = N_d - \left(2 \sum_{i=1}^{k} \delta_i(a) - k \right)$.

This motivates the following approach, similar to the breakpoint median algorithm of [8]. Build a graph G where the vertices are defined as the extremities (head and tail) of a unique copy for each gene family in the considered genomes D_i (so a gene family a induces two vertices a_h and a_t), and weighted edges are defined as follows: for any edge $e = (x, y)$ such that x and y form an adjacency in at least one of the genomes D_i, the weight of e is $w(e) = 2 \sum_{i=1}^{k} \delta_i(e) - k$. Any matching M on G defines a trivial genome A_M, having the adjacencies corresponding to the edges in the matching M. Also, if $W(M)$ denotes the weight of the matching M, that is the sum of the weights of the edges in M we have that

$$s(A_M) = N_d - \left(2 \sum_{i=1}^{k} |A_M \cap D_i| - k|A_M| \right)$$

$$= N_d - \sum_{e \in M} \left(2 \sum_{i=1}^{k} \delta_i(e) - k \right)$$

$$= N_d - W(M)$$

Therefore, solving a maximum weight matching problem on G solves the directed median problem. To handle the case when some D_i is not reduced, we can rely on Lemma 2 that implies that the genomes can be reduced first without impacting the optimality of a trivial genome obtained by a maximum weight matching. Combined with the tractability of computing a maximum weight matching [25], this proves our last theorem.

Theorem 3. *Let $k \geq 2$ and $D_1, \ldots, D_k$ be k genomes on the same set of n gene families, having respectively $n_1, \ldots, n_k$ adjacencies. The directed SCJ-TD-FD median problem for these genomes can be solved in time and space $O(n(n_1 + \cdots + n_k) \log(n_1 + \cdots + n_k))$.*

Remark 2. In the case of the median of two genomes D_1 and D_2, note that the only edges with strictly positive weight in the graph are defined by adjacencies that appear in both D_1 and D_2, while edges appearing just once have weight 0. So a median genome can be defined as a maximum matching over the unweighted graph defined only by adjacencies that appear in both genomes, and given such a median, it can be augmented by any subset of edges appearing just once that do not re-use a gene extremity already used in the matching.

4   Experimental Results

We ran experiments on simulated instances with the aim to evaluate the ability of the d-SCJ-TD-FD distance to correlate with the true number of syntenic events. We followed a simulation protocol inspired from [16]. The code itself was programmed in Python and is available via github[2]. We first describe the simulation protocol, followed by the results we obtained.

---

[2] https://github.com/acme92/SCJTDFD.

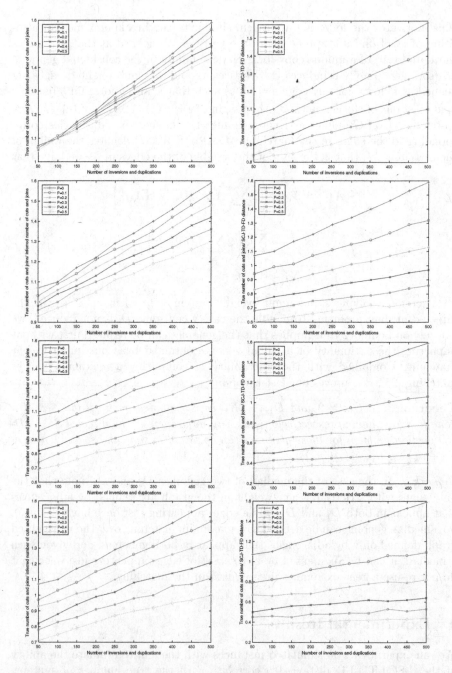

Fig. 6. Experimental results, for four duplications parameters – single-gene segmental duplication (top row), two-genes segmental duplication (second row), five-genes segmental duplications (third row), variable length segmental duplications (bottom row) – and two measured quantities – inferred cuts and joins (left column) and SCJ-TD-FD distance (right column).

We started from a genome A composed of a single linear chromosome containing 1000 single-copy genes. Then, we transformed A genome into a genome D through a sequence of random segmental duplications and inversions. We fixed the number N of evolutionary events (from 50 to 500 by steps of 50) and the probability P that a given event is a segmental duplication (from 0 to 0.5 by steps of 0.1). A segmental duplication is defined by three parameters: the position of the first gene of the duplicated segment, the length of the duplicated segment, and the breakpoint where the duplicated segment is transposed into; we considered two models of segmental duplications, one with fixed segment length L (with L taking values in $\{1, 2, 5\}$) and one where for each segment, L is picked randomly (under the uniform distribution) in $\{1, 2, 5, 10\}$. Inversion breakpoints were chosen randomly, again under the uniform distribution. For each array of parameters, we ran 50 replicates.

For each instance, we compared two quantities to the true number of cuts and joins in the scenario transforming A into D, which is roughly four times the number of inversions plus three times the number of segmental duplications: first we compared the full SCJ-TD-FD distance, defined as stated in Theorem 1 and the number of cuts and joins ($|A - D| + |D - A|$). Figure 6 illustrates the results we obtained.

We can make several observations from these results. The first one is a general trend that both measured quantities (the number of cuts and joins and the full SCJ-TD-FD distances) scale linearly with the true number of cuts and joins. The second observation is that, as expected, the slope and a y-intercept of the graphs depend from both the frequency of duplications and the length of the duplicated segments. This leaves open the question of using the SCJ-TD-FD distance as an estimator of the number of cuts and joins in an evolutionary model where the probability of duplication compared to rearrangements (that can be estimated for example from reconciled gene trees and adjacency forests [24]) is given and the length of duplicated segments is expected to follow a well defined distribution.

5   Conclusion

In this work, we introduced a simple variant of the SCJ model that accounts for duplications, and showed that, in this model, computing a directed parsimonious genomic distance from a trivial ancestral genome to a non-trivial descendant genome can be done in linear time and that a directed median can be computed in polynomial time. The tractability stems mostly from the combination of assuming that one genome is trivial and of a simplified model of duplication where gene duplication are single-gene events. However we believe it is interesting to push the tractability boundaries of the SCJ models toward augmented models of evolution (here accounting for duplications). Moreover, our work is motivated by the increasing performance of ancestral gene order reconstruction methods, that can now account for complex gene histories using reconciled gene trees and motivate the directed distance approach, and provides an additional positive result along the line of the research program outlined in [21].

For example, our algorithm will allow to extend the small parsimony algorithm PhySca introduced in [26] to a duplication-aware framework by allowing to score exactly and quickly an ancestral gene order configuration within a species phylogeny (work in progress).

There are several avenues to extend the results we presented in this paper. It will likely be easy to modify our algorithm to work in an extended the evolution model to integrate the loss of gene families and de-novo creation of genes. Our main result provides a simple algorithm that computes a parsimonious scenario, however it is likely one among a large number of parsimonious scenarios, and it is open to see if the results of [27] about counting and sampling SCJ parsimonious scenarios can be extended to our model. An important open question toward a more realistic model of evolution concerns the possibility to include larger scale duplications as unit-cost events. The case of a single whole genome duplication and of whole-chromosome duplications have been shown to be tractable [5,20], but to the best of our knowledge there is no known result including segmental duplications in which a contiguous segment of genes is duplicated either in tandem or appearing as a single chromosome. It also remains to be seen if the directed SCJ-TD-FD distance can be used toward the computation of an estimated distance in a more realistic evolutionary distance, similarly to the use of the breakpoint distance to estimate the true DCJ distance [28]; our experimental results suggest this is a promising avenue, although it might be difficult to obtain analytical results in models mixing rearrangements and duplications. Finally, the question of the tractability of the small parsimony problem in our model is also, to the best of our knowledge, still open. It is known to be tractable in the pure SCJ model (*i.e.* with no duplications) due to the independence of adjacencies; this assumption does not hold anymore here and the small parsimony problem is thus likely more difficult in our model. Our tractability result for the directed median is a first step toward this goal as it already provides a building block for a bottom-up ancestral reconstruction algorithm.

Acknowledgments. CC is supported by a Discovery Grant from the Natural Sciences and Engineering Research Council of Canada. PF is supported by the Genome Canada grant PathoGiST.

References

1. Fertin, G., Labarre, A., Rusu, I., Tannier, E., Vialette, S.: Combinatorics of Genome Rearrangements. MIT Press, Cambridge (2009)
2. Wang, D., Li, D., Ning, K., Wang, L.: Core-genome scaffold comparison reveals the prevalence that inversion events are associated with pairs of inverted repeats. BMC Genom. **18**(1), 268 (2017)
3. Neafsey, D., Waterhouse, R., et al.: Mosquito genomics. Highly evolvable malaria vectors: the genomes of 16 Anopheles mosquitoes. Science **347**(6217), 1258522 (2015)
4. Hannenhalli, S., Pevzner, P.: Transforming cabbage into turnip (polynomial algorithm for sorting signed permutations by reversals). In: 27th Annual ACM Symposium on the Theory of Computing (STOC 1995), pp. 178–189 (1995)

5. Feijão, P., Meidanis, J.: SCJ: a breakpoint-like distance that simplifies several rearrangement problems. IEEE/ACM Trans. Comput. Biol. Bioinform. **8**(5), 1318–1329 (2011)
6. da Silva, P., Machado, R., Dantas, S., Braga, M.: DCJ-indel and DCJ-substitution distances with distinct operation costs. Algorithms Mol. Biol. **8**(1), 21 (2013)
7. Braga, M., Willing, E., Stoye, J.: Double cut and join with insertions and deletions. J. Comput. Biol. **18**(9), 1167–1184 (2011)
8. Tannier, E., Zheng, C., Sankoff, D.: Multichromosomal median and halving problems under various different genomic distances. BMC Bioinform. **10**, 120 (2009)
9. Shao, M., Lin, Y., Moret, B.: An exact algorithm to compute the double-cut-and-join distance for genomes with duplicate genes. J. Comput. Biol. **22**(5), 425–435 (2015)
10. Compeau, P.E.C.: DCJ-Indel sorting revisited. Algorithms Mol. Biol. **8**, 6 (2013)
11. Rubert, D., Feijão, P., Braga, M., Stoye, J., Martinez, F.: Approximating the DCJ distance of balanced genomes in linear time. Algorithms Mol. Biol. **12**, 3 (2017)
12. Bryant, D.: The complexity of calculating exemplar distances. In: Sankoff, D., Nadeau, J.H. (eds.) Comparative Genomics: Empirical and Analytical Approaches to Gene Order Dynamics, Map Alignment and Evolution of Gene Families, vol. 1, pp. 207–211. Springer, Dordrecht (2000). doi:10.1007/978-94-011-4309-7_19
13. Blin, G., Chauve, C., Fertin, G.: The breakpoint distance for signed sequences. In: Algorithms and Computational Methods for Biochemical and Evolutionary Networks (CompBioNets 2004). Text in Algorithms, vol. 3, pp. 3–16 (2004)
14. Angibaud, S., Fertin, G., Rusu, I., Thevenin, A., Vialette, S.: On the approximability of comparing genomes with duplicates. J. Graph Algorithms Appl. **13**(1), 19–53 (2009)
15. Blin, G., Fertin, G., Sikora, F., Vialette, S.: The EXEMPLARBREAKPOINTDISTANCE for non-trivial genomes cannot be approximated. In: Das, S., Uehara, R. (eds.) WALCOM 2009. LNCS, vol. 5431, pp. 357–368. Springer, Heidelberg (2009). doi:10.1007/978-3-642-00202-1_31
16. Shao, M., Moret, B.: A fast and exact algorithm for the exemplar breakpoint distance. J. Comput. Biol. **23**(5), 337–346 (2016)
17. Shao, M., Moret, B.: On computing breakpoint distances for genomes with duplicate genes. J. Comput. Biol. (2016, ahead of print). doi:10.1089/cmb.2016.0149
18. Wei, Z., Zhu, D., Wang, L.: A dynamic programming algorithm for (1,2)-exemplar breakpoint distance. J. Comput. Biol. **22**(7), 666–676 (2014)
19. Angibaud, S., Fertin, G., Rusu, I., Thévenin, A., Vialette, S.: Efficient tools for computing the number of breakpoints and the number of adjacencies between two genomes with duplicate genes. J. Comput. Biol. **15**(8), 1093–1115 (2008)
20. Zeira, R., Shamir, R.: Sorting by cuts, joins, and whole chromosome duplications. J. Comput. Biol. **24**(2), 127–137 (2017)
21. Sankoff, D., El-Mabrouk, N.: Duplication, rearrangement and reconciliation. In: Sankoff, D., Nadeau, J.H. (eds.) Comparative Genomics: Empirical and Analytical Approaches to Gene Order Dynamics Map, Alignment and Evolution of Gene Families, vol. 1, pp. 537–550. Springer, Dordrecht (2000). doi:10.1007/978-94-011-4309-7_46
22. Sankoff, D.: Genome rearrangement with gene families. Bioinformatics **15**(11), 909–917 (1999)
23. Chauve, C., El-Mabrouk, N., Guéguen, L., Semeria, M., Tannier, E.: Duplication, rearrangement and reconciliation a follow-up 13 years later. In: Chauve, C., El-Mabrouk, N., Tannier, E. (eds.) Models and Algorithms for Genome Evolution, vol. 19, pp. 47–62. Springer, London (2013). doi:10.1007/978-1-4471-5298-9_4

24. Duchemin, W., Anselmetti, Y., Patterson, M., Ponty, Y., Bérard, S., Chauve, C., Scornavacca, C., Daubin, V., Tannier, E.: DeCoSTAR: reconstructing the ancestral organization of genes or genomes using reconciled phylogenies. Genome Biol. Evol. **9**(5), 1312–1319 (2017)
25. Plummer, M.D., Lovász, L.: Matching Theory. Elsevier, Amsterdam (1986)
26. Luhmann N., Lafond M., Thevenin A., Ouangraoua A., Wittler R., Chauve C.: The SCJ small parsimony problem for weighted gene adjacencies. IEEE/ACM Trans. Comput. Biol. Bioinform. (2017, ahead of print). doi:10.1109/TCBB.2017.2661761
27. Miklós, I., Kiss, S., Tannier, E.: Counting and sampling SCJ small parsimony solutions. Theoret. Comput. Sci. **552**, 83–98 (2014)
28. Biller, P., Guéguen, L., Tannier, E.: Moments of genome evolution by Double Cut-and-Join. BMC Bioinform. **16**(Suppl 14), S7 (2015)

Generation and Comparative Genomics
of Synthetic Dengue Viruses

Eli Goz[1,2], Yael Tsalenchuck[2], Rony Oren Benaroya[2], Shimshi Atar[1],
Tahel Altman[2], Justin Julander[3], and Tamir Tuller[1,2,4(✉)]

[1] Department of Biomedical Engineering, Tel-Aviv University,
Ramat Aviv, Israel
tamirtul@post.tau.ac.il
[2] SynVaccine Ltd., Ramat Hachayal, Tel Aviv, Israel
[3] Institute for Antiviral Research, Utah State University, Logan, UT, USA
[4] Sagol School of Neuroscience, Tel-Aviv University, Ramat Aviv, Israel

Abstract. Synthetic virology is an important multidisciplinary scientific field,
with emerging applications in biotechnology and medicine, aiming at developing methods to generate and engineer synthetic viruses. Here we demonstrate
a full multidisciplinary pipeline for generation and analysis of synthetic RNA
viruses and specifically apply it to Dengue virus type 2 (DENV-2). The major
steps of the pipeline include comparative genomics of endogenous and synthetic
viral strains. In particular, we show that although the synthetic DENV-2 viruses
were found to have lower nucleotide variability, their phenotype, as reflected in
the study of the AG129 mouse model morbidity, RNA levels, and neutralization
antibodies, is similar or even more pathogenic in comparison to the wildtype
master strain. These results may suggest that synthetic DENV-2 may enhance
virulence if the correct sequence is selected. The approach reported here can be
used for understanding the functionality and the fitness effects of any set of
mutations in viral RNA. It can be also used for editing RNA viruses for various
target applications.

1 Background

The ability to synthesize and engineer RNA viruses has important applications to
biotechnology and human health [1–5]. Although some previously published studies
involved generation and analysis of various synthetic RNA viruses (e.g., [6–12]), more
studies are required to understand their evolution and population dynamics together
with developing efficient synthetic biology approaches for their research [13, 14].

Dengue virus (DENV) is an important RNA flavivirus with 4 genomically diverse
serotypes. Its genome is a positive polarity single stranded RNA of approximately

---

E. Goz and Y. Tsalenchuck—Equal contribution.

© Springer International Publishing AG 2017
J. Meidanis and L. Nakhleh (Eds.): RECOMB-CG 2017, LNBI 10562, pp. 31–52, 2017.
DOI: 10.1007/978-3-319-67979-2_3

11 kb composed of 3 parts: the unique coding sequence (CDS) that produces 10 mature viral proteins via a single precursor polyprotein, and two flanking untranslated regions (UTRs) that contain important structural and functional elements [15–17]. Dengue is an important human pathogen, widely recognized as a major public health concern [18]; therefore new insights into biological and evolutionary properties of its wildtype and synthetically generated strains may have significant influence on the development of new antiviral vaccines and therapies.

Here we demonstrate, for the first time, a comprehensive analysis of DENV-2 (Dengue virus, serotype 2) synthetic variants using a multidisciplinary pipeline which combines comparative genomics, synthetic biology, next generation sequencing (NGS), and experiments with animal models of viral infection. Specifically, basing on DENV-2 New Guinea C wildtype strain (WT), two synthetic variants were generated; comparative genomics as well as their replication and pathogenesis analyses in AG129 mouse model were performed. The suggested approach can be also directly applied to other RNA viruses.

2   Results

2.1   Study Outline

A full comparative genomics pipeline for synthetic biology of RNA viruses demonstrated in this study is described in Fig. 1 (see Sect. 4 for more details): DENV-2 New Guinea C wild type viral strain (Fig. 1.I) was sequenced (Sanger) (Fig. 1.II). The resulting sequence was used as a master sequence to generate synthetic variants. In order to demonstrate the robustness of the reported results two different approaches were examined: (1) *Incellulo* construction of subgenomic infectious amplicons (Fig. 1. III), and (2) PCR *assembly* of full-length infectious DNA (Fig. 1.IV). After three passages of the wildtype and synthetic DENV-2 viruses in Vero E6 cells (Fig. 1.V), viral RNA was extracted and next generation sequencing was performed (Fig. 1.VI). The resulting sequences were verified and analysis of genomic variants (variants calling) with respect to single nucleotide variations (SNV) and inserts was performed (Fig. 1.VII). The results were compared to a population sample of 618 available DENV-2 genomes (Fig. 1.VIII-IV) with respect to several genomic features: nucleotide variability, secondary structure, codon and codon pair bias, CpG and GC contents. In addition, the wildtype and synthetic (generated by the *assembly* method) viruses were compared in vivo in the AG129 mouse model (Fig. 1.X); the comparison included: weight change, survival rates, RNA levels, and neutralizing antibodies (Ab). In the following we describe the results of our analyses.

2.2   Next Generation Sequencing and Comparative Genomics Analysis of the Synthetic and Wildtype Viruses

The NGS and comparative genomics analysis of the populations of synthetic and WT viruses in the cell line is summarized in Figs. 2 and 3 (further details can be found in the Methods section).

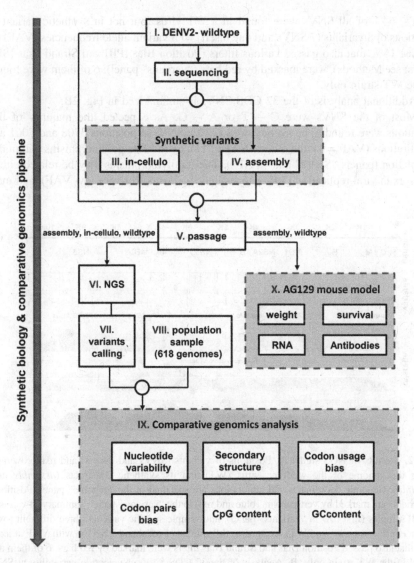

Fig. 1. A full comparative genomics pipeline for synthetic biology of RNA viruses. Details can be found in the main text.

The positions of the discovered variants (SNVs and inserts) and read coverage along the genome for the WT virus and its synthetic constructs (*assembly* and *incellulo*) are shown in **Fig. 2A**. Genomic coordinates and annotations are specified in the "genome" panel. In total, 38 variants were found: 35 in the CDS and 3 in the UTRs; positions of these variants were marked by vertical bars (blue and red) in the "variants" panel. 32 (91%) of the variants in the CDS were due to SNVs, and the rest 3 (9%) were due to inserts; 23 (72%) of SNVs in the CDS were synonymous. In the UTRs all variants were due to inserts. All the inserts were due to an additional A within a stretch of A's. 64% of all SNVs were found in a WT virus (but not in synthetic variants). Positions of 8 variants (7 SNVs and one insert) with variant allele frequencies (VAF) of at least 15%, that also passed various filters (Position Bias [PB] and Strand Bias [SB] filters; see Methods), were marked by red bars ("variants" panel); 6 of them were found in the WT strain only.

Additional analysis of the 32 CDS SNVs is summarized in Fig. 2B:

Most of the SNVs were C → T or A → G. As expected, the majority of the variations were found to be synonymous (72%). SNVs at positions 3708 and 5061 are substitutions (VAF = 1), the others (VAF < 1) may indicate polymorphisms within the population (panels "SNVs" and "VAF"). These polymorphisms may be related either to errors in viral replication [19] or to sequencing errors (for very low VAF), and may

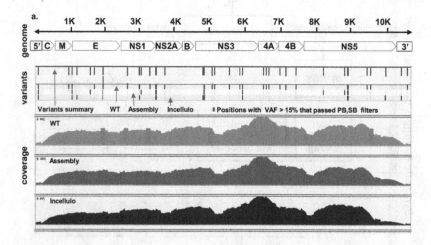

Fig. 2. Results of NGS study. **A.** Positions of variants (SNVs and inserts) and read coverage along the genome of the wildtype virus (WT) and its synthetic constructs (*assembly* and *incellulo*). Genomic coordinates and annotations are specified in the "genome" panel. Positions of variants are marked by vertical bars (blue and red) in the "variants" panel; summary of variants for all samples (first row in "variants" panel), and sample-specific variants (three different rows for *incellulo*, *assembly*, and WT) are shown. 8 Positions (1 indel and 7 SNVs) with VAF at least 0.15 that also passed Position Bias and Strand Bias filters were marked by red bars; 6 of them are found in the WT strain only. **B.** Analysis of the 32 CDS SNVs. Nucleotide variability at SNV positions based on a set of 618 aligned DENV2 genomes is represented by the corresponding sequence logo in the "variability" panel. The coordinates of SNVs are specified in the x-axis; corresponding genes are specified in the bottom row. The variability at each position is represented by a stack of letters. The relative sizes of the letters indicate their frequency in the alignment at specific position. The total height of the letters depicts the information content of the position, in bits. The SNVs at each position are depicted in the "SNVs" panel in the following format: reference codon (reference AA) → variant codon (variant AA). The altered nucleotides are marked in red. The Variant allele frequencies (VAF) at each position were specified in the "VAF" panel. Non synonymous SNVs (18%) were marked by "N". Positions that overlap with regions that undergo a conserved selection for strong or weak mRNA folding (based on [16]) were marked by "s" or "w" respectively. 7 CDS SNVs with VAF of at least 0.15 that also passed Position Bias and Strand Bias filters were marked by red asterisks. (Color figure online)

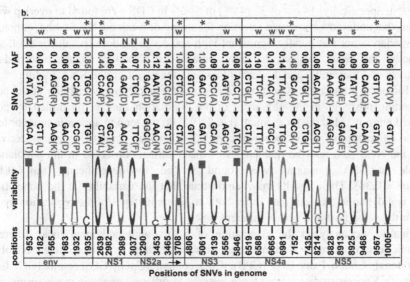

Positions of SNVs in genome

*   Positions with VAF > 15% that passed PB,SB filters

s/w   Positions that overlap with regions with significantly strong (s) / weak (w) RNA structures

N   Positions with non-synonymous CDS SNVs

Fig. 2. (*continued*)

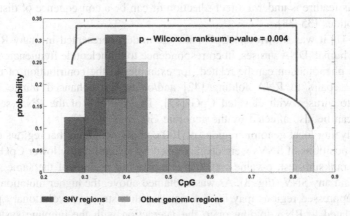

Fig. 3. CpG dinucleotides are significantly suppressed in regions of 100 codons around SNVs as compared to genomic regions of the same size that do not contain any SNV.

play a role in improving the fitness of the WT virus [19–21]. 7 significant SNVs (VAF ≥ 15% that also passed PB and SB filters) were marked by red asterisks.

Further, the positions of the CDS SNVs were compared to the population sample consisting of 618 available DENV-2 genomes (see Methods for additional details). In particular relations to regions with significantly strong/weak RNA structures ([16] and Methods) and to position-wise nucleotide variability (panel "variability") were analyzed. Specifically: 14 out of 32 (44%) of SNVs in CDS were found to overlap with regions that undergo a conserved selection for strong or weak mRNA folding, although this overlap is not statistically significant with respect to random (p-value = 0.14, see Methods); this result suggests that only some of the SNVs are related to secondary structures and may influence the viral fitness via the effect on the RNA folding.

Moreover, no accord between the SNVs and the nucleotide variability at the corresponding positions in the alignment of the population sample (quantified by normalized entropy, Methods) was found (i.e. SNV's do not tend to appear in more or less conserved positions). Also, Spearman correlation between variant frequencies derived from the NGS data and the frequencies of these variants observed in the population sample at SNV positions was found to be low/insignificant (rho = 0.16, p-value = 0.38). These results may/demonstrate that the variability at the population level is a dynamic trait and may differ both at different stages of the replication cycle within the same host and across different hosts.

Finally, we analyzed the association of SNVs with additional local different genomic features, namely: Codons Usage Bias as measured by ENC (Efficient Number of Codons), Codon Pairs Bias (CPB), CpG content (CpG), GC content (GC) (see Methods). Various studies related these features to different genomic mechanisms that take part in the viral replication cycle and are related to the viral fitness. For example, it was suggested that viral codons may be under selection to improve the translation efficiency, among others, via adaptation to the host tRNA pool (or other translation resources)(e.g., [22–34]). It was also suggested that translation efficiency is affected not only by single codons, but also by distribution of codon pairs (although it is debated whether this feature is under a direct selection or can be a consequence of distribution of dinucleotides [35–38].

In [39–41] it was shown that CpG pairs are under-represented in many RNA and most small human DNA viruses, in correspondence to dinucleotide frequencies of their hosts. This phenomenon can be related, for example, to the contribution of the CpG stacking basepairs to RNA folding [42] and/or to the enhanced innate immune responses to viruses with elevated CpG [43]. The stability of the RNA secondary structures can be also affected by the genomic GC content [44].

By analyzing local genomic regions of 100 codons we found that regions centered around the positions of SNVs were characterized by a significantly lower CpG content (Wilcoxon ranksum test, p-value = 0.004) compared to regions of the same size that didn't contain any SNV (Fig. 3). As was explained above, the higher mutation rates in the CpG suppressed regions may be associated with a stronger evolutionary pressure possibly related to RNA folding or to the interaction with the immune system. We didn't find significant differences between the SNV and non-SNV regions with respect to the rest of the analyzed features (supplementary material, Sect. 1).

2.3   AG129 Mouse Model Study

An AG129 mouse model was used to characterize the pathogenicity and immunization capabilities of synthetic DENV-2 generated by the Assembly method. Infection with the WT virus resulted in around 50% mortality (Fig. 4A, Table 1), which was consistent with intraperitoneal infection of AG129 mice in previous studies in our lab.

Synthetic wild-type virus injection resulted in 75% mortality and had a similar mortality curve to mice infected with the WT virus. The mortality rates caused by both synthetic and WT infection were significantly higher (p-value < 0.001) than in sham-infected controls (Fig. 4A).

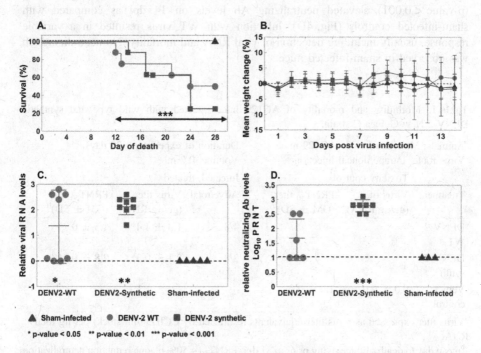

Fig. 4. Results of AG129 mouse model study. A. Percentage of survival of AG129 mice infected with wild-type and synthetic DENV-2 (p-value < 0.001 for 12 dpi and furhter, as compared with sham controls). B. Mean weight change (%) of AG129 mice infected with different DENV-2 variants. C. Virus titers from serum collected 3 dpi from infected AG129 mice (**p-value < 0.01, *p-value < 0.05, as compared with sham controls). D. 50% plaque reduction neutralization titers (PRNT$_{50}$) from serum taken 14 dpi (***p-value < 0.001, as compared with sham controls). Note that there were not sufficient quantities of serum for 5 samples: 2 from group 1(infected with WT variant), 1 from group 2 (infected with synthetic variant) and 2 from control group for quantification of neutralizing Ab.

Some weight loss was observed at various times over the course of infection in individual animals infected with WT or synthetic DENV-2, although the general trend was for little average weight change beyond baseline within all groups (Fig. 4B). For some reason, vehicle inoculated control animals (6–7 weeks old) did not gain weight as expected (Fig. 4B), although reasons for this are unknown.

Viral RNA was detected in all animals infected with synthetic DENV-2 in serum collected 3 days post infection (dpi), which was significantly (p-value < 0.01) higher than sham-infected controls (Fig. 4C). Half of the animals inoculated with the WT virus had detectable titers, while the other half had RNA levels below that of assay detection (Fig. 4C). Despite the lower rate of detection of viral RNA in these samples, the mean viral RNA level in this group was significantly (p-value < 0.05) elevated as compared with controls. No viral RNA was detected in the serum of sham-infected mice.

As anticipated, animals infected with synthetic DENV-2 elicited significantly (p-value < 0.001) elevated neutralizing Ab levels on 14 dpi as compared with sham-infected controls (Fig. 4D). Infection with WT virus resulted in a variable response, usually including detection of viral RNA and mortality. No neutralizing Ab was observed in sham-infected mice.

Table 1. Morbidity and mortality of AG129 mice infected with wild type and synthetic DENV-2 New Guinea C strain.

Animals: Male and female AG129 mice Virus route: Intraperitoneal injection			Duration of experiment: 28 days Volume: 0.2 ml		
	Toxicity controls		Infected, treated		
Treatment	Virus titer (ge/μg RNA)[a]	PRNT$_{50}$ titer (GM ± SD)[b]	Alive/total	Virus titer (ge/μg RNA)	PRNT$_{50}$ titer (GM ± SD)[b]
DENV-2 WT	–	–	4/8	1.3 ± 1.4*	1.5 ± 0.7
DENV-2 synth	–	–	2/8	2.1 ± 0.3**	2.8 ± 0.2***
Vehicle control	0.0 ± 0.0	1.0 ± 0.0	–	–	–

[a]Virus titer expressed as virus titer equivalents (estimated by CCID50/ml stock) per μg total RNA.
[b]mean day to death of mice dying prior to 31 dpi. PRNT$_{50}$ – 50% plaque reduction neutralization titers. GM – geometric mean; SD – standard deviation
(***p-value < 0.001, **p < 0.01, *p < 0.05, as compared with sham-infected [vehicle] controls)

3 Discussion

This study demonstrates a multidisciplinary pipeline for comparing and analyzing synthetic and wildtype viruses. Central aspects/steps of this pipeline includes comparative genomics: First, synthetic viruses can be designed based on comparative genomics analysis of WT viral genomes (e.g., [16, 45]). Second, the synthetic variants can be analyzed based on comparative genomics tools; this includes the comparisons of the variants to each other and to WT genomes. Among others, this study demonstrates the interactions between comparative genomics and other research tools in synthetic virology.

Specifically, we report here for the first time a set of SNVs that appear in a master wildtype DENV-2 virus in comparison to its synthetic constructs. In general, the wildtype virus was found to have a higher genomic variability (i.e. more nucleotide variations) than its synthetic strains. It is known that the population of viral variants is important for its fitness [20, 21]; thus, this analysis demonstrates an approach for connecting viral fitness to the structure of its population.

In addition, our study demonstrates that the synthetic variants may cause more morbidity and mortality than the wildtype virus. Also, infection with the synthetic virus resulted in higher serum viral RNA titers (as measured 3 dpi) and more consistent and high levels of neutralizing Ab (as measured 14 dpi) as compared to infection with the wildtype DENV-2. These findings may be related to the already mentioned fact that the synthetic virus includes lower levels of genomic sequence variability. RNA viruses are often defined as quasispecies [46], collections of closely related mutant spectra or mutant clouds. The fact that a virus exists as a 'cloud' of closely related genomes constitutes an evolutionary advantage, among others by enabling the virus to efficiently adjust to different hosts or cells. Therefore, higher level of variability in the wildtype virus may improve its replication rate in hosts such as Vero cells (the origin of the investigated wildtype virus) and/or humans/*A. aegypti*, but on the other hand, decreas its fitness/replication rate in mice. It is also possible that the wildtype virus undergoes a direct selection for reduced virulence (among others this may be related to its variability) which actually increases its transmissibility and fitness [47].

The approach described here can be performed on any RNA virus and can be used as a platform technology for rational design and synthesis of live attenuated vaccines for RNA viruses. Specifically, in the near future we plan to use this approach for studying the effect of synonymous mutations on the fitness of various types of viruses (e.g. different serotypes of DENV and Zika). To this end, we will generate large libraries of viral genomes with different type of synonymous mutations. The RNA and protein levels of these libraries will be measured in cell lines and model organisms and the mutations will be connected to the observed measurements and to other viral phenotypes (e.g. cytopathic effect).

4   Methods

4.1   Virology

Reagents: Minimum Essential Medium Eagle- Earle`s (EMEM medium), Fetal Bovine Serum (FBS), Penicillin-Streptomycin-Nystatin Solution, and Dulbecco's Phosphate Buffered Saline (DPBS) were obtained from Biological Industries (Bet-Haemek, IL). Opti-MEM medium was obtained from Gibco-ThermoFisher (Grand Island, NY). TransIT-X2 Mirus reagent was also used (MC-MIR-6003; Madison, WI).

Cell Lines and Virus: Vero E6 (ATCC CRL-1586) cells were cultured in EMEM supplemented with 10% Fetal Calf Serum, streptomycin 0.1 mg/ml, penicillin 100 u/ml, nystatin 12.5 U/ml, 0.29 mg/ml at 37 °C in a 5% CO_2 incubator. DENV-2, New Guinea-C strain was prepared in Vero E6 cells grown in EMEM medium containing 2% inactive FBS, Strep-Pen-Nys solution and incubated at 37 °C for 10 to 12 days until the appearance of cytopathic effects (CPE). The supernatant was then harvested and divided into aliquots and stored at −80 °C. Viral RNA was quantified using RT-PCR and infectious tissue culture assay.

Cell Transfection and Infection: Vero E6 cells 20,000 cell/well were seeded in 96-well tissue culture plates and incubated overnight at 37 °C. A final amount of 2000 ng of an equimolar mix of all cDNA was amplified by PCR and incubated with 2.4 ml of TransIT-X2 Mirus reagent in 400 ml of Opti-MEM medium. Following the manufacturer's instructions, the mix was incubated for 15 min at room temperature, then divided into 8 wells (of a 96-well plate) with Vero E6 cells and incubated for 4 h. After incubation the cell supernatant was removed and cells were washed three times with D-PBS, and 100 ml of EMEM medium supplemented with 2% inactive FBS was added. The supernatant was harvested 10 to 12 days post transfection. The supernatant was centrifuged and then passaged two more times as follows: 0.8 to 1 ml supernatant was added to 25 cm^2 culture flask of Vero E6 cells (400,000 cells incubated overnight), and incubated for 4 h. After incubation the cell supernatant was removed and washed three times with D-PBS, and 6 ml of EMEM medium supplemented with 2% inactive FBS was added. The cell supernatant was incubated for 10 to 12 days until the appearance of cytopathic effects (CPE). The supernatant was then harvested divided to aliquots and stored in at 80 °C.

4.2 Molecular Biology

Dengue Virus Sequencing and Construction of Subgenomic Amplicons' Plasmids: Dengue virus type 2, New Guinea C (Culture collection) RNA was extracted from aliquoted culture supernatants using the QIAamp Viral RNA mini kit (Qiagen). cDNA was synthesized using Maxima H Minus First Strand DNA synthesis kit (Thermo Scientific) with random hexamer primers, according to manufacturer's protocol. cDNA was amplified by PCR in 500–3000 bp overlapping amplicons using PfuUltra II Hotstart PCR Master Mix(Agilent Technologies). The 500–1000 bp amplicons were sequenced using the Sanger method. The 2500–3500 bp amplicons were cloned into pJET1.2/blunt vectors using the CloneJET PCR Cloning Kit (Thermo Scientific). Cytomegalovirus promoter was inserted before the first fragment (0–2.5 kb). The last fragment (7–10.7 kb) was synthesized using the solid-phase DNA synthesis method (BioBasic), followed by hepatitis delta and simian polyadenylation signal.

Subgenomic Infectious Amplicons: Dengue subgenomic amplicons were generated as previously described in [6]. Entire viral genome was amplified by PCR in 4 DNA fragments of approximately 3 kb, each with 75–80 bp overlapping arms. The first fragment was flanked at 5′ by the cytomegalovirus promoter. The last fragment was flanked at 3′ by the hepatitis delta ribozyme followed by the simian polyadenylation signal. The PCR products were purified using MinElute PCR purification kit (Qiagen) and 1000 ng of equimolar mix was transfected to Vero E6 cell line (ATCC).

Assembly PCR of Full-Length Infectious DNA: Dengue viruses' genomes were amplified by PCR in 4 DNA fragments of approximately 3 kb, each with 100–500 bp overlapping ends. The first fragment was flanked at 5′ by the cytomegalovirus promoter. The last fragment was flanked at 3′ by the hepatitis delta ribozyme followed by the simian polyadenylation signal. The overlapping fragments were extended using Q5

High-Fidelity DNA Polymerase (New England Biolabs), using the neighboring fragment as a primer.

NGS Sequencing: Dengue virus type 2, New Guinea C (Culture collection) RNA was extracted from aliquoted culture supernatants using QIAamp Viral RNA mini kit (Qiagen). cDNA was synthesized using Maxima H Minus First Strand DNA synthesis kit (Thermo Scientific) with random hexamer primers, according to manufacturer's instructions. The cDNA was amplified in seven 1700–2500 bp overlapping amplicons using PfuUltra II Hotstart PCR Master Mix (Agilent Technologies). PCR products were separated in a 1% agarose TAE gel. After confirming the bands of interest were present without non-specific products, the PCR product was purified using the MIniElute PCR purification kit (Qiagen).

Sequencing libraries were prepared using an in-house (INCPM) DNA-Seq protocol, and sequenced 2 × 150 on an Illumina MiSeq nano v2.

4.3   Computational Analysis

NGS Analysis and Variants Calling. Three DENV cDNA samples were sequenced and analyzed: one of the samples was the wildtype (WT) strain, with the WT (Sanger) sequence used as a reference genome for NGS analysis. The two (*assembly* and *incellulo*) synthetic variants of the virus synthesized using two different methods (full length assembly PCR and subgenomic infectious amplicons).

Sequencing libraries were prepared using the INCPM DNA-seq protocol, and sequenced 2 × 150 on an Illumina MiSeq nano v2 PE150. Sequenced reads were mapped to a reference genome (genome of the wildtype virus) using BWA MEM v0.75 [48]. Among all read-pairs with the same alignment, a single representative read was chosen using Picard MarkDuplicates (http://picard.sourceforge.net).

Variants were called using Freebayes v1.0.2 [49] with the setting -f 0.05–ploidy 5 (in order to increase sensitivity to low allele fraction variants). Freebayes is a haplotype based Bayesian genetic variant detector designed to find small polymorphisms and complex events (composite insertion and substitution events) smaller than the length of a short-read sequencing alignment. It uses short-read alignments (BAM files with Phred+33 encoded quality scores) for individuals from a population and a reference genome to determine the most-likely combination of genotypes for the population at each position in the reference.

Positions of putatively polymorphic variants were identified and characterized by Variant Allele Frequency (VAF) - the fraction of variant alleles, out of all observed reads.

Two types of biases were tested for each variant:

Strand Bias (SB) is a type of sequencing bias in which one (positive/negative) strand is favored over the other, which can result in incorrect evaluation of the amount of evidence observed for one allele vs. the other. We used GATK Strand-Odds-Ratio (SOR) test [50] to determine if there is strand bias between forward and reverse strands for the reference or alternate alleles. Positions with SOR values > 4 may be considered as biased.

Position Bias (PB) is a composite measure composed of (phred scaled) probabilities for observing positional bias within supporting reads for a variant. Positions with PB values > 40 may be considered as biased.

Variants with VAF > 15% that passed SB ($\leq$ 4) and PB ($\leq$ 40) filters were marked as especially significant/strong.

Comparative Genomics Analysis of SNVs. Multiple alignments of 618 DENV2 coding regions (population sample), and regions with conserved selection for strong or weak mRNA folding were computed as in [16].

Briefly, the translated amino acid sequences corresponding to the coding regions were aligned by Clustal Omega package [51] with default parameters and then mapped backed to the nucleotide sequences basing on the original nucleotide composition of each coding region. Multiple alignment conservation scores can be found in the supplementary material, Sect. 2.

Coordinates of different genomic features (ORFs/UTRs) were obtained from DENV2 reference strain NC_001474.2 and aligned to the genomes used in this study.

Local minimum free folding energy (MFE) profiles were constructed by applying a 39 nt length sliding window to each genomic sequence (robustness to the sliding window size and to multiple sequence alignment inaccuracies was discussed in our previous paper [16]): at each step the MFE of a local subsequence enclosed by the corresponding window was calculated by the Vienna (v. 2.1.9) package RNAfold function with default parameters [52]. These profiles were compared to folding profiles of corresponding randomized variants preserving different genomic features of the wildtype sequences which are not necessarily directly related to mRNA folding: amino acid order and content, ("vertical") distribution of synonymous codons at each position in the alignment, frequency of di-nucleotides. As a result, regions with positions that undergo a significantly conserved (across different genomes) selection at the synonymous/silent level for weak/strong mRNA folding (MFE-selected regions) were identified. More details, including the coordinates of the identified regions with significant folding signals, can be found in [16], and in the supplementary material, Sect. 3.

Statistical analysis of the overlap between the positions of SNVs and regions selected for strong/weak RNA structure was performed as follows: from the analysis in [16] the total length of MFE - selected regions (for weak and strong folding) occupies $\sim$ 1/3 of the entire coding sequence (3368/10176), i.e. a randomly chosen point lies in a MFE-selected region with a probability p $\sim$ 1/3. On the other hand, in our case, 14 out of 32 SNV positions overlap with some MFE-selected region. In order to compare this number with what is expected in random under a Uniform distribution hypothesis we calculate the Hyper-geometric p-value: Let N (Population size) = 10176 (total DENV2 CDS length); Successes in population (K) = 3368 (The total length of MFE - selected regions); Sample Size (n) = 32 (number of CDS SNV positions); Successes in Sample (k) = 14 (number of SNV positions that overlap with MFE-selected regions). Then, the probability to get at least 14 positions overlapping with MFE selected positions can be calculated as:

$$p = 1 - \sum_{k=1}^{13} \frac{\binom{K}{k}\binom{N-k}{n-k}}{\binom{N}{n}} \sim 0.14$$

For a specific genomic position i, the nucleotide variability with respect to the population sample was quantified by a normalized Shannon entropy of a distribution on nucleotides normalized by the maximal entropy value at this position in the corresponding alignment:

$$V_i = \frac{\sum_{j=1}^{n} p_j \log_2(p_j)}{\log_2 n}$$

Where $n = 4$, i.e. the number of different possible nucleotides. This variability measure takes values between 0 and 1, and describes how dispersed the distribution of the alphabet elements is: higher values correspond to more uniform nucleotide usage; lower values correspond to more biased nucleotide, indicating that some nucleotides are preferred/positions are conserved.

Effective Number of Codons (ENC) is a measure that quantifies how far the codon usage of a coding sequence departs from equal usage of synonymous codons [53]. For each amino acid (AA) let us define x_i to be the number of its synonymous codons of each type in the sequence, and n to be the number of times this AA appears in the sequence:

$$n = \sum_{i}^{d} x_i$$

The frequency of each codon is therefore:

$$p_i = x_i/n$$

The ENC for a specific AA is:

$$ENC_A = 1/F_A, \text{ where } F_A = \sum_{i}^{d} p_i^2$$

ENC for the group of AAs with degeneracy $d(A_d)$:

$$ENC_{A_d} = 1/F_{A_d}, \text{ where } F_{A_d} = \frac{1}{|A_d|}\sum_{A \in A_d} F_A$$

(when an AA is missing, the corresponding effective number of codons is defined as an average over the given AAs of the same degeneracy).

Finally ENC for a gene is defined as an average of the group ENCs over all degeneracy AA groups weighted by the number of AAs in each group computed over the entire coding sequence.

$$ENC = 2 + \frac{9}{F_{A_2}} + \frac{1}{F_{A_3}} + \frac{5}{F_{A_4}} + \frac{3}{F_{A_6}}$$

ENC can take values from 20, in the case of extreme bias where one codon is exclusively used for each amino acid (AA), to 61 when the use of alternative synonymous codons is equally likely. Therefore smaller ENC values correspond to a higher bias in synonymous codons usage; consequently, a negative correlation with ENC values means is equivalent to a positive correlation with synonymous codons usage.

Codon Pairs Bias (CPB). To quantify codon pair bias, we follow [36] and define a codon pair score (CPS) as the log ratio of the observed over the expected number of occurrences of this codon pair in the coding sequence. To achieve independence from amino acid and codon bias, the expected frequency is calculated based on the relative proportion of the number of times an amino acid is encoded by a specific codon:

$$CPS = \log\left(\frac{F(AB)}{\frac{F(A) \times F(B)}{F(X) \times F(Y)} \times F(XY)}\right),$$

where the codon pair AB encodes for amino acid pair XY and F denotes the number of occurrences. The codon pair bias (CPB) of a virus is them defined as an average codon pair scores over all codon pairs comprising all viral coding sequences:

$$CPB = \frac{1}{k-1} \sum_{I=1}^{k-1} CPS[i]$$

CpG Content. Following [54] we compute a dinucleotide score (DNTS) for a pair of nucleotides XY as an odds ratio of observed over expected frequencies:

$$DNTS = \frac{F(XY)}{F(X)F(Y)},$$

where F denotes the frequency of occurrences.

Specifically, the CpG score is equal to the DNTS corresponding to the CG nucleotide.

The Dinucleotide Pair Bias (DNTB) of a virus is defined as an average of dinucleotide scores over all dinucleotides comprising all viral sequences:

$$DNTB = \frac{1}{k-1} \sum_{i=1}^{k-1} DNTS[i]$$

GC content is defined as:

$$GC\% = \frac{F(G) + F(C)}{F(A) + F(G) + F(C) + F(T)}$$

Where F() is a number of occurrences of each one of nucleotides A,G,C, and T.

4.4 AG129 Study

Animals. Male and female AG129 6–7 weeks old mice with an average weight of 20 g were used. Because of the immunocompromised nature of this strain, AG129 mice (produced in house) were housed in pre-sterilized bonneted (HEPA-filtered) cages in a ventilated cage rack during the duration of the experiment. Animals were randomly assigned to cages and individually marked with eartags. All instruments, eartags, gloved hands, or other surfaces that came in contact with the mice were cold-sterilized with a 10% betadine solution and/or 75% EtOH.

Facilities. Experiments were conducted in the BSL-2 animal suite at the Utah State University Laboratory Animal Research Center (LARC). All personnel receive continual special training on blood-borne pathogen handling by this university's Environmental Health and Safety Office. Standard operating procedures for BSL-2 were used.

Quantification of Viremia. The virus titers in plasma were assayed using the Brilliant II QRT-PCR Master Mix 1-step kit with samples run on the Mic (Magnetic Induction Cycler) real time PCR machine (Bio Molecular Systems, Inc). Breifly, RNA was extracted from serum samples collected 3 dpi using the QIAamp MinElute Virus Spin Kit (QIAGEN, cat# 57704), and was eluted with 25 μl of elution buffer. A volume of 5 μl of the RNA preparation was added to the appropriate mixture of PCR reagents following manufactures protocol. A virus stock of known titer was also extracted in parallel for use in quantification. Samples were subjected to 40 cycles of 15 s at 95 °C and 60 s at 60 °C following an initial single cycle of 30 min at 50 °C and 10 min at 95 °C. Samples of unknown quantity were quantified by extrapolation of C(t) values using a curve generated from serial dilutions of standard samples.

Experimental Design. Animals were block-randomized by cage to two groups, with 8 included in each. Groups of 8 mice were challenged intraperitoneally (i.p.) with wild-type and synthetic viral variant. A third group of mice was also included as normal controls, and were sham-infected with vehicle to monitor handling, inoculation and caging techniques for effects on the immunocompromised AG129 mice. Mice were i.p. infected with 0.2 ml of each virus preparation. Mortality was observed daily for 28 days. Mice were weighed on day 0 and every other day through 11 dpi. Serum was collected from all animals on 3 dpi for quantification of viremia. Serum was also collected on 14 dpi for measurement of neutralizing antibody titer.

Statistical Analysis of Mice Data. Survival data were analyzed using the Wilcoxon log-rank survival analysis, and all other statistical analyses were performed using one-way ANOVA with a Bonferroni group comparison.

Acknowledgment. E.G. is supported, in part, by a fellowship from the Edmond J. Safra Center for Bioinformatics at Tel-Aviv University. T.T. is partially supported by the Minerva ARCHES award.

Appendix: Supplementary Material

1. **No difference in various genomic features in 100 codon regions around SNV compared to regions that do not contains SNVs**

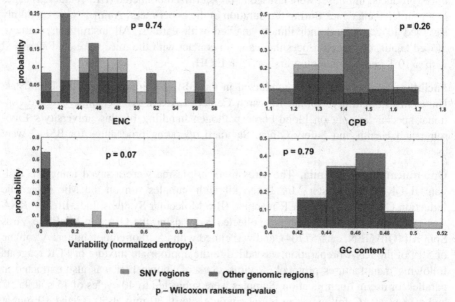

2. Multiple alignment of 618 DENV-2 genomes analyzed in this study - conservation scores

Multiple alignment conservation score was defined by us as an average sum-of-pair score (SP). For the i-th column in the alignment we define $P_{ijk} = 1$ for every pair A_{ij} and A_{ik} of elements (either nucleotides of amino acids, depending on the type of the aligned sequences) which are equal to each other and $P_{ijk} = 0$ otherwise. The score Si for the ith column is

$$S_i = \frac{1}{N(N-1)/2} \sum_{j=1}^{N} \sum_{k=j+1}^{N} P_{ijk}$$

and the SP for the alignment is:

$$SP = \frac{1}{M} \sum_{i=1}^{M} S_i$$

The following values summarize the SP scores for the multiple alignment of 618 DENV-2 coding sequences analyzed in this study: SP(amino acids) = 0.97, SP (nucleotides) = 0.94

3. List of regions selected for strong/weak folding energy used in this study

Coordinates of regions predicted to be selected for strong/weak folding energy can be found in the following tables (see details in reference [16] in main text):

Each row in a file corresponds to one region (number of rows = number of regions) and contains 3 comma separated values x, y, z in the following order:

Region start coordinate, region end coordinate, maximum folding selection conservation index (FSCI) in the cluster.

The coordinates are given with respect to the start of the polyprotein coding sequence in the reference genome NC_001474.2. E.g., coordinates x, y for some region correspond to the nucleotides at x-th and y-th positions in the coding sequence of NC_001474.2

weak folding			strong folding		
start	end	FSCI	119	165	0.21
96	153	0.47	266	353	0.29
186	247	0.33	387	615	0.74
332	416	0.43	695	745	0.44
441	490	0.23	820	872	0.68
530	686	0.93	1010	1142	0.48
781	830	0.29	1179	1272	0.31
867	917	0.31	1287	1332	0.21
1098	1191	0.95	1353	1520	0.74
1263	1311	0.49	1521	1583	0.54
1333	1381	0.33	1585	1635	0.57
1506	1563	0.3	1662	1739	0.85
1617	1673	0.4	1964	2078	0.48
1705	1840	0.66	2157	2231	0.43
1879	1968	0.82	2244	2313	0.36
2020	2086	0.66	2345	2413	0.37
2123	2171	0.54	2523	2688	0.7
2219	2291	0.49	2689	2805	0.48
2374	2432	0.28	2839	2903	0.38
2817	2865	0.33	2932	2980	0.26
2885	2933	0.33	2991	3038	0.2
2998	3061	0.38	3263	3312	0.51
3062	3107	0.22	3333	3406	0.8
3258	3328	0.32	3419	3464	0.27
3378	3438	0.38	3481	3545	0.41
3564	3613	0.56	3572	3631	0.35
3717	3766	0.35	3793	3838	0.2
3844	3904	0.34	3845	3941	0.21
4046	4106	0.74	3972	4020	0.33
4204	4292	0.53	4029	4080	0.35
4355	4403	0.32	4122	4171	0.29
4465	4563	0.65	4197	4251	0.32
4576	4625	0.25	4347	4399	0.35
4629	4682	0.41	4407	4525	0.52
4758	4806	0.28	4680	4732	0.44
5028	5077	0.26	4869	4914	0.21

5239	5290	0.21		5083	5131	0.24
5299	5387	0.36		5142	5235	0.87
5458	5557	0.45		5347	5396	0.22
5626	5674	0.23		5567	5618	0.5
5772	5823	0.68		5940	6000	0.48
5844	5940	0.62		6080	6150	0.78
6129	6179	0.3		6230	6282	0.31
6186	6231	0.71		6328	6376	0.26
6338	6386	0.24		6712	6776	0.48
6475	6589	0.8		6803	6871	0.42
6649	6718	0.38		6946	6994	0.37
6752	6815	0.29		7137	7187	0.34
6895	6983	0.45		7192	7271	0.52
6995	7183	0.93		7286	7434	0.68
7240	7299	0.65		7608	7668	0.63
7332	7379	0.26		7679	7727	0.25
7412	7461	0.3		7799	7851	0.25
7658	7709	0.32		7871	7928	0.45
7731	7780	0.3		8073	8166	0.49
7785	7835	0.5		8229	8296	0.41
7913	8004	0.29		8345	8393	0.27
8022	8079	0.5		8696	8754	0.41
8207	8266	0.3		8873	8951	0.98
8279	8331	0.51		9258	9309	0.47
8409	8463	0.67		9460	9508	0.23
8470	8556	0.63		9585	9636	0.67
8580	8624	0.25		9666	9848	0.95
8747	8821	0.49		9880	9944	0.57
8920	8969	0.34		9950	10021	0.51
9046	9102	0.6				
9156	9213	0.53				
9214	9292	0.94				
9344	9385	0.2				
9395	9532	0.78				
9578	9691	0.74				
9829	9887	0.56				
10019	10069	0.3				
10109	10167	0.46				

References

1. Sainsbury, F., Cañizares, M.C., Lomonossoff, G.P.: Cowpea mosaic virus: the plant virus-based biotechnology workhorse. Annu. Rev. Phytopathol. **48**, 437–455 (2010)
2. Kaufman, H.L., Kohlhapp, F.J., Zloza, A.: Oncolytic viruses: a new class of immunotherapy drugs. Nat. Rev. Drug Discov. **14**, 642–662 (2015)
3. Reyes, A., Semenkovich, N.P., Whiteson, K., Rohwer, F., Gordon, J.I.: Going viral: next-generation sequencing applied to phage populations in the human gut. Nat. Rev. Microbiol. **10**, 607–617 (2012)
4. Kay, M.A., Glorioso, J.C., Naldini, L.: Viral vectors for gene therapy: the art of turning infectious agents into vehicles of therapeutics. Nat. Med. **7**, 33–40 (2001)
5. Soto, C.M., Ratna, B.R.: Virus hybrids as nanomaterials for biotechnology. Curr. Opin. Biotechnol. **21**, 426–438 (2010)
6. Aubry, F., Nougairede, A., de Fabritus, L., Querat, G., Gould, E.A., de Lamballerie, X.: Single-stranded positive-sense RNA viruses generated in days using infectious subgenomic amplicons. J. Gen. Virol. **95**, 2462–2467 (2014)
7. Gadea, G., Bos, S., Krejbich-Trotot, P., Clain, E., Viranaicken, W., El-Kalamouni, C., Mavingui, P., Desprs, P.: A robust method for the rapid generation of recombinant Zika virus expressing the GFP reporter gene. Virology **497**, 157–162 (2016)
8. Aubry, F., Nougair'de, A., de Fabritus, L., Piorkowski, G., Gould, E.A., de Lamballerie, X.: ISA-Lation of single-stranded positive-sense RNA viruses from non-infectious clinical/animal samples. PLoS ONE **10**, e0138703 (2015)
9. Du, R., Wang, M., Hu, Z., Wang, H., Deng, F.: An in vitro recombination-based reverse genetic system for rapid mutagenesis of structural genes of the Japanese encephalitis virus. Virol. Sin. **30**, 354–362 (2015)
10. Pu, S.-Y., Wu, R.-H., Yang, C.-C., Jao, T.-M., Tsai, M.-H., Wang, J.-C., Lin, H.-M., Chao, Y.-S., Yueh, A.: Successful propagation of flavivirus infectious cDNAs by a novel method to reduce the cryptic bacterial promoter activity of virus genomes. J. Virol. **85**, 2927–2941 (2011)
11. Santos, J.J., Cordeiro, M.T., Bertani, G.R., Marques, E.T., Gil, L.H.: Construction and characterisation of a complete reverse genetics system of dengue virus type. Memórias do Inst. Oswaldo Cruz **108**, 983–991 (2013)
12. Siridechadilok, B., Gomutsukhavadee, M., Sawaengpol, T., Sangiambut, S., Puttikhunt, C., Chin-inmanu, K., Suriyaphol, P., Malasit, P., Screaton, G., Mongkolsapaya, J.: A simplified positive-sense-RNA virus construction approach that enhances analysis throughput. J. Virol. **87**, 12667–12674 (2013)
13. Steinhauer, D.A., Holland, J.J.: Rapid evolution of RNA viruses. Annu. Rev. Microbiol. **41**, 409–431 (1987)
14. Jenkins, G.M., Rambaut, A., Pybus, O.G., Holmes, E.C.: Rates of molecular evolution in RNA Viruses: a quantitative phylogenetic analysis. J. Mol. Evol. **54**, 156–165 (2002)
15. Goodfellow, I., Chaudhry, Y., Richardson, A., Meredith, J., Almond, J.W., Barclay, W., Evans, D.J.: Identification of a cis-acting replication element within the poliovirus coding region. J. Virol. **74**, 4590–4600 (2000)
16. Goz, E., Tuller, T.: Widespread signatures of local mRNA folding structure selection in four Dengue virus serotypes. BMC Genomics. **16**(Suppl 1), S4 (2015)
17. Alvarez, D.E., Lodeiro, M.F., Ludueña, S.J., Pietrasanta, L.I., Gamarnik, A.V.: Long-range RNA-RNA interactions circularize the dengue virus genome. J. Virol. **79**, 6631–6643 (2005)

18. Guzman, M.G., Halstead, S.B., Artsob, H., Buchy, P., Farrar, J., Gubler, D.J., Hunsperger, E., Kroeger, A., Margolis, H.S., Martínez, E., Nathan, M.B., Pelegrino, J.L., Simmons, C., Yoksan, S., Peeling, R.W.: Dengue: a continuing global threat. Nat. Rev. Microbiol. **8**, S7–S16 (2010)

19. Lauring, A.S., Frydman, J., Andino, R.: The role of mutational robustness in RNA virus evolution. Nat. Rev. Microbiol. **11**, 327–336 (2013)

20. Lauring, A.S., Andino, R., Boone, C., Holden, D., Liu, T.: Quasispecies theory and the behavior of RNA viruses. PLoS Pathog. **6**, e1001005 (2010)

21. Eigen, M.: Viral quasispecies. Sci. Am. **269**, 42–49 (1993)

22. Jenkins, G.M., Pagel, M., Gould, E.A., de A Zanotto, P.M., Holmes, E.C.: Evolution of base composition and codon usage bias in the genus Flavivirus. J. Mol. Evol. **52**, 383–390 (2001)

23. Burns, C.C., Shaw, J., Campagnoli, R., Jorba, J., Vincent, A., Quay, J., Kew, O.: Modulation of poliovirus replicative fitness in hela cells by deoptimization of synonymous codon usage in the capsid region. J. Virol. **80**, 3259–3272 (2006)

24. Gu, W., Zhou, T., Ma, J., Sun, X., Lu, Z.: Analysis of synonymous codon usage in SARS coronavirus and other viruses in the Nidovirales. Virus Res. **101**, 155–161 (2004)

25. Tao, P., Dai, L., Luo, M., Tang, F., Tien, P., Pan, Z.: Analysis of synonymous codon usage in classical swine fever virus. Virus Genes **38**, 104–112 (2009)

26. Jia, R., Cheng, A., Wang, M., Xin, H., Guo, Y., Zhu, D., Qi, X., Zhao, L., Ge, H., Chen, X.: Analysis of synonymous codon usage in the UL24 gene of duck enteritis virus. Virus Genes **38**, 96–103 (2009)

27. Zhou, J.-H., Zhang, J., Chen, H.-T., Ma, L.-N., Liu, Y.-S.: Analysis of synonymous codon usage in foot-and-mouth disease virus. Vet. Res. Commun. **34**, 393–404 (2010)

28. Liu, Y., Zhou, J., Chen, H., Ma, L., Pejsak, Z., Ding, Y., Zhang, J.: The characteristics of the synonymous codon usage in enterovirus 71 virus and the effects of host on the virus in codon usage pattern. Infect. Genet. Evol. **11**, 1168–1173 (2011)

29. Das, S., Paul, S., Dutta, C.: Synonymous codon usage in adenoviruses: influence of mutation, selection and protein hydropathy. Virus Res. **117**, 227–236 (2006)

30. Aragonès, L., Guix, S., Ribes, E., Bosch, A., Pintó, R.M.: Fine-Tuning translation kinetics selection as the driving force of codon usage bias in the hepatitis a virus capsid. PLoS Pathog. **6**, e1000797 (2010)

31. Bull, J.J., Molineux, I.J., Wilke, C.O.: Slow fitness recovery in a codon-modified viral genome. Mol. Biol. Evol. **29**, 2997–3004 (2012)

32. Rocha, E.P.C.: Codon usage bias from tRNA's point of view: redundancy, specialization, and efficient decoding for translation optimization. Genome Res. **14**, 2279–2286 (2004)

33. Dana, A., Tuller, T.: The effect of tRNA levels on decoding times of mRNA codons. Nucleic Acids Res. **42**, 9171–9181 (2014)

34. Goz, E., Mioduser, O., Diament, A., Tuller, T.: Evidence of translation efficiency adaptation of the coding regions of the bacteriophage lambda. DNA Res. **24**(4), 333–342 (2017)

35. Martrus, G., Nevot, M., Andres, C., Clotet, B., Martinez, M.A.: Changes in codon-pair bias of human immunodeficiency virus type 1 have profound effects on virus replication in cell culture. Retrovirology. **10**, 78 (2013)

36. Coleman, J.R., Papamichail, D., Skiena, S., Futcher, B., Wimmer, E., Mueller, S.: Virus attenuation by genome-scale changes in codon pair bias. Science **320**, 1784–1787 (2008)

37. Mueller, S., Coleman, J.R., Papamichail, D., Ward, C.B., Nimnual, A., Futcher, B., Skiena, S., Wimmer, E.: Live attenuated influenza virus vaccines by computer-aided rational design. Nat. Biotechnol. **28**, 723–726 (2010)

38. Tulloch, F., Atkinson, N.J., Evans, D.J., Ryan, M.D., Simmonds, P.: RNA virus attenuation by codon pair deoptimisation is an artefact of increases in CpG/UpA dinucleotide frequencies. Elife. **3**, e04531 (2014)

39. Greenbaum, B.D., Levine, A.J., Bhanot, G., Rabadan, R.: Patterns of evolution and host gene mimicry in influenza and other RNA viruses. PLoS Pathog. **4**, e1000079 (2008)
40. Rima, B.K., McFerran, N.: V: Dinucleotide and stop codon frequencies in single-stranded RNA viruses. J. Gen. Virol. **78**, 2859–2870 (1997)
41. Karlin, S., Doerfler, W., Cardon, L.R.: Why is CpG suppressed in the genomes of virtually all small eukaryotic viruses but not in those of large eukaryotic viruses? J. Virol. **68**, 2889–2897 (1994)
42. Yakovchuk, P., Protozanova, E., Frank-Kamenetskii, M.D.: Base-stacking and base-pairing contributions into thermal stability of the DNA double helix. Nucleic Acids Res. **34**, 564–574 (2006)
43. Cheng, X., Virk, N., Chen, W., Ji, S., Ji, S., Sun, Y., Wu, X.: CpG usage in RNA viruses: data and hypotheses. PLoS ONE **8**, e74109 (2013)
44. Wang, A.H., Hakoshima, T., van der Marel, G., van Boom, J.H., Rich, A.: AT base pairs are less stable than GC base pairs in Z-DNA: the crystal structure of d(m5CGTAm 5CG). Cell **37**, 321–331 (1984)
45. Goz, E., Tuller, T.: Evidence of a direct evolutionary selection for strong folding and mutational robustness within HIV coding regions. J. Comput. Biol. **23**, 641–650 (2016)
46. Domingo, E., Sheldon, J., Perales, C.: Viral quasispecies evolution. Microbiol. Mol. Biol. Rev. **76**, 159–216 (2012)
47. Read, A.F.: The evolution of virulence. Trends Microbiol. **2**, 73–76 (1994)
48. Li, H., Durbin, R.: Fast and accurate short read alignment with Burrows-Wheeler transform. Bioinformatics **25**, 1754–1760 (2009)
49. Garrison, E., Marth, G.: Haplotype-based variant detection from short-read sequencing (2012)
50. McKenna, A., Hanna, M., Banks, E., Sivachenko, A., Cibulskis, K., Kernytsky, A., Garimella, K., Altshuler, D., Gabriel, S., Daly, M., DePristo, M.A.: The genome analysis toolkit: a MapReduce framework for analyzing next-generation DNA sequencing data. Genome Res. **20**, 1297–1303 (2010)
51. Sievers, F., Wilm, A., Dineen, D., Gibson, T.J., Karplus, K., Li, W., Lopez, R., McWilliam, H., Remmert, M., Söding, J., Thompson, J.D., Higgins, D.G.: Fast, scalable generation of high-quality protein multiple sequence alignments using Clustal Omega. Mol. Syst. Biol. **7**, 539 (2011)
52. Lorenz, R., Bernhart, S.H., Höner Zu Siederdissen, C.: ViennaRNA Package 2.0. Algorithms Mol. Biol. **6**, 26 (2011)
53. Wright, F.: The effective number of codons used in a gene. Gene **87**, 23–29 (1990)
54. Kariin, S., Burge, C.: Dinucleotide relative abundance extremes: a genomic signature. Trends Genet. **11**, 283–290 (1995)

ASTRAL-III: Increased Scalability and Impacts of Contracting Low Support Branches

Chao Zhang, Erfan Sayyari, and Siavash Mirarab[(✉)]

Department of Electrical and Computer Engineering,
University of California at San Diego, San Diego, USA
{chz069,esayyari,smirarab}@ucsd.edu

Abstract. Discordances between species trees and gene trees can complicate phylogenetics reconstruction. ASTRAL is a leading method for inferring species trees given gene trees while accounting for incomplete lineage sorting. It finds the tree that shares the maximum number of quartets with input trees, drawing bipartitions from a predefined set of bipartitions X. In this paper, we introduce ASTRAL-III, which substantially improves on ASTRAL-II in terms of running time by handling polytomies more efficiently, exploiting similarities between gene trees, and trimming unnecessary parts of the search space. The asymptotic running time in the presence of polytomies is reduced from $O(n^3 k|X|^{1.726})$ for n species and k genes to $O(D|X|^{1.726})$ where $D = O(nk)$ is the sum of degrees of all *unique* nodes in input trees. ASTRAL-III enables us to test whether contracting low support branches in gene trees improves the accuracy by reducing noise. In extensive simulations and on real data, we show that removing branches with *very* low support improves accuracy while overly aggressive filtering is harmful.

Keywords: Phylogenomics · Incomplete lineage sorting. ASTRAL

1 Introduction

Reconstructing species phylogenies from a collection of input trees each inferred from a different part of the genome is becoming the standard practice in phylogenomics (e.g., [1–5]). This two-step approach stands in contrast to concatenation [6], where all of the sequences are combined into a supermatrix and analyzed in one maximum likelihood analysis. The two-step approach promises to effectively account for discordances between gene trees and the species tree [7] (but see recent literature for ongoing debates [8–11]) and is more efficient than statistical co-estimation of gene trees and the species tree [12] or site-based estimation of the species tree [13]. Among several causes of gene tree discordance [14], incomplete lineage sorting (ILS) is believed to be ubiquitous [15] and is extensively studied. ILS is typically modeled by the multi-species coalescent model (MSCM) [16,17], where branches of the species tree represent populations, and lineages are allowed to coalesce inside each branch; lineages that fail to coalesce at the root of each branch are moved to the parent branch.

© Springer International Publishing AG 2017
J. Meidanis and L. Nakhleh (Eds.): RECOMB CG 2017, LNBI 10562, pp. 53–75, 2017.
DOI: 10.1007/978-3-319-67979-2_4

Several methods are proposed to infer a species tree from a collection of input trees (even though these trees need not be inferred from functional genes, following the conventions of the field, we will call them "gene trees"). Examples of summary methods include MP-EST [18], NJst [19], DISTIQUE [20], and STAR [21], which only use the topology of the input gene trees, and GLASS [22] and STEAC [21], which also uses the input branch lengths. While most methods need rooted gene trees as input, NJst and DISTIQUE can take unrooted input. These methods are all proved statistically consistent under the MSCM when the input gene trees are error-free, but no summary method is proved consistent when input trees are inferred from sequence data [23].

One of the statistically consistent methods under the MSCM is ASTRAL [24], which takes as input a collection of unrooted gene tree topologies and produces an unrooted species tree. ASTRAL uses dynamic programming to find the tree that shares the maximum number of induced quartet topologies with the collection of input gene trees. Since this problem is NP-Hard [25], ASTRAL solves a constrained version of the problem exactly, where the search space is limited to a predefined set of bipartitions X. In ASTRAL-I, the set X is the collection of all bipartitions in input gene trees. Showing that this space is not always large enough, ASTRAL-II [26] uses several heuristics to further augment the search space. Using the fact that for unrooted quartet trees the species tree always matches the most likely gene tree [27], ASTRAL is proved statistically consistent, even when solving the constrained problem, and its accuracy has been established in simulations [20,24,26,28,29]. ASTRAL-II has running time $O(nk|X|^2)$ for n species and k binary genes. Finally, ASTRAL has the ability to compute branch lengths in coalescent units [14] and a measure of branch support called local posterior probability [30]. Perhaps most importantly, ASTRAL and ASTRAL-II have been adopted by the community as one of the main methods of performing phylogenomics, and many biological analyses have adopted them.

ASTRAL-II has several shortcomings, some of which we address here by introducing ASTRAL-III. While ASTRAL-II can analyze datasets of 1,000 species and 1,000 genes on average in a day, ASTRAL-II has trouble scaling to many tens of thousands of input trees. Datasets with more than ten thousands genomic loci are already available (e.g., [3]) and with the increase in genome sequencing, more will be available in future. Moreover, being able to handle large numbers of input trees enables using multiple trees per locus (e.g., a Bayesian sample) as input to ASTRAL. The limited scalability of ASTRAL with k is because of a $\Theta(nk)$ factor in the running time that corresponds to scoring a potential node in the species tree against all nodes of the input gene trees. This computation does not exploit similarities between gene trees, a shortcoming that we fix in ASTRAL-III. Moreover, while ASTRAL-II can handle polytomies in input gene trees, in the presence of polytomies of maximum degree d_m, its running time inflates to $O(d_m^3 k|X|^2) = O(n^3 k|X|^2)$, which quickly becomes prohibitive for input trees with polytomies of large degrees. ASTRAL-III uses a mathematical trick to enable scoring of gene tree polytomies in time similar to binary nodes.

The ability to handle large polytomies in input gene trees is important for two reasons. On the one hand, some of the conditions that are conducive to ILS, namely shallow trees with many short branches, are also likely to produce gene sequence data that are identical between two species. A sensible gene tree (e.g., those produced by FastTree [31]) would leave the relationship between identical sequences unresolved (tools such as RAxML that output a random resolution take care to warn the user about such input data). On the other hand, all summary methods, including ASTRAL, are sensitive to gene tree estimation error [26,32–36]. One way of dealing with gene tree error, previously studied in the context of minimizing deep coalescence [37], is to contract low support branches in gene trees and use these unresolved trees as input to the summary method. While earlier studies found no evidence that this approach helps ASTRAL when the support is judged by SH-like FastTree support [26], no study has tested this approach with bootstrap support values. We will for the first time evaluate the effectiveness of contracting low support branches and show that conservative filtering of very low support branches does, in fact, help the accuracy. We note that the main competitors to ASTRAL, namely NJst [19] and its fast implementation, ASTRID [38], are not able to handle polytomies in input gene trees. ASTRAL-III makes it efficient to use unresolved gene trees as input to the species tree. Empirically, we observe that ASTRAL-III improves the running time compared to ASTRAL-II by a factor of 3X-4X for binary trees with large numbers of genes. Moreover, ASTRAL-III finishes on a dataset of 5,000 species and 500 genes in 18–30 h (24 on average). The ASTRAL-III software is publicly available at https://github.com/smirarab/ASTRAL.

2   Background and Notation

2.1   Notations and Definitions

We denote the set of species by L and let $n = |L|$. Let G be the set of k input gene trees. The set of quartet trees induced by any tree t is denoted by $Q(t)$. We refer to any subset of L as a cluster and refer to clusters with cardinality one as singletons. We define a partition as a set of clusters that are pairwise mutually exclusive (note that we abuse the term here, as the union of all clusters in a partition need not give the complete set). A bipartition (tripartition) is a partition with cardinality two (three); a partition with cardinality at least four corresponds to a polytomy and is referred to as a polytomy in this paper. Let X (the constraint bipartition set) be a set of clusters such that for each $A \in X$, we also have $L - A \in X$. We use Y to represent the set of all tripartitions examined in the ASTRAL dynamic programming:

$$Y = \{(A', A - A', L - A) | A' \subset A, A \in X, A' \in X, A - A' \in X\}.$$

We use $N(g)$ to represent the set of partitions correspondent to internal nodes in the gene tree g. We use E to denote the set of unique partitions and the number of times they appear in G:

$$E = \{(M, \sum_{g \in G} |N(g) \cap \{M\}|)|M \in N(g), g \in G\} \tag{1}$$

and we define D as the sum of the cardinalities of unique partitions in gene trees:

$$D = \sum_{(M,c) \in E} |M|. \tag{2}$$

Finally, we use $[d]$ to represent the set $\{1, 2 \ldots, d\}$.

2.2 Background on ASTRAL-I and ASTRAL-II

The problem addressed by ASTRAL is:

Given: a set G of input gene trees
Find: find the species tree t that maximizes $\sum_{g \in G} |Q(g) \cap Q(t)|$.

Lanford and Scornavacca recently proved this problem is NP-hard [25]. ASTRAL solves a constrained version of this problem where a set of clusters X restricts bipartitions that the output species tree may include (note $\forall A \in X : L - A \in X$).

To solve the constrained version, ASTRAL uses a dynamic programming method with the following recursive relation to obtain the optimal tree.

$$V(A) = \max_{(A'|A-A'|L-A) \in Y} V(A') + V(A - A') + w(A'|A - A'|L - A)$$

where the function $w(T)$ scores each tripartition $T = (A|B|C)$ against each node in each input gene tree. Let partition $M = (M_1|M_2|\ldots|M_d)$ represent an internal node of degree d in a gene tree. The overall contribution of T to the score of any species tree that includes T is:

$$w(T) = \sum_{g \in G} \sum_{M \in N(g)} \frac{1}{2} QI(T, M) \tag{3}$$

where, defining $a_i = |A \cap M_i|$, $b_i = |B \cap M_i|$, and $c_i = |C \cap M_i|$, we have:

$$QI((A|B|C), M) = \sum_{i \in [d]} \sum_{j \in [d] - \{i\}} \sum_{k \in [d] - \{i,j\}} \frac{a_i + b_j + c_k - 3}{2} a_i b_j c_k. \tag{4}$$

As previously proved [24], $QI(T, M)$ computes twice the number of quartet trees that are going to be shared between any two trees if one includes only T and the other includes only M. ASTRAL-II requires $\Theta(d^3)$ time for computing $QI(.)$, making the overall running time $O(n^3 k|Y|)$ with polytomies of unbounded degrees or $O(nk|Y|)$ in the absence of polytomies.

3 ASTRAL-III Algorithmic Improvements

Noting trivially that $|Y| < |X|^2$, the previously published running time analysis of ASTRAL-II was $O(nk|X|^2)$ for binary gene trees and $O(n^3 k|X|^2)$ for input trees with polytomies. A recent result by Kane and Tao [39] (motivated by the question raised in analyzing the ASTRAL algorithm) indicates that $|Y| \leq |X|^{3/\log_3(27/4)}$. This result immediately gives a better upper bound:

Corollary 1. *ASTRAL-II runs in $O(nk|X|^{1.726})$ and $O(n^3k|X|^{1.726})$, respectively, with and without polytomies in gene trees.*

ASTRAL-III further improves this running time using three new features:

1. A new way of handling polytomies is introduced to reduce the running time for scoring a gene tree to $O(n)$, instead of $O(n^3)$, in the presence of polytomies, which reduces the total running time to $O(nk|X|^{1.726})$ irrespective of the gene tree resolution.
2. A polytree is used to represent gene trees, and this enables an algorithm that reduce the overall running time from $O(nk|X|^{1.726})$ to $O(D|X|^{1.726})$.
3. An A*-like algorithm is used to trim parts of the dynamic programming DAG.

In addition to these running time improvements, ASTRAL-III changes parameters of heuristics described in ASTRAL-II [26] to expand the size of set X for gene trees that include polytomies. We next describe each improvement in detail.

3.1 Efficient Handling of Polytomies

Recall that ASTRAL-II uses Eq. 4 to score a tripartition against a polytomy of size d in $\Theta(d^3)$ time. We now show this can be improved.

Lemma 1. *Let $QI(T, M)$ be twice the number of quartet tree topologies shared between an unrooted tree that only includes a node corresponding to the tripartition $T = (A|B|C)$ and another tree that includes only a node corresponding to a partition $M = (M_1|M_2|\ldots|M_d)$ of degree d; then, $QI(T, M)$ can be computed in time $\Theta(d)$.*

Proof. In $\Theta(d)$ time, we can compute:

$$S_a = \sum_{i \in [d]} a_i \quad \text{and} \quad S_{a,b} = \sum_{i \in [d]} a_i b_i \tag{5}$$

where $a_i = |A \cap M_i|$ and $b_i = |B \cap M_i|$; ditto for S_b, S_c, $S_{a,c}$ and $S_{b,c}$. Equation 4, as proved before [26], computes twice the number of quartet tree topologies shared between an unrooted tree with internal node T and another tree with one internal node M. Equation 4 can be rewritten using these intermediate sums as:

$$QI((A|B|C), M) = \sum_{i \in [d]} \binom{a_i}{2}((S_b - b_i)(S_c - c_i) - S_{b,c} + b_i c_i)$$

$$+ \sum_{i \in [d]} \binom{b_i}{2}((S_a - a_i)(S_c - c_i) - S_{a,c} + a_i c_i) \tag{6}$$

$$+ \sum_{i \in [d]} \binom{c_i}{2}((S_a - a_i)(S_b - b_i) - S_{a,b} + a_i b_i)$$

(the derivation is given in the Appendix A). Computing Eq. 6 instead of Eq. 4 clearly reduces the running time to $\Theta(d)$ instead of $\Theta(d^3)$. □

ASTRAL needs to score each of the $|Y|$ tripartitions considered in the dynamic programming against each internal node of each input gene tree. The sum of degrees of k trees on n leaves is $O(nk)$ (since that sum can never exceed the number of bipartitions in gene trees) and thus:

Corollary 2. *Scoring a tripartition (i.e., computing w) can be done in $O(nk)$.*

3.2 Gene Trees as a Polytree

ASTRAL-II scores each dynamic programming tripartition against each individual node of each gene tree. However, nodes that are repeated in several genes need not be recomputed. Recalling the definitions of E and D (Eqs. 1 and 2),

Lemma 2. *The score of a tripartition $T = (A|B|C)$ against all gene trees (i.e., the $w(T)$ score) can be computed in $\Theta(D)$.*

Proof. In ASTRAL-III, we keep track of nodes that appear in multiple trees. This enables us to reduce the total calculation by using multiplicities:

$$w(T) = \sum_{(M,c)\in E} c \times QI(T, M). \tag{7}$$

We achieve this in two steps. In the first step, for each distinct gene tree cluster W, we compute the cardinality of the intersection of W and sets A, B, and C once using a depth first search with memoization. Let $children(W)$ denote the set of children of W in an arbitrarily chosen tree $g \in G$ containing W. Then, we have the following recursive relation:

$$|W \cap A| = \sum_{Z \in children(W)} |Z \cap A| \tag{8}$$

(ditto for $|W \cap B|$ and $|W \cap C|$). All such intersection values can be computed in a post-order traversal of a polytree. In this polytree, all unique clusters in the gene trees are represented as vertices and parent-child relations are represented as edges; note that when a cluster has different children in two different input trees, we arbitrary choose one set of children and ignore the others. The polytree will include no more than D edges; thus, the time complexity of traversing this polytree and computing Eq. 8 for all nodes is $\Theta(D)$. Once all intersections are computed, in the second step, we simply compute the sum in Eq. 7. Each $QI(.)$ computation requires $\Theta(d)$ by Lemma 1. Recalling that $D = \sum_{(M,c)\in E} |M|$, it is clear that computing Eq. 7 requires $\Theta(D)$. Therefore, both steps can be performed in $\Theta(D)$.   □

Theorem 1. *The ASTRAL-III running time is $O(D|X|^{1.726})$ for both binary and unresolved gene trees.*

Proof. By results of Kane and Tao [39], the size of the set Y is $O(|X|^{1.726})$, and for each element in Y, by Lemma 2, we require $O(D)$ to compute the weights, regardless of the presence or absence of polytomies. The running time of ASTRAL is dominated by computing the weights [26]. Thus, the overall running time is $O(D|Y|) = O(D|X|^{1.726})$.

3.3   Trimming of the Dynamic Programming

Our last feature does not improve theoretical running time but can result in some improvements in the experimental running time. Our main insight is that $U(A) = \frac{w(A|A|L)}{2} - \frac{w(A|A|A)}{3}$ is an upperbound of $V(A)$ (see the Appendix A for proof). Since $V(A) \leq U(A)$, for any $(A'|A-A'|L-A') \in Y$ and $(A''|A-A''|L-A'') \in Y$, we no longer need to recursively compute $V(A'')$ and $V(A - A'')$ when:

$$
\begin{aligned}
U(A'') + U(A - A'') + w(A''|A - A''|L - A) \leq \\
V(A') + V(A - A') + w(A'|A - A'|L - A)
\end{aligned}
\tag{9}
$$

Thus, in ASTRAL-III we trim the DAG of the memoized recursive dynamic programming when this calculation indicates that a path has no chance of improving the final score. To heuristically improve the efficiency of this approach, we order all $(A'|A - A'|L - A) \in Y$ according to $U(A') + U(A - A') + w(A'|A - A'|L - A)$.

4   Experimental Setup

Using simulation studies and on real data, we study two research questions:

RQ1: Can contracting low support branches improve the accuracy of ASTRAL?
RQ2: How does ASTRAL-III running time compare to ASTRAL-II for large
  polytomies and many gene trees?

Note that addressing RQ1 in a scalable fashion is made possible only through the running time improvements of ASTRAL-III.

4.1   Datasets

Avian Biological Dataset: Neoaves have gone through a rapid radiation, and therefore, have extremely high levels of ILS [3]. A dataset of 48 whole-genomes was used to resolve this rapid radiation [3]. MP-EST run on 14,446 gene trees (exons, introns, and UCEs) produced a tree that conflicted with strong evidence from the literature and other analyses on the dataset (e.g., the Passerimorphae/-Falcons/Seriemas grade was not recovered). This motivated the development of the statistical binning method to reduce the impacts of gene tree error [32,33]. MP-EST run on binned gene trees produced results that were largely congruent with the concatenation and other analyses. Here, we test if simply contracting low support branches of gene trees produces a tree that is congruent with other analyses and the literature. This analysis is made possible because ASTRAL-III can handle datasets with a large number of polytomies and large k efficiently.

Simulated Avian-Like Dataset: We use a simulated dataset that was previously used in the statistical binning paper [32] to emulate the biological avian dataset. Since estimating the true branch lengths in coalescent units are hard, three versions of this dataset are available: 1X is the default version, whereas

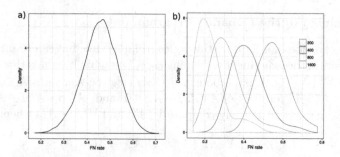

Fig. 1. Properties of the S100 dataset. (a) The density plot of the amount of true gene discordance measured by the FN rate between the true species tree and the true gene trees. (b) The density plot of gene tree estimation error measured by FN rate between true gene trees and estimated gene trees for different set of sequence lengths.

0.5X divides each branch length in half (increasing ILS) and 2X multiplies them by 2 (reducing ILS). The amount of true discordance (ILS), measured by the average RF distances between true species tree and true gene trees, is moderate at 0.35 for 2X, high at 0.47 for 1X, and very high at 0.59 for 0.5X. Moreover, to study the impact of gene tree estimation error, sequence lengths were varied to create four conditions: very high error with 250 bp alignments (0.67 RF distance between true gene trees and estimated gene trees), high error with 500 bp (0.54 RF), medium error with 1000 bp (0.39 RF) and moderate error with 1500 bp (0.30 RF). We use 1000 gene trees, and 20 replicates per condition. Gene trees are estimated using RAxML [40] with 200 replicates of bootstrapping.

SimPhy-Homogeneous (S100): We simulated 50 replicates of a 101-taxon dataset using SimPhy [41] under the MSCM, where each replicate has a different species tree. In order to generate the species trees, we used the birth-only process with birth rate 10^{-7}, fixed haploid effective population size of 400K, and the number of generations sampled from a log-normal distribution with mean 2.5 M. For each replicate, 1000 true gene trees are simulated under the MSCM (the exact simulation commands are given in Appendix B and parameters are shown in Table 2). The amount of ILS, measured by the false-negative (FN) rate between true species trees and true gene trees, mostly ranged between 0.3 and 0.6 with an average of 0.46 (Fig. 1). We use Indelible [42] to simulate the nucleotide sequences along the gene trees using the GTR evolutionary model [43] with 4 different fixed sequence lengths: 1600, 800, 400, and 200 bp (Table 2). We then use FastTree2 [31] to estimate both ML and 100 bootstrapped gene trees under the GTR model for each gene of each replicate (> 2000200 runs in total). Gene tree estimation error, measured by the FN rate between the true gene trees and the estimated gene trees, depended on the sequence length as shown in Fig. 1 (0.55, 0.42, 0.31, and 0.23 on average for 200 bp, 400 bp, 800 bp, and 1600 bp, respectively). We sample 1000, 500, 200, or 50 genes to generate datasets with varying numbers of gene trees.

4.2 Methods and Evaluation

We compare ASTRAL-III (version 5.2.5) to ASTRAL-II (version 4.11.1) in terms of running time. To address RQ1, we draw the bootstrap support values on the ML gene trees using the newick utility package [44]. We then contract branches with bootstrap support up to a threshold (0, 3, 5, 7, 10, 20, 33, 50, and 75%,) using the newick utility and use these contracted gene trees as input to ASTRAL. Together with the original set, this creates 10 different ways to run ASTRAL.

To measure the accuracy of estimated species trees, we use False Negative (FN) rate. Note that in all our species tree comparisons, FN rate is equivalent to normalized Robinson-Foulds (RF) [45] metric, since the ASTRAL species trees are fully resolved. All running times are measured on a cluster with servers with Intel(R) Xeon(R) CPU E5-2680 v3 @ 2.50 GHz; each run was assigned to a single process, sharing cache and memory with other jobs.

5 Results

5.1 Impact of Contracting Low Support Branches on Accuracy

We investigate the impact of contracting branches with low support (RQ1) on our two simulated datasets (avian and S100) and on the real avian dataset.

S100: On this dataset, contracting *very* low support branches in most cases improves the accuracy (Table 1 and Fig. 2); however, the excessive removal of

Table 1. Species tree error (FN ratio) for all model conditions of the S100 dataset, with true gene trees (*true*), no filtering (*non*), and all filtering thresholds (*columns*).

Genes	Alignment	true	non	0	3	5	7	10	20	33	50	75
50	200 bp	7.0	18.0	**16.1**	16.5	16.3	16.5	16.9	17.3	19.9	23.5	31.6
50	400 bp		14.2	13.5	13.3	13.3	13.4	**13.2**	14.1	15.0	16.6	21.0
50	800 bp		12.3	11.9	11.7	11.6	11.3	11.4	**11.2**	12.2	12.9	15.9
50	1600 bp		10.6	10.4	10.4	10.3	**10.1**	10.3	10.4	10.5	10.8	12.4
200	200 bp	3.7	11.6	10.5	**10.3**	10.8	10.6	10.8	10.7	12.5	15.3	21.5
200	400 bp		9.2	8.4	**8.3**	8.3	8.3	8.4	8.6	9.2	10.1	13.6
200	800 bp		7.4	7.3	7.2	7.1	**7.0**	**7.0**	7.1	7.4	7.7	9.1
200	1600 bp		6.1	6.4	6.3	6.3	6.3	6.2	**6.1**	6.6	6.5	7.4
500	200 bp	2.4	9.9	8.8	8.8	8.7	**8.5**	**8.5**	8.6	9.8	11.8	16.7
500	400 bp		7.3	7.1	6.6	6.6	6.7	**6.5**	6.6	7.0	8.0	10.8
500	800 bp		5.6	5.5	5.5	5.6	5.4	5.4	**5.3**	**5.3**	5.7	6.6
500	1600 bp		4.8	4.5	4.6	4.6	4.5	4.6	4.6	**4.4**	4.8	5.3
1000	200 bp	1.5	9.1	8.0	7.6	7.3	7.3	**7.1**	7.6	8.6	10.2	13.6
1000	400 bp		6.9	6.0	5.8	5.9	5.7	**5.6**	**5.6**	5.8	6.8	8.5
1000	800 bp		5.5	5.0	5.0	4.9	5.0	4.9	4.7	**4.6**	4.8	5.8
1000	1600 bp		4.1	4.2	4.1	4.0	4.0	**3.9**	4.1	4.1	4.1	4.5

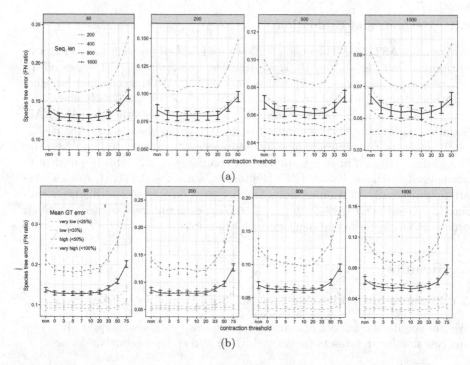

(a)

(b)

Fig. 2. The error in species trees estimated by ASTRAL-III on the S100 dataset given $k = 50, 200, 500,$ or 1000 genes (*boxes*) and with full FastTree gene trees (*non*) or trees with branches with $\leq \{0, 3, 5, 7, 10, 20, 33, 50\}\%$ support contracted (*x-axis*). Average FN error and standard error bars are shown for all 50 replicates with the four alignment lengths combined (*black solid line*); average FN error broken down by alignment length (a) or gene tree error (b) is also shown (*dashed colored lines*). In (b) we divide the replicates based on their average gene tree error (normalized RF) into four categories: $[0, \frac{1}{4}], (\frac{1}{4}, \frac{1}{3}], (\frac{1}{3}, \frac{1}{2}], (\frac{1}{2}, 1]$. (Color figure online)

branches with high, moderate or occasionally even low support degrades the accuracy. The threshold where contracting starts to become detrimental depends on the condition, especially the number of gene trees and the alignment length.

As the number of genes increases, the optimal threshold for contracting also tends to increase. Combining all model conditions, the error continues to drop until a 10% contracting threshold with 1000 genes, whereas no substantial improvement is observed after contracting branches with 0% support for 50 genes. The alignment length and gene tree error also impact the effect of contraction. For short alignments (200 bp) and 1000 genes, contracting branches with up to 10% support reduces the species tree error by 21% (from 8.9% with no contraction to 7.0%). As alignment length grows (and gene tree error decreases), benefits of gene tree contraction diminish, so that with 1600 bp genes, the reduction in error is merely from 4.1% to 3.7%. While aggressive filtering at 33% or higher sometimes increases the error compared to no filtering, filtering at 10% is either neutral or beneficial on average for all conditions in this dataset.

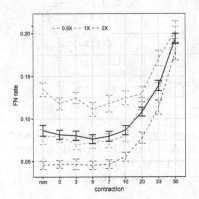 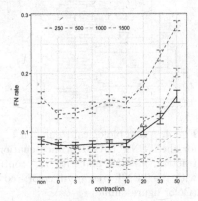

Fig. 3. Species trees error for ASTRAL-III on the avian dataset given $k = 1000$ genes with (*left*) fixed sequence lengths $= 500$ and varying levels of ILS, or (*right*) fixed ILS (1X) and varying sequence length, in each case both with full FastTree gene trees (*non*) or trees with branches with $\leq \{0, 3, 5, 7, 10, 20, 33, 50\}\%$ support contracted (*x-axis*). Average FN error and standard error bars are shown for all 20 replicates (*black solid line*) and also for each model condition separately (*dashed color lines*). (Color figure online)

Avian-Like Simulations: On the avian dataset, overall, contracting low support branches helps accuracy marginally, but the extent of improvements depends on the model condition (Fig. 3). We first fix the sequence length to 500 bp and allow the amount of ILS to change (e.g., from moderate with 2X to very high with 0.5X). With moderate ILS (2X), we see no improvements as a result of contracting low support branches, perhaps because the average error is below 5% even with no contraction. Going to high and very high ILS, we start to see improvements with contracted gene trees. For example, removing branches of up to 5% support reduces the error from 13% to 11% with 0.5X, and from 8% to 7% for the 1X condition. Just like the S100 dataset, aggressive filtering reduces the accuracy, but here, thresholds of 20% and higher seem to be detrimental. When ILS is fixed to 1X and sequence length is varied (Fig. 3), contracting is helpful mostly with short sequences (e.g., 250 bp). With longer sequences, where gene tree estimation error is low, little or no improvement in accuracy is obtained. The best accuracy is typically observed by contracting at 0–5%. Overall, the gains in accuracy comparing no contraction to contraction at 0% are statistically significant with p-value 0.042 (are close to significant with p-values 0.087 and 0.058 for 3% and 5% thresholds) according to a two-way ANOVA test with the sequence length and ILS levels as extra variables. Irrespective of significance, the improvements are not large on this dataset.

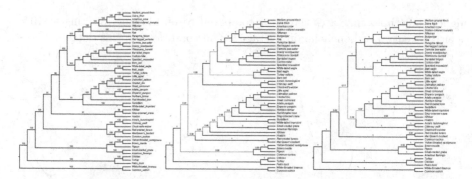

Fig. 4. Avian biological results on 14,446 unbinned gene trees. Species trees are shown for the TENT RAxML concatenation tree [3] (*left*), ASTRAL-III tree with no contraction (*middle*), and with 33% contraction (*right*). ASTRAL-III branches conflicting with the TENT tree are in color; red indicates disagreement with strong evidence [3]. (Color figure online)

Avian Biological Dataset: On the avian dataset with 14 K genes, ASTRAL-III managed to complete with 0%, 33%, 50%, and 75% thresholds in 48 h. Results on the runs that did finish are very interesting. The ASTRAL-III tree with no contraction had 11 and 9 branch differences respectively with the TENT and the binned MP-EST analyses from the original paper [3]. Some of these differences were on strong results (Fig. 4) from the avian dataset (e.g., the Columbea group). After contracting branches below 33% BS, the ASTRAL-III tree had only 4 and 3 branch differences, respectively with TENT and the binned MP-EST trees; these differences among the branches that were deemed unresolved by Jarvis *et al.* and changed among their analyses as well [3]. ASTRAL-III obtained on collapsed gene trees agreed with all major new findings of Jarvis et al. [3]. For example, at 33% filtering, the novel Columbea group was corroborated, whereas the unresolved tree completely missed this important clade (Fig. 4).

5.2   Running Time Improvements

The Impact of the Number of Genes (k): We evaluate the improvement of ASTRAL-III compared to ASTRAL-II on the avian simulated dataset, changing the number of genes from 2^8 to 2^{16}. We allow each replicate run to take up to two days. ASTRAL-II is not able to finish on the dataset with $k = 2^{16}$, while ASTRAL-III finishes on all conditions. ASTRAL-III improves the running time over ASTRAL-II and the extent of the improvement depends on k (Fig. 5). With 1000 genes or more, there is at least a 2.7X improvement. With 2^{13} genes, the largest value where both versions could run, ASTRAL-III finishes on average four times faster than ASTRAL-II (190 versus 756 min). Moreover, fitting a line to the average running time in the log-log scale graph reveals that on this dataset, the running time of ASTRAL-III on average grows as $O(k^{2.04})$, which is better than that of ASTRAL-II at $O(k^{2.27})$, and both are better than the theoretical worst case, which is $O(k^{2.726})$.

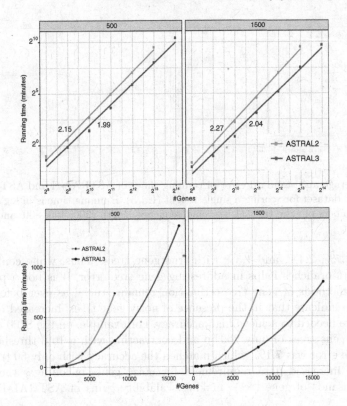

Fig. 5. Average running time of ASTRAL-II versus ASTRAL-III on the avian dataset with 500 bp or 1500 bp alignments with varying numbers of gens (k), shown both in log-log scale (*top*) and normal scale (*bottom*). A line (*top*) or a LOESS curve (*bottom*) is fit to the data points. ASTRAL-II could not finish on 2^{14} genes in the allotted 48 h time. Line slopes are shown for the log-log plot. Averages are over 4 runs.

The Impact of Polytomies: ASTRAL-III has a clear advantage compared to ASTRAL-II with respect to the running time when gene trees include polytomies (Fig. 6). Since ASTRAL-II and ASTRAL-III can potentially weight different numbers of tripartitions, we show the running time per weight calculation (i.e., Eq. 3). As we contract low support branches and hence increase the prevalence of polytomies, the weight calculation time quickly grows for ASTRAL-II, whereas, in ASTRAL-III, the weight calculation time remains flat, or even decreases.

6 Discussion

Comparison to True Gene Trees: Although we observed improvements in the tree accuracy with contracting low support branches, the gap between performance on true gene trees and estimated gene trees remains wide (Table 1). On the S100 dataset, respectively for 50, 200, 500, and 1000 genes, the average error with 1600 bp gene trees were 10.1%, 6.1%, 4.4%, and 3.9% compared

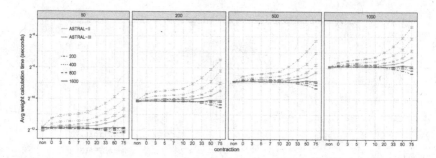

Fig. 6. Average and standard error of running times of ASTRAL-II and ASTRAL-III on the S100 dataset for scoring a single weight (Eq. 3). Running time is in log scale for varying numbers of gene trees (*boxes*) and sequence length (200, 400, 800, and 1600).

to 7.0%, 3.7%, 2.4%, and 1.5% with true gene trees. Thus, while contracting low support branches helps in addressing gene tree error, it is not a panacea. Improved methods of gene tree estimation remain crucial as ever before. Our results also indicate that in the presence of noisy gene trees, increased numbers of genes are needed to achieve high accuracy. For example, on the S100 dataset, with 1000 gene trees of only 200 bp and contracting with a 10% threshold, the species tree error was 7.1%, which matched the accuracy with only 50 true gene trees. The increased data requirement for noisy genes encourages the use of many thousands of gene trees, making scalability gains of ASTRAL-III more relevant.

Arbitrary Resolutions and 0% Filtering: Interestingly, in most datasets, the most substantial improvements were observed when only 0% BS branches were removed, and one can assume that such branches are essentially resolved randomly. As the use of Ultra Conserved Elements [9] continues to gain in popularity, instances where two or more taxa have identical sequences in some genes will become more prevalent. Many tree estimation methods generate binary trees even under such conditions. Removing branches that are arbitrarily resolved make sense and, as our results indicate, will improve accuracy.

Statistical Consistency: While removing branches with low support can improve the accuracy empirically, its theoretical justification is less clear. In principle, branches that have low support are not necessarily expected to have a random distribution among gene trees. Thus, while the empirical results could support the use of (conservative) filtering, it must be understood that the resulting procedure may introduce small biases. Whether ASTRAL remains statistically consistent given contracted gene trees should be studied in future work.

Other Strategies: Beyond removing branches with low bootstrap support, several other strategies could be employed. Our previous results [24] indicate that simply using a concatenation of all bootstrap gene trees as input to ASTRAL

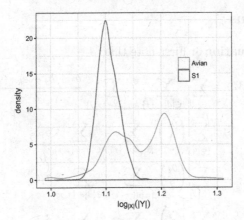

Fig. 7. The density plots of $\log_X |Y|$ across all ASTRAL-III runs reported in this paper. Size of the dynamic programming space Y is never above $|X|^{1.312}$ here.

increases error, perhaps because of the increased noise in bootstrap replicates [30,34]. However, it is possible that a fixed sized sample from a Bayesian estimate of each gene tree distribution would improve accuracy.

$|Y|$: The ASTRAL-III running time analysis of $O(|X|^{1.726})$ is based on the fact that $|Y| \leq |X|^{1.726}$ [39]. However, this upper bound is for specialized formations of the set X. Empirically, as the size of set X increases, the size of $|Y|$ in ASTRAL-III does not increase as fast as the worst-case scenario implies. Across all of our ASTRAL-III runs in this paper, $|Y|$ ranged mostly between $|X|^{1.05}$ and $|X|^{1.25}$, with an average of $|X|^{1.11}$ (Fig. 7). Thus, the average running time of ASTRAL seems closer to $O(D|X|^{1.1})$, though, the exact value depends on the dataset. The size of X is not currently controlled to be a polynomial function of n and k, but such constraints can be imposed in future versions of ASTRAL.

Large n: To assess scalability limits of ASTRAL-III, we tested it on 4 replicates of a dataset with 5,000 species and 500 true gene trees (simulation procedure described in Appendix B). ASTRAL-III took between 18 and 30 h to run on this dataset (24 h on average). We also ran ASTRAL-III on similar datasets with 1,000 and 2,000 species. Average over our four replicates, ASTRAL running time increases as $O(n^{1.9})$. Our attempts to analyze 10K species within 72 h failed. Future work should reproduce these results with more replicates.

Acknowledgments. This work was supported by the NSF grant IIS-1565862 to SM and ES. Computations were performed on the San Diego Supercomputer Center (SDSC) through XSEDE allocations, which is supported by the NSF grant ACI-1053575.

A Derivations

Derivation of Equation 6: First note that:

$$
QI((A|B|C), M) = \sum_{i\in[d]} \sum_{j\in[d]-\{i\}} \sum_{k\in[d]-\{i,j\}} \frac{a_i + b_j + c_k - 3}{2} a_i b_j c_k
$$

$$
= \sum_{i\in[d]} \binom{a_i}{2} \sum_{j\in[d]-\{i\}} \sum_{k\in[d]-\{i,j\}} b_j c_k
$$

$$
+ \sum_{i\in[d]} \binom{b_i}{2} \sum_{j\in[d]-\{i\}} \sum_{k\in[d]-\{i,j\}} a_j c_k \tag{10}
$$

$$
+ \sum_{i\in[d]} \binom{c_i}{2} \sum_{j\in[d]-\{i\}} \sum_{k\in[d]-\{i,j\}} a_j b_k .
$$

Now, we note that:

$$
\sum_{j\in[d]-\{i\}} \sum_{k\in[d]-\{i,j\}} b_j c_k
$$

$$
= \sum_{j\in[d]-\{i\}} b_j \sum_{k\in[d]-\{i,j\}} c_k
$$

$$
= \sum_{j\in[d]-\{i\}} b_j (S_c - c_i - c_j) \tag{11}
$$

$$
= -b_i(S_c - c_i - c_i) + \sum_{j\in[d]} b_j(S_c - c_i - c_j)
$$

$$
= 2b_i c_i - S_c b_i + S_b S_c - S_b c_i - S_{b,c}
$$

$$
= (S_b - b_i)(S_c - c_i) - S_{b,c} + b_i c_i
$$

Replacing this (and similar calculations for other terms) in Eq. 10 directly gives us the Eq. 6:

$$
QI((A|B|C), M) = \sum_{i\in[d]} \binom{a_i}{2} ((S_b - b_i)(S_c - c_i) - S_{b,c} + b_i c_i)
$$

$$
+ \sum_{i\in[d]} \binom{b_i}{2} ((S_a - a_i)(S_c - c_i) - S_{a,c} + a_i c_i) \tag{12}
$$

$$
+ \sum_{i\in[d]} \binom{c_i}{2} ((S_a - a_i)(S_b - b_i) - S_{a,b} + a_i b_i)
$$

Derivation of the Upperbound $U(Z)$: In ASTRAL, $V(Z)$ denotes the total contribution to the support of the best rooted tree T_Z on taxon set Z, where each quartet tree in the set of input gene trees contributes 0 if it conflicts with T_Z or only intersects it with one leaf, and otherwise contributes 1 or 2, depending on the number of nodes in T_Z it maps to. Let $U(Z)$ be the sum of max possible support of each quartet tree in the gene trees with respect to any resolution T_Z of set Z, allowing the resolution to change for each gene tree. In other words, let $Q(Z)$ be the set of quartets that would be resolved one way or another in any resolution of Z, and note that these are quartets that include two or leaves in Z; then, $U(Z)$ is the number of resolved gene tree quartets that would match *some* resolution of Z and are included in $Q(Z)$. More formally,

$$U(Z) = \sum_{g \in G} \sum_{M \in N(g)} \sum_{T \in Q(Z)} QI(T, M) \,,$$

where

$$Q_1(Z) = \{\{\{v, w\}, \{x\}, \{y\}\} | \{x, y\} \subset Z, \{v, w\} \subset L - \{x, y\}\} \,,$$
$$Q_2(Z) = \{\{\{v, w\}, \{x\}, \{y\}\} | \{v, w, x\} \subset Z, y \in L - Z\} \,, \text{ and}$$
$$Q(Z) = Q_1(Z) \cup Q_2(Z) \,, Q_1(Z) \cap Q_2(Z) = \emptyset \,.$$

Clearly, $V(Z) \leq U(Z)$ (equality can be achieved only if all gene trees are compatible with some resolution of Z). Then, letting $d = |M|$ and defining $z_i = |Z \cap M_i|$ and $l_i = |L \cap M_i| = |M_i|$, we have

$$\sum_{\{A,B,C\} \in Q(Z)} QI((A|B|C), M)$$

$$= \sum_{\{A,B,C\} \in Q_1(Z)} QI((A|B|C), M) + \sum_{\{A,B,C\} \in Q_2(Z)} QI((A|B|C), M)$$

$$= \sum_{i \in [d]} \sum_{j \in [d]-\{i\}} \sum_{k \in [d]-\{i\}-[j]} \binom{l_i}{2} z_j z_k$$

$$+ \sum_{i \in [d]} \sum_{j \in [d]-\{i\}} \sum_{k \in [d]-\{i\}-[j]} \binom{z_i}{2} (z_j(l_k - z_k) + (l_j - z_j)z_k)$$

$$= \sum_{i \in [d]} \sum_{j \in [d]-\{i\}} \sum_{k \in [d]-\{i,j\}} \binom{l_i}{2} \frac{z_j z_k}{2}$$

$$+ \sum_{i \in [d]} \sum_{j \in [d]-\{i\}} \sum_{k \in [d]-\{i,j\}} \binom{z_i}{2} \frac{z_j(l_k - z_k) + (l_j - z_j)z_k}{2}$$

$$= \sum_{i \in [d]} \sum_{j \in [d]-\{i\}} \sum_{k \in [d]-\{i,j\}} \binom{l_i}{2} \frac{z_j z_k}{2}$$

$$+ \sum_{i \in [d]} \sum_{j \in [d]-\{i\}} \sum_{k \in [d]-\{i,j\}} \binom{z_i}{2} z_j(l_k - z_k) \,.$$

(13)

Notice that based on Eq. 4,

$$\frac{QI((Z|Z|L),M)}{2} - \frac{QI((Z|Z|Z),M)}{3} =$$

$$\frac{1}{2}\sum_{i\in[d]}\sum_{j\in[d]-\{i\}}\sum_{k\in[d]-\{i,j\}} z_i z_j l_k \frac{z_i+z_j+l_k-3}{2} =$$

$$-\frac{1}{3}\sum_{i\in[d]}\sum_{j\in[d]-\{i\}}\sum_{k\in[d]-\{i,j\}} z_i z_j z_k \frac{z_i+z_j+z_k-3}{2} =$$

$$\frac{1}{2}\sum_{i\in[d]}\sum_{j\in[d]-\{i\}}\sum_{k\in[d]-\{i,j\}} (\binom{z_i}{2}z_j l_k + z_i \binom{z_j}{2}l_k + z_i z_j \binom{l_k}{2})$$

$$-\frac{1}{3}\sum_{i\in[d]}\sum_{j\in[d]-\{i\}}\sum_{k\in[d]-\{i,j\}} (\binom{z_i}{2}z_j z_k + z_i \binom{z_j}{2}z_k + z_i z_j \binom{z_k}{2}) =$$

$$\frac{1}{2}\sum_{i\in[d]}\sum_{j\in[d]-\{i\}}\sum_{k\in[d]-\{i,j\}} (\binom{z_i}{2}z_j l_k + \binom{z_i}{2}z_j l_k + \binom{l_i}{2}z_j z_k) \qquad (14)$$

$$-\frac{1}{3}\sum_{i\in[d]}\sum_{j\in[d]-\{i\}}\sum_{k\in[d]-\{i,j\}} (\binom{z_i}{2}z_j z_k + \binom{z_i}{2}z_j z_k + \binom{z_i}{2}z_j z_k) =$$

$$\frac{1}{2}\sum_{i\in[d]}\sum_{j\in[d]-\{i\}}\sum_{k\in[d]-\{i,j\}} (\binom{l_i}{2}z_j z_k + 2\binom{z_i}{2}z_j l_k)$$

$$-\frac{1}{3}\sum_{i\in[d]}\sum_{j\in[d]-\{i\}}\sum_{k\in[d]-\{i,j\}} 3\binom{z_i}{2}z_j z_k =$$

$$\sum_{A,B,C\in Q(Z)} QI((A|B|C),M) .$$

(going from the fourth term to the fifth is accomplished by changing the order of sums). Therefore,

$$U(Z) = \sum_{g\in G}\sum_{M\in N(g)} (\frac{QI((Z|Z|L),M)}{2} - \frac{QI((Z|Z|Z),M)}{3})$$

$$= \frac{w(Z|Z|L)}{2} - \frac{w(Z|Z|Z)}{3} . \qquad (15)$$

B   Simulations and Commands

Simulation Setup

S100: In order to generate the gene trees and species trees using the Simphy we use this command:

Table 2. Species tree and gene tree generation parameters used for simphy [41], and sequence evolution parameters for the GTR model used for Indelible [42] for the S100 dataset.

Parameter name	Parameter value
Speciation rate	0.0000001
Extinsion rate	0
Number of leaves	100
Ingroup divergence to the ingroup ratio	1.0
Generations	$LogN(1.470055e+01, 2.500000e\text{-}01)$
Haploid effective population size	400000
Global substitution rate	$LogN(-1.727461e+01, 6.931472e\text{-}01)$
Lineage specific rate gamma shape	$LogN(1.500000e+00, 1)$
Gene family specific rate gamma shape	$LogN(1.551533e+00, 6.931472e\text{-}01)$
Gene tree branch specific rate gamma shape	$LogN(1.500000e+00, 1)$
Seed	9644
Sequence length	1600, 800, 400, 200
Sequence base frequencies	$Dirichlet(A=36, C=26, G=28, T=32)$
Sequence transition rates	$Dirichlet(TC=16, TA=3, TG=5, CA=5, CG=6, AG=15)$

```
simphy −rs 50 −rl f:1000 −rg 1 −sb f:0.0000001 −sd f:0
−st ln:14.70055,0.25 −sl f:100 −so f:1 −si f:1 −sp
f:400000 −su ln:−17.27461,0.6931472 −hh f:1 −hs ln:1.5,1
−hl ln:1.551533,0.6931472 −hg ln:1.5,1 −cs 9644 −v 3
−o ASTRALIII −ot 0 −op 1 −od 1
```

Larege-n Simulated Dataset: In order to compare running time performances of ASTRAL-II and ASTRAL-III, we created another dataset with very large numbers of species using Simphy and under the MSCM. Since we are only comparing running times, we only use true gene trees to infer the ASTRAL species trees. We have three sub-datasets with 5000, 2000, and 1000 species (plus one outgroup). Each sub-dataset has 4 replicates, and each replicate has a different species tree with 500 gene trees. Species trees are generated based on the birth-death process with birth and date rates from log uniform distributions. We sampled the number of generations and effective population size from log normal and uniform distributions respectively such that we have medium amounts of ILS. The average FN rates between the true gene trees and the species tree ranges between 4% and 23% for 1K, between 21% and 58% for 2K, and between 21% and 33% for 5K.

In order to generate the gene trees and true species trees using the Simphy we use parameters given in Table 3 and the following command.

1K:

```
simphy −rs 20 −rl f:1000 −rg 1 −sb lu:0.0000001,0.000001 −sd
lu:0.0000001,sb −st ln:16,1 −sl f:1000 −so f:1 −si f:1 −sp
u:10000,1000000 −su ln:−17.27461,0.6931472 −hh f:1 −hs ln:1.5,1 −hl
ln:1.551533,0.6931472 −hg ln:1.5,1 −cs 9644 −v 3 −o 5k.species −ot 0
−op 1 −od 1
```

Table 3. Species tree and gene tree generation parameters in Simphy [41] for 1K-taxon, 2K-taxon and 5K-taxon datasets

Parameter Name	Parameter value
Speciation rate	LogU[1.000000e-07, 1.000000e-06)
Extinsion rate	LogU[1.000000e-07, SB)
Locus trees	1000
Gene trees	1
Number of leaves	1000, 2000, or 5000
Ingroup divergence to the ingroup ratio	1.0
Generations	LogN(16, 1)
Haploid effective population size	Uniform[10000, 1000000]
Global substitution rate	LogN(−1.727461e + 01, 6.931472e-01)
Lineage specific rate gamma shape	LogN(1.500000e + 00,1)
Gene family specific rate gamma shape	LogN(1.551533e + 00, 6.931472e-01)
Gene tree branch specific rate gamma shape	LogN(1.500000e + 00, 1)
Seed	9644

2K:

```
simphy −rs 20 −rl f:1000 −rg 1 −sb lu:0.0000001,0.000001 −sd
lu:0.0000001,sb −st ln:16,1 −sl f:2000 −so f:1 −si f:1 −sp
u:10000,1000000 −su ln:−17.27461,0.6931472 −hh f:1 −hs ln:1.5,1 −hl
ln:1.551533,0.6931472 −hg ln:1.5,1 −cs 9644 −v 3 −o 5k.species −ot 0
−op 1 −od 1
```

5K:

```
simphy −rs 20 −rl f:1000 −rg 1 −sb lu:0.0000001,0.000001 −sd
lu:0.0000001,sb −st ln:16,1 −sl f:5000 −so f:1 −si f:1 −sp
u:10000,1000000 −su ln:−17.27461,0.6931472 −hh f:1 −hs ln:1.5,1 −hl
ln:1.551533,0.6931472 −hg ln:1.5,1 −cs 9644 −v 3 −o 5k.species −ot 0
−op 1 −od 1
```

Commands

Contracting Branches: In order to contract gene tree branches with bootstrap up to a certain threshold we used this command:

```
nw_ed genetree 'i & (b<=$threshold)' o
```

Drawing Bootstrap Support on ML Gene Trees: In order to draw bootstrap support on best ML gene trees we first reroot both best ML gene tree, and the bootstrap gene trees using this command:

```
nw_support   bootstrapgenetrees taxon > bootstrapgenetrees.rerooted
nw_support      bestMLgenetree taxon > bestMLgenetree.rerooted
```

Then we draw bootstrap supports on the branches:

```
nw_support -p
bestMLgenetree.rerooted bootstrapgenetrees.rerooted >
bestMLgenetree.rerooted.final
```

Gene Tree Estimation: We used FastTree version 2.1.9 Double precision. In order to estimated best ML gene trees we used the following command:

```
fasttree -nt -gtr -nopr -gamma -n <num> <all-genes.phylip>
```

where we have all the alignments in the PHYLIP format in the file all-genes.phylip for each replicate, and $< num >$ is the number of alignments in this file.

For bootstrapping analysis, we first generate bootstrapped sequences using RAxML version 8.2.9 with the following command:

```
raxmlHPC -s alignment.phylip -f j
         -b <seed number> -n BS -m GTRGAMMA -# 100
```

and then we Fasttree to perform the actual ML analyses; for FastTree bootstrap runs, we use the same command and models that we used for best ML gene trees.

Running ASTRAL: ASTRAL-II in this paper refers to ASTRAL version 4.11.1 and ASTRAL-III refers to ASTRAL version 5.2.5. Both versions can be found in the link below:

```
https://github.com/chaoszhang/ASTRAL/releases/tag/paper
```

Both versions of ASTRAL program were run with following command:

```
java -jar <program> -t 0 -i <input> -o <output> &> <log>
```

References

1. Song, S., Liu, L., Edwards, S.V., Wu, S.: Resolving conflict in eutherian mammal phylogeny using phylogenomics and the multispecies coalescent model. Proc. Nat. Acad. Sci. **109**(37), 14942–14947 (2012)
2. Wickett, N.J., Mirarab, S., Nguyen, N., et al.: Phylotranscriptomic analysis of the origin and early diversification of land plants. Proc. Nat. Acad. Sci. **111**(45), 4859–4868 (2014)
3. Jarvis, E.D., Mirarab, S., Aberer, A.J., et al.: Whole-genome analyses resolve early branches in the tree of life of modern birds. Science **346**(6215), 1320–1331 (2014)
4. Laumer, C.E., Hejnol, A., Giribet, G.: Nuclear genomic signals of the 'microturbellarian' roots of platyhelminth evolutionary innovation. eLife 4 (2015)
5. Tarver, J.E., dos Reis, M., Mirarab, S., et al.: The interrelationships of placental mammals and the limits of phylogenetic inference. Genome Biol. Evol. **8**(2), 330–344 (2016)
6. Rokas, A., Williams, B.L., King, N., Carroll, S.B.: Genome-scale approaches to resolving incongruence in molecular phylogenies. Nature **425**(6960), 798–804 (2003)

7. Maddison, W.P.: Gene trees in species trees. Syst. Biol. **46**(3), 523–536 (1997)
8. Springer, M.S., Gatesy, J.: The gene tree delusion. Mol. Phylogenet. Evol. **94**(Part A), 1–33 (2016)
9. Meiklejohn, K.A., Faircloth, B.C., Glenn, T.C., Kimball, R.T., Braun, E.L.: Analysis of a rapid evolutionary radiation using ultraconserved elements: evidence for a bias in some multispecies coalescent methods. Syst. Biol. **65**(4), 612–627 (2016)
10. Edwards, S.V., Xi, Z., Janke, A., et al.: Implementing and testing the multispecies coalescent model: a valuable paradigm for phylogenomics. Mol. Phylogenet. Evol. **94**, 447–462 (2016)
11. Shen, X.X., Hittinger, C.T., Rokas, A.: Studies can be driven by a handful of genes. Nature **1**, 1–10 (2017)
12. Heled, J., Drummond, A.J.: Bayesian inference of species trees from multilocus data. Mol. Biol. Evol. **27**(3), 570–580 (2010)
13. Chifman, J., Kubatko, L.S.: Quartet inference from SNP data under the coalescent model. Bioinformatics **30**(23), 3317–3324 (2014)
14. Degnan, J.H., Rosenberg, N.A.: Gene tree discordance, phylogenetic inference and the multispecies coalescent. Trends Ecol. Evol. **24**(6), 332–340 (2009)
15. Edwards, S.V.: Is a new and general theory of molecular systematics emerging? Evolution **63**(1), 1–19 (2009)
16. Pamilo, P., Nei, M.: Relationships between gene trees and species trees. Mol. Biol. Evol. **5**(5), 568–583 (1988)
17. Rannala, B., Yang, Z.: Bayes estimation of species divergence times and ancestral population sizes using DNA sequences from multiple loci. Genetics **164**(4), 1645–1656 (2003)
18. Liu, L., Yu, L., Edwards, S.V.: A maximum pseudo-likelihood approach for estimating species trees under the coalescent model. BMC Evol. Biol. **10**(1), 302 (2010)
19. Liu, L., Yu, L.: Estimating species trees from unrooted gene trees. Syst. Biol. **60**, 661–667 (2011)
20. Sayyari, E., Mirarab, S.: Anchoring quartet-based phylogenetic distances and applications to species tree reconstruction. BMC Genomics **17**(S10), 101–113 (2016)
21. Liu, L., Yu, L., Pearl, D.K., Edwards, S.V.: Estimating species phylogenies using coalescence times among sequences. Syst. Biol. **58**(5), 468–477 (2009)
22. Mossel, E., Roch, S.: Incomplete lineage sorting: consistent phylogeny estimation from multiple loci. IEEE/ACM Trans. Comput. Biol. Bioinform. (TCBB) **7**(1), 166–171 (2010)
23. Roch, S., Warnow, T.: On the robustness to gene tree estimation error (or lack thereof) of coalescent-based species tree methods. Syst. Biol. **64**(4), 663–676 (2015)
24. Mirarab, S., Reaz, R., Bayzid, M.S., et al.: ASTRAL: genome-scale coalescent-based species tree estimation. Bioinformatics **30**(17), i541–i548 (2014)
25. Lafond, M., Scornavacca, C.: On the Weighted Quartet Consensus problem. arxiv: 1610.00505 (2016)
26. Mirarab, S., Warnow, T.: ASTRAL-II: coalescent-based species tree estimation with many hundreds of taxa and thousands of genes. Bioinformatics **31**(12), i44–i52 (2015)
27. Allman, E.S., Degnan, J.H., Rhodes, J.A.: Determining species tree topologies from clade probabilities under the coalescent. J. Theor. Biol. **289**(1), 96–106 (2011)
28. Shekhar, S., Roch, S., Mirarab, S.: Species tree estimation using ASTRAL: how many genes are enough? In: Proceedings of International Conference on Research in Computational Molecular Biology (RECOMB) (to appear) (2017)

29. Davidson, R., Vachaspati, P., Mirarab, S., Warnow, T.: Phylogenomic species tree estimation in the presence of incomplete lineage sorting and horizontal gene transfer. BMC Genomics **16**(Suppl 10), S1 (2015)
30. Sayyari, E., Mirarab, S.: Fast coalescent-based computation of local branch support from quartet frequencies. Mol. Biol. Evol. **33**(7), 1654–1668 (2016)
31. Price, M.N., Dehal, P.S., Arkin, A.P.: FastTree-2 - approximately maximum-likelihood trees for large alignments. PLoS ONE **5**(3), e9490 (2010)
32. Mirarab, S., Bayzid, M.S., Boussau, B., Warnow, T.: Statistical binning enables an accurate coalescent-based estimation of the avian tree. Science **346**(6215), 1250463–1250463 (2014)
33. Bayzid, M.S., Mirarab, S., Boussau, B., Warnow, T.: Weighted statistical binning: enabling statistically consistent genome-scale phylogenetic analyses. PLoS ONE **10**(6), e0129183 (2015)
34. Mirarab, S., Bayzid, M.S., Warnow, T.: Evaluating summary methods for multi-locus species tree estimation in the presence of incomplete lineage sorting. Syst. Biol. **65**(3), 366–380 (2016)
35. Patel, S., Kimball, R., Braun, E.: Error in phylogenetic estimation for bushes in the tree of life. Phylogenet. Evol. Biol. **1**(2), 2 (2013)
36. Gatesy, J., Springer, M.S.: Phylogenetic analysis at deep timescales: unreliable gene trees, bypassed hidden support, and the coalescence/concatalescence conundrum. Mol. Phylogenet. Evol. **80**, 231–266 (2014)
37. Yu, Y., Warnow, T., Nakhleh, L.: Algorithms for MDC-based multi-locus phylogeny inference: beyond rooted binary gene trees on single alleles. J. Comput. Biol. **18**(11), 1543–1559 (2011)
38. Vachaspati, P., Warnow, T.: ASTRID: accurate species trees from internode distances. BMC genomics **16**(Suppl 10), S3 (2015)
39. Kane, D., Tao, T.: A bound on partitioning clusters (2017). arXiv:11702.00912
40. Stamatakis, A.: RAxML version 8: a tool for phylogenetic analysis and post-analysis of large phylogenies. Bioinformatics **30**(9), 1312–1313 (2014)
41. Mallo, D., De Oliveira Martins, L., Posada, D.: SimPhy: Phylogenomic simulation of gene, locus and species trees. Syst. Biol. **65**(2), syv082 (2016)
42. Fletcher, W., Yang, Z.: INDELible: a flexible simulator of biological sequence evolution. Mol. Biol. Evol. **26**(8), 1879–1888 (2009)
43. Tavaré, S.: Some probabilistic and statistical problems in the analysis of DNA sequences. Lect. Math. Life Sci. **17**, 57–86 (1986)
44. Junier, T., Zdobnov, E.M.: The Newick utilities: high-throughput phylogenetic tree processing in the UNIX shell. Bioinformatics **26**(13), 1669–1670 (2010)
45. Robinson, D., Foulds, L.: Comparison of phylogenetic trees. Math. Biosci. **53**(1–2), 131–147 (1981)

Algorithms for Computing the Family-Free Genomic Similarity Under DCJ

Diego P. Rubert[1], Gabriel L. Medeiros[1], Edna A. Hoshino[1],
Marília D.V. Braga[2], Jens Stoye[2], and Fábio V. Martinez[1(✉)]

[1] Faculdade de Computação, Universidade Federal de Mato Grosso do Sul,
Campo Grande, MS, Brazil
{diego,gabriel_medeiros,eah,fhvm}@facom.ufms.br
[2] Faculty of Technology and Center for Biotechnology (CeBiTec),
Bielefeld University, Bielefeld, Germany
{marilia.braga,jens.stoye}@uni-bielefeld.de

Abstract. The genomic similarity is a large-scale measure for comparing two given genomes. In this work we study the (NP-hard) problem of computing the genomic similarity under the DCJ model in a setting that does not assume that the genes of the compared genomes are grouped into gene families. This problem is called family-free DCJ similarity. Here we propose an exact ILP algorithm to solve it, we show its APX-hardness, and we present three combinatorial heuristics, with computational experiments comparing their results to the ILP. Experiments on simulated datasets show that the proposed heuristics are very fast and even competitive with respect to the ILP algorithm for some instances.

Keywords: Genome rearrangement · Double-cut-and-join · Family-free genomic similarity

1 Introduction

A central question in comparative genomics is the elucidation of similarities and differences between genomes. Local and global measures can be employed. A popular set of global measures is based on the number of genome rearrangements necessary to transform one genome into another one [23]. Genome rearrangements are large scale mutations, changing the number of chromosomes and/or the positions and orientations of DNA segments. Examples of such rearrangements are inversions, translocations, fusions, and fissions.

As a first step before such a comparison can be performed, some preprocessing is required. The most common method, adopted for about 20 years [23,24], is to base the analysis on the order of conserved syntenic DNA segments across different genomes and group homologous segments into *families*. This setting is said to be *family-based*. Without duplicate segments, i.e., with the additional restriction that at most one representative of each family occurs in any genome, several polynomial time algorithms have been proposed to compute genomic

J. Meidanis and L. Nakhleh (Eds.): RECOMB CG 2017, LNBI 10562, pp. 76–100, 2017.
DOI: 10.1007/978-3-319-67979-2_5

distances and similarities [5,6,9,16,27]. However, when duplicates are allowed, problems become more intricate and many presented approaches are NP-hard [1–3,11,12,24,26].

Although family information can be obtained by accessing public databases or by direct computing, data can be incorrect, and inaccurate families could be providing support to erroneous assumptions of homology between segments [15]. Thus, it is not always possible to classify each segment unambiguously into a single family, and an alternative to the family-based setting was proposed recently [10]. It consists of studying genome rearrangements without prior family assignment, by directly accessing the pairwise similarities between DNA segments of the compared genomes. This approach is said to be *family-free* (FF).

The *double cut and join* (DCJ) operation, that consists of cutting a genome in two distinct positions and joining the four resultant open ends in a different way, represents most of large-scale rearrangements that modify genomes [27]. In this work we are interested in the problem of computing the overall similarity of two given genomes in a family-free setting under the DCJ model. This problem is called FFDCJ similarity. The complexity of computing the FFDCJ similarity was proven to be NP-hard [20], while the counterpart problem of computing the FFDCJ distance was already proven to be APX-hard [20]. In the remainder of this paper, after preliminaries and a formal definition of the FFDCJ similarity problem, we first present an exact ILP algorithm to solve it. We then show the APX-hardness of the FFDCJ similarity problem and present three combinatorial heuristics, with computational experiments comparing their results to the ILP for datasets simulated by a framework for genome evolution.

2   Preliminaries

Each segment (often called *gene*) g of a genome is an oriented DNA fragment and its two distinct *extremities* are called *tail* and *head*, denoted by g^t and g^h, respectively. A genome is composed of a set of chromosomes, each of which can be circular or linear and is a sequence of genes. Each one of the two extremities of a linear chromosome is called a *telomere*, represented by the symbol ∘. An *adjacency* in a chromosome is then either the extremity of a gene that is adjacent to a telomere, or a pair of consecutive gene extremities. As an example, observe that the adjacencies 5^h, 5^t2^t, 2^h4^t, 4^h3^t, 3^h6^t, 6^h1^h and 1^t can define a linear chromosome. Another representation of the same linear chromosome, flanked by parentheses for the sake of clarity, would be (∘ −5 2 4 3 6 −1 ∘), in which the genes preceded by the minus sign (−) have reverse orientation.

A *double cut and join* or DCJ operation applied to a genome A is the operation that cuts two adjacencies of A and joins the separated extremities in a different way, creating two new adjacencies. For example, a DCJ acting on two adjacencies pq and rs would create either the adjacencies pr and qs, or the adjacencies ps and qr (this could correspond to an inversion, a reciprocal translocation between two linear chromosomes, a fusion of two circular chromosomes, or an excision of a circular chromosome). In the same way, a DCJ acting on two adjacencies pq and r would create either pr and q, or p and qr (in this case, the operation could correspond to an inversion, a translocation, or a fusion of a circular and a linear chromosome). For the cases described so far we can notice that for each pair of cuts there are two possibilities of joining. There are two special cases of a DCJ operation, in which there is only one possibility of joining. The first is a DCJ acting on two adjacencies p and q, that would create only one new adjacency pq (that could represent a circularization of one or a fusion of two linear chromosomes). Conversely, a DCJ can act on only one adjacency pq and create the two adjacencies p and q (representing a linearization of a circular or a fission of a linear chromosome).

In the remainder of this section we extend the notation introduced in [20]. In general we consider the comparison of two distinct genomes, that will be denoted by A and B. Respectively, we denote by $\mathcal{A}$ the set of genes in genome A, and by $\mathcal{B}$ the set of genes in genome B.

2.1   Adjacency Graph and Family-Based DCJ Similarity

In most versions of the family-based setting the two genomes A and B have the same content, that is, $\mathcal{A} = \mathcal{B}$. When in addition there are no duplicates, that is, when there is exactly one representative of each family in each genome, we can easily build the *adjacency graph* of genomes A and B, denoted by $AG(A, B)$ [6]. It is a bipartite multigraph such that each partition corresponds to the set of adjacencies of one of the two input genomes, and an edge connects the same extremities of genes in both genomes. In other words, there is a one-to-one correspondence between the set of edges in $AG(A, B)$ and the set of gene extremities. Since the graph is bipartite and vertices have degree one or two, the adjacency graph is a collection of paths and even cycles. An example of an adjacency graph is presented in Fig. 1.

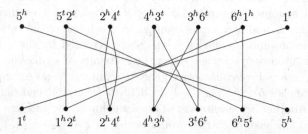

Fig. 1. The adjacency graph for the genomes $A = \{(\circ\ -5\ 2\ 4\ 3\ 6\ -1\ \circ)\}$ and $B = \{(\circ\ 1\ 2\ 4\ -3\ 6\ 5\ \circ)\}$.

It is well known that a DCJ operation that modifies $AG(A, B)$ by increasing either the number of even cycles by one or the number of odd paths by two decreases the DCJ distance between genomes A and B [6]. This type of DCJ operation is said to be *optimal*. Conversely, if we are interested in a DCJ similarity measure between A and B, rather than a distance measure, then it should be increased by such an optimal DCJ operation. This suggests that a formula for a DCJ similarity between two genomes should correlate to the number of connected components (in the following just *components*) of the corresponding adjacency graph.

Moreover, when the genomes A and B are identical, their corresponding adjacency graph is a collection of 2-cycles and 1-paths [6]. This should correspond to the maximum value of our DCJ similarity measure. We know that an optimal operation can always be applied to adjacencies that belong to one of the two genomes and to one single component of $AG(A, B)$, until the graph becomes a collection of 2-cycles and 1-paths. In other words, each component of the graph can be *sorted*, that is, converted into a collection of 2-cycles and 1-paths independently of the other components. Furthermore, it is known that each of the following components – an even cycle with $2d + 2$ edges, or an odd path with $2d + 1$ edges, or an even path with $2d$ edges – can be sorted with exactly d optimal DCJ operations. This suggests that the three listed components should have equivalent weights in the DCJ similarity formula. However, we should also take into consideration that, for the same d, components with more edges should actually have a higher weight.

Let $\mathcal{P}, \mathcal{I}$, and $\mathcal{C}$ represent the sets of components in $AG(A, B)$ that are even paths, odd paths and cycles, respectively. We have the following formula for the family-based DCJ similarity:

$$s_{\text{DCJ}}(A, B) = \sum_{C \in \mathcal{P}} \left(\frac{|C|}{|C|+2} \right) + \sum_{C \in \mathcal{I}} \left(\frac{|C|}{|C|+1} \right) + \sum_{C \in \mathcal{C}} \left(\frac{|C|}{|C|} \right) \qquad (1)$$

$$= \sum_{C \in \mathcal{P}} \left(\frac{|C|}{|C|+2} \right) + \sum_{C \in \mathcal{I}} \left(\frac{|C|}{|C|+1} \right) + c,$$

where c is the number of cycles in $AG(A, B)$. In Fig. 1 the DCJ similarity is $s_{\text{DCJ}}(A, B) = 2 \left(\frac{1}{2} \right) + 3 = 4$. Observe that $s_{\text{DCJ}}(A, B)$ is a positive value, upper bounded by n, where $n = |\mathcal{A}| = |\mathcal{B}|$.

The formula to compute $s_{\text{DCJ}}(A, B)$ in Eq. (1) is actually the family-based version of the family-free DCJ similarity defined in [20], as we will see in the following subsections.

2.2   Gene Similarity Graph

In the family-free setting, each gene in each genome is represented by a unique (signed) symbol, thus $\mathcal{A} \cap \mathcal{B} = \emptyset$ and the cardinalities $|\mathcal{A}|$ and $|\mathcal{B}|$ may be distinct. Let a be a gene in A and b be a gene in B, then their *normalized gene similarity* is given by some value $\sigma(a, b)$ such that $0 \leq \sigma(a, b) \leq 1$.

We can represent the gene similarities between the genes of genome A and the genes of genome B with respect to σ in the so called *gene similarity graph* [10], denoted by $GS_\sigma(A, B)$. This is a weighted bipartite graph whose partitions

$\mathcal{A}$ and $\mathcal{B}$ are the sets of (signed) genes in genomes A and B, respectively. Furthermore, for each pair of genes (a, b) such that $a \in \mathcal{A}$ and $b \in \mathcal{B}$, if $\sigma(a, b) > 0$ then there is an edge e connecting a and b in $GS_\sigma(A, B)$ whose weight is $\sigma(e) := \sigma(a, b)$. An example of a gene similarity graph is given in Fig. 2.

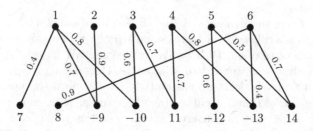

Fig. 2. Representation of a gene similarity graph for two unichromosomal linear genomes $A = \{(\circ\ 1\ 2\ 3\ 4\ 5\ 6\ \circ)\}$ and $B = \{(\circ\ 7\ 8\ -9\ -10\ 11\ -12\ -13\ 14\ \circ)\}$.

2.3 Weighted Adjacency Graph

The *weighted adjacency graph* $AG_\sigma(A, B)$ of two genomes A and B has a vertex for each adjacency in A and a vertex for each adjacency in B. For a gene a in A and a gene b in B with gene similarity $\sigma(a, b) > 0$ there is one edge e^h connecting the vertices containing the two heads a^h and b^h and one edge e^t connecting the vertices containing the two tails a^t and b^t. The weight of each of these edges is $\sigma(e^h) = \sigma(e^t) = \sigma(a, b)$. Differently from the simple adjacency graph, the weighted adjacency graph cannot be easily decomposed into cycles and paths, since its vertices can have degree greater than 2. As an example, the weighted adjacency graph corresponding to the gene similarity graph of Fig. 2 is given in Fig. 3.

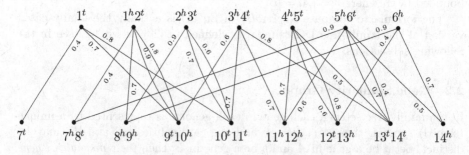

Fig. 3. The weighted adjacency graph $AG_\sigma(A, B)$ for two unichromosomal linear genomes $A = \{(\circ\ 1\ 2\ 3\ 4\ 5\ 6\ \circ)\}$ and $B = \{(\circ\ 7\ 8\ -9\ -10\ 11\ -12\ -13\ 14\ \circ)\}$.

2.4 Reduced Genomes

Let A and B be two genomes and let $GS_\sigma(A, B)$ be their gene similarity graph. Now let $M = \{e_1, e_2, \ldots, e_n\}$ be a matching in $GS_\sigma(A, B)$ and denote by $w(M) = \sum_{e_i \in M} \sigma(e_i)$ the weight of M, that is the sum of its edge weights. Since the endpoints of each edge $e_i = (a, b)$ in M are not saturated by any other edge of M, we can unambiguously define the function $\ell^M(a) = \ell^M(b) = i$ to relabel each vertex in A and B [20]. The *reduced genome* A^M is obtained by deleting from A all genes not saturated by M, and renaming each saturated gene a to $\ell^M(a)$, preserving its orientation (sign). Similarly, the reduced genome B^M is obtained by deleting from B all genes that are not saturated by M, and renaming each saturated gene b to $\ell^M(b)$, preserving its orientation. Observe that the set of genes in A^M and in B^M is $\mathcal{G}(M) = \{\ell^M(g) : g$ is saturated by the matching $M\} = \{1, 2, \ldots, n\}$.

2.5 Weighted Adjacency Graph of Reduced Genomes

Let A^M and B^M be the reduced genomes for a given matching M of $GS_\sigma(A, B)$. The weighted adjacency graph $AG_\sigma(A^M, B^M)$ can be obtained from $AG_\sigma(A, B)$ by deleting all edges that are not elements of M and relabeling the adjacencies according to ℓ^M. Vertices that have no connections are then also deleted from the graph. Another way to obtain the same graph is building the adjacency graph of A^M and B^M and adding weights to the edges as follows. For each gene i in $\mathcal{G}(M)$, both edges $i^t i^t$ and $i^h i^h$ inherit the weight of edge e_i in M, that is, $\sigma(i^t i^t) = \sigma(i^h i^h) = \sigma(e_i)$. Consequently, the graph $AG_\sigma(A^M, B^M)$ is also a collection of paths and even cycles and differs from $AG(A^M, B^M)$ only by the edge weights.

Observe that, for each edge $e \in M$, we have two edges of weight $\sigma(e)$ in $AG_\sigma(A^M, B^M)$, thus $w(AG_\sigma(A^M, B^M)) = 2w(M)$, where $w(C) = \sum_{e \in C} \sigma(e)$ is the sum of the weights of all edges in a subgraph C. Examples of weighted adjacency graphs of reduced genomes are shown in Fig. 4.

2.6 The Family-Free DCJ Similarity

For a given matching M in $GS_\sigma(A, B)$, a first formula for the weighted DCJ (wDCJ) similarity s_σ of the reduced genomes A^M and B^M was proposed in [10] only considering the cycles of $AG_\sigma(A^M, B^M)$. Following, this definition was modified and extended in [20], in order to consider the normalized total weight of all components of the weighted adjacency graph. Let $\mathcal{P}$, $\mathcal{I}$, and $\mathcal{C}$ represent the sets of components in $AG_\sigma(A^M, B^M)$ that are even paths, odd paths and cycles, respectively. Then the wDCJ similarity s_σ is given by the following formula [20]:

$$s_\sigma(A^M, B^M) = \sum_{C \in \mathcal{P}} \left(\frac{w(C)}{|C|+2} \right) + \sum_{C \in \mathcal{I}} \left(\frac{w(C)}{|C|+1} \right) + \sum_{C \in \mathcal{C}} \left(\frac{w(C)}{|C|} \right) \qquad (2)$$

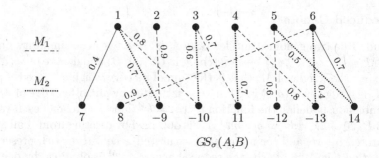

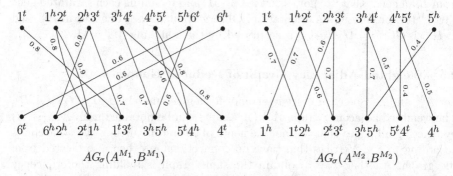

Fig. 4. Considering, as in Fig. 2, the genomes $A = \{(\circ\ 1\ 2\ 3\ 4\ 5\ 6\ \circ)\}$ and $B = \{(\circ\ 7\ 8\ -9\ -10\ 11\ -12\ -13\ 14\ \circ)\}$, let M_1 (dashed edges) and M_2 (dotted edges) be two distinct maximal matchings in $GS_\sigma(A, B)$, shown in the upper part. The two resulting weighted adjacency graphs $AG_\sigma(A^{M_1}, B^{M_1})$, that has two cycles and two even paths, and $AG_\sigma(A^{M_2}, B^{M_2})$, that has two odd paths, are shown in the lower part.

Observe that, when the weights of all edges in M are equal to 1, this formula is equivalent to the one in Eq. (1).

The goal now is to compute the family-free DCJ similarity, i.e., to find a matching in $GS_\sigma(A, B)$ that maximizes s_σ. However, although $s_\sigma(A^M, B^M)$ is a positive value upper bounded by $|M|$, the behaviour of the wDCJ similarity does not correlate with the size of the matching, since smaller matchings, that possibly discard gene assignments, can lead to higher wDCJ similarities [20]. For this reason, the wDCJ similarity function is restricted to *maximal matchings* only, ensuring that no pair of genes with positive gene similarity score is simply discarded, even though it might decrease the overall wDCJ similarity. We then have the following optimization problem:

Problem FFDCJ-SIMILARITY(A, B): Given genomes A and B and their gene similarities σ, calculate their family-free DCJ similarity

$$s_{\text{FFDCJ}}(A, B) = \max_{M \in \mathbb{M}}\{s_\sigma(A^M, B^M)\}, \tag{3}$$

where $\mathbb{M}$ is the set of all maximal matchings in $GS_\sigma(A, B)$.

Problem FFDCJ-SIMILARITY is NP-hard [20]. Moreover, one can directly correlate the problem to the adjacency similarity problem, where the goal is to maximize the number of preserved adjacencies between two given genomes [1]. However, since there the objective is to maximize the number of cycles of length 2, even an approximation for the adjacency similarity problem is not a good algorithm for the FFDCJ-SIMILARITY problem, where cycles of higher lengths are possible in the solution [22].

2.7 Capping Telomeres

A very useful preprocessing to $AG_\sigma(A, B)$ is the *capping* of telomeres, a general technique for simplifying algorithms that handle genomes with linear chromosomes, commonly used in the context of family-based settings [16, 25, 27]. Given two genomes A and B with i and j linear chromosomes, respectively, for each vertex representing only one extremity we add a *null extremity* τ to it (e.g., 1^t of Fig. 4 becomes $\tau 1^t$). Furthermore, in order to add the same number of null extremities to both genomes, $|2j - 2i|$ *null adjacencies* $\tau\tau$ (composed of two null extremities) are added to genome A, if $i < j$, or to genome B, if $j < i$. Finally, for each null extremity of a vertex in A we add to $AG_\sigma(A, B)$ a *null edge* with weight 0 to each null extremity of vertices in B. Consequently, after capping of telomeres the graph $AG_\sigma(A, B)$ has no vertex of degree one. Notice that, if before the capping p was a path of weight w connecting telomeres in $AG_\sigma(A, B)$, then after the capping p will be part of a cycle closed by null extremities with normalized weight $\frac{w}{|p|+1}$ if p is an odd path, or of normalized weight $\frac{w}{|p|+2}$ if p is an even path. In any of the two cases, the normalized weight is consistent with the wDCJ similarity formula in Eq. (2).

3 An Exact Algorithm

In order to exactly compute the family-free DCJ similarity between two given genomes, we propose an integer linear program (ILP) formulation that is similar to the one for the family-free DCJ distance given in [20]. It adopts the same notation and also uses an approach to solve the maximum cycle decomposition problem as in [26].

   Let A and B be two genomes, let $G = GS_\sigma(A, B)$ be their gene similarity graph, and let X_A and X_B be the extremity sets (including null extremities) with respect to A and B for the capped adjacency graph $AG_\sigma(A, B)$, respectively. The weight $w(e)$ of an edge e in G is also denoted by w_e. For the ILP formulation, an extension $H = (V_H, E_H)$ of the capped weighted adjacency graph $AG_\sigma(A, B)$ is defined such that $V_H = X_A \cup X_B$ and $E_H = E_m \cup E_a \cup E_s$ has three types of edges: (i) *matching edges* that connect two extremities in different extremity sets, one in X_A and the other in X_B, if they are null extremities or there exists an edge connecting these genes in G; the set of matching edges is denoted by E_m; (ii) *adjacency edges* that connect two extremities in the same extremity set if they form an adjacency; the set of adjacency edges is denoted by E_a; and (*iii*) *self edges* that connect two extremities of the same gene in an extremity set; the set of self edges is denoted by E_s. Matching edges have weights defined by the normalized gene similarity σ, all adjacency and self edges have weight 0. Notice that any edge in G corresponds to two matching edges in H.

The description of the ILP follows. For each edge in H, we create a binary variable x_e to indicate whether e will be in the final solution. We require first that each adjacency edge be chosen:

$$x_e = 1, \qquad \forall\, e \in E_a.$$

Now we rename each vertex in H such that $V_H = \{v_1, v_2, \ldots, v_k\}$ with $k = |V_H|$. We require that each of these vertices be adjacent to exactly one matching or self edge:

$$\sum_{e=v_r v_t \in E_m \cup E_s} x_e = 1, \forall\, v_r \in X_A, \quad \text{and} \quad \sum_{e=v_r v_t \in E_m \cup E_s} x_e = 1, \forall\, v_t \in X_B.$$

Then, we require that the final solution be valid, meaning that if one extremity of a gene in A is assigned to an extremity of a gene in B, then the other extremities of these two genes have to be assigned as well:

$$x_{a^h b^h} = x_{a^t b^t}, \qquad \forall\, ab \in E_G.$$

We also require that the matching be maximal. This can easily be ensured if we guarantee that at least one of the vertices connected by an edge in the gene similarity graph be chosen, which is equivalent to not allowing both of the corresponding self edges in the weighted adjacency graph be chosen:

$$x_{a^h a^t} + x_{b^h b^t} \leq 1, \qquad \forall\, ab \in E_G.$$

To count the number of cycles, we use the same strategy as described in [26]. For each vertex v_i we define a variable y_i that labels v_i such that

$$0 \leq y_i \leq i, \qquad 1 \leq i \leq k.$$

We also require that adjacent vertices have the same label, forcing all vertices in the same cycle to have the same label:

$$y_i \leq y_j + i \cdot (1 - x_e), \ \forall\, e = v_i v_j \in E_H,$$
$$y_j \leq y_i + j \cdot (1 - x_e), \ \forall\, e = v_i v_j \in E_H.$$

We create a binary variable z_i, for each vertex v_i, to verify whether y_i is equal to its upper bound i:

$$i \cdot z_i \leq y_i, \qquad 1 \leq i \leq k.$$

Since all variables y_i in the same cycle have the same label but a different upper bound, only one of the y_i can be equal to its upper bound i. This means that z_i is 1 if the cycle with vertex i as representative is used in a solution.

Now, let $L = \{2j : j = 1, \ldots, n\}$ be the set of possible cycle lengths in H, where $n := \min(|A|, |B|)$. We create the binary variable x_{ei} to indicate whether e is in i, for each $e \in E_H$ and each cycle i. We also create the binary variable x_{ei}^ℓ to indicate whether e belongs to i and the length of cycle i is ℓ, for each $e \in E_H$, each cycle i, and each $\ell \in L$.

We require that if an edge e belongs to a cycle i, then it can be true for only one length $\ell \in L$. Thus,

$$\sum_{\ell \in L} x_{ei}^\ell \leq x_{ei}, \quad \forall\, e \in E_H \text{ and } 1 \leq i \leq k. \tag{4}$$

We create another binary variable z_i^ℓ to indicate whether cycle i has length ℓ. Then $\ell \cdot z_i^\ell$ is an upper bound for the total number of edges in cycle i of length ℓ:

$$\sum_{e \in E_M} x_{ei}^\ell \leq \ell \cdot z_i^\ell, \quad \forall\, \ell \in L \text{ and } 1 \leq i \leq k.$$

The length of a cycle i is given by $\ell \cdot z_i^\ell$, for $i = 1, \ldots, k$ and $\ell \in L$. On the other hand, it is the total amount of matching edges e in cycle i. That is,

$$\sum_{\ell \in L} \ell \cdot z_i^\ell = \sum_{e \in E_m} x_{ei}, \quad 1 \leq i \leq k.$$

We have to ensure that each cycle i must have just one length:

$$\sum_{\ell \in L} z_i^\ell = z_i, \quad 1 \leq i \leq k.$$

Now we create the binary variable y_{ri} to indicate whether the vertex v_r is in cycle i. Thus, if $x_{ei} = 1$, i.e., if the edge $e = v_r v_t$ in H is chosen in cycle i, then $y_{ri} = 1 = y_{ti}$ (and $x_e = 1$ as well). Hence,

$$\left.\begin{aligned} x_{ei} &\leq x_e, \\ x_{ei} &\leq y_{ri}, \\ x_{ei} &\leq y_{ti}, \\ x_{ei} &\geq x_e + y_{ri} + y_{ti} - 2, \end{aligned}\right\} \quad \forall\, e = v_r v_t \in E_H \text{ and } 1 \leq i \leq k. \tag{5}$$

Since y_r is an integer variable, we associate y_r to the corresponding binary variable y_{ri}, for any vertex v_r belonging to cycle i:

$$y_r = \sum_{i=1}^{r} i \cdot y_{ri}, \quad \forall\, v_r \in V_H.$$

Furthermore, we must ensure that each vertex v_r may belong to at most one cycle:

$$\sum_{i=1}^{r} y_{ri} \leq 1, \quad \forall\, v_r \in V_H.$$

Finally, we set the objective function as follows:

$$\text{maximize} \quad \sum_{i=1}^{k} \sum_{\ell \in L} \sum_{e \in E_m} \frac{w_e x_{ei}^{\ell}}{\ell}.$$

Note that, with this formulation, we do not have any path as a component. Therefore, the objective function above is exactly the family-free DCJ similarity $s_{\text{FFDCJ}}(A, B)$ as defined in Eqs. (2) and (3).

Notice that the ILP formulation has $O(N^4)$ variables and $O(N^3)$ constraints, where $N = |A| + |B|$. The number of variables is proportional to the number of variables x_{ei}^{ℓ}, and the number of constraints is upper bounded by (4) and (5).

4   APX-hardness and Heuristics

In this section we first state that problem FFDCJ-SIMILARITY is APX-hard and provide a lower bound for the approximation ratio.

Theorem 1. FFDCJ-SIMILARITY *is APX-hard and cannot be approximated with approximation ratio better than* $22/21 = 1.0476\ldots$, *unless* $P = NP$.

Proof. See Appendix A.

We now propose three heuristic algorithms to compute the family-free DCJ similarity of two given genomes: a greedy-like heuristic collecting the best density cycles in the weighted adjacency graph, a greedy-like heuristic collecting cycles of increasing lengths in the weighted adjacency graph, and a heuristic that tries to collect sets of cycles of increasing lengths with maximum total density in the weighted adjacency graph by a weighted maximum independent set (WMIS) algorithm.

The algorithms select disjoint cycles in the capped $AG_{\sigma}(A, B)$, inducing a matching M in $GS_{\sigma}(A, B)$. In addition to being disjoint, the selected cycles must also be consistent: we say that two edges in $AG_{\sigma}(A, B)$ are *consistent* if one connects the head and the other connects the tail of the same pair of genes, or if they connect extremities of distinct genes in both genomes. Otherwise they are *inconsistent*. A set of edges, in particular a cycle, is consistent if it has no pair of inconsistent edges. A set of cycles is consistent if the union of all of their edges is consistent.

All three heuristics have a common adjustment step: the deletion of blocking genes, that works as follows. After some iterations of cycle selection that increase the matching M, one or more genes may become blocking and thus must be deleted. This happens when the algorithms either find no cycle, or find some cycles but they are all inconsistent with previous selections, having however genes in $GS_{\sigma}(A, B)$ unsaturated by M. We call them *blocking genes*. Whenever this occurs, we can find and delete blocking genes by (i) finding genes in $GS_{\sigma}(A, B)$ having all neighbors saturated by M and thus blocking or (ii) finding a vertex set $S \subseteq A$ or $S \subseteq B$ such that, for the set of neighbors $N(S)$ of vertices in S, we have $|S| > |N(S)|$ (Hall's theorem), then choosing $|S| - |N(S)|$ genes in S as blocking. After deleting a blocking gene g, $AG_\sigma(A, B)$ must be adjusted accordingly, removing edges corresponding to extremities g^t or g^h of g and "merging" the two vertices that represented these extremities. At the end of the three algorithms we have a maximal matching M, and the union of selected cycles is equivalent to $AG_\sigma(A^M, B^M)$.

The three heuristics have an initial step where all cycles of the weighted adjacency graph are generated (see Step 4 of each one). Thus, the running time of the heuristics is potentially exponential in the number of vertices of the weighted adjacency graph. In Sect. 5, these three heuristics will be compared to the exact ILP algorithm from Sect. 3 regarding quality and running time.

4.1 Best Density

The best density heuristic is shown in Algorithm 1 (GREEDY-DENSITY). Its first step is to generate all cycles in the weighted adjacency graph based on the gene similarity graph of two given genomes. Cycles are arranged in order decreasing by their densities, i.e., the weight divided by the squared length. Then, consistent cycles are collected following this criterion and the family-free DCJ similarity is computed on these collected cycles and remaining components.

Algorithm 1. GREEDY-DENSITY(A, B, σ)

Input: genomes A and B, gene similarity function σ
Output: a family-free DCJ similarity between A and B
1: $M := \emptyset$; $C := \emptyset$.
2: Build the gene similarity graph $GS_\sigma(A, B)$.
3: Build the capped weighted adjacency graph $AG_\sigma(A, B)$.
4: List all cycles C of $AG_\sigma(A, B)$ in decreasing order of their density $w(C)/|C|^2$.
5: While it is possible, select the best density consistent cycle C that is also consistent with all cycles in C and add it to C, let $AG_\sigma(A, B) := AG_\sigma(A, B) \setminus C$, update M by adding the new gene connections induced by C.
6: Find and delete blocking genes, returning to Step 4 if there are genes in $GS_\sigma(A, B)$ unsaturated by M.
7: Return $s_\sigma(A, B) = \sum_{C \in C} \left(\frac{w(C)}{|C|} \right)$

Step 4 is the core of GREEDY-DENSITY, where cycles are obtained by a procedure based on Johnson's algorithm [18, 19]. Although the number of cycles may be exponential in the size of the input graph, the implemented algorithm restricts the size of cycles found and is good enough for graphs and for many experiments, as presented in Sect. 5.

4.2 Best Length

The best length heuristic is shown in Algorithm 2 (GREEDY-LENGTH). As in the best density heuristic, the first step is to generate all cycles in the weighted adjacency graph based on the gene similarity graph of the two given genomes. However, cycles are arranged in order increasing by their lengths, where ties are broken by selecting cycles with greater density. Similar to above, consistent cycles are collected following this criterion, and the family-free DCJ similarity is computed on these collected cycles and remaining components.

Algorithm 2. GREEDY-LENGTH(A, B, σ)

Input: genomes A and B, gene similarity function σ
Output: a family-free DCJ similarity between A and B
1: $M := \emptyset; \mathcal{C} := \emptyset$.
2: Build the gene similarity graph $GS_\sigma(A, B)$.
3: Build the capped weighted adjacency graph $AG_\sigma(A, B)$.
4: List all cycles C of $AG_\sigma(A, B)$ in increasing order of their lengths $|C|$.
5: Iterate over the list of cycles as follows. Select consistent cycles of length 2, while its is possible. Then, select consistent cycles of length 4. And so on, until there are no more cycles. Let C be the cycle selected at each iteration. Add C to $\mathcal{C}$, let $AG_\sigma(A, B) := AG_\sigma(A, B) \setminus C$, update M by adding the new gene connections induced by C.
6: Find and delete blocking genes, returning to Step 4 if there are genes in $GS_\sigma(A, B)$ unsaturated by M.
7: Return $s_\sigma(A, B) = \sum_{C \in \mathcal{C}} \left(\frac{w(C)}{|C|} \right)$

Again, Step 4 is the core of the algorithm and it is implemented based on Johnson's algorithm [18, 19], but we first find and select cycles of length 2, then of length 4, and so on.

4.3   Best Length with Weighted Maximum Independent Set

The best length heuristic with WMIS is shown in Algorithm 3 (GREEDY-WMIS) and is a variation of GREEDY-LENGTH. Instead of selecting cycles of greater density for a fixed length, this algorithm selects the greatest amount of cycles for a fixed length by a WMIS algorithm. The heuristic builds a *cycle graph* where each vertex is a cycle of $AG_\sigma(A, B)$, the weight of a vertex is the density of the cycle it represents and two vertices are adjacent if the cycles they represent are inconsistent. The heuristic tries to find next an independent set with the greatest weight in the cycle graph. Since this graph is not d-claw-free for any fixed d, the WMIS algorithm [7] does not guarantee any fixed ratio.

5   Experimental Results

Experiments for the ILP and our heuristics were conducted on an Intel i7-4770 3.40 GHz machine with 16 GB of memory. In order to do so, we produced simulated datasets by the Artificial Life Simulator (ALF) [14]. Gurobi Optimizer 7.0

Algorithm 3. GREEDY-WMIS(A, B, σ)

Input: genomes A and B, gene similarity function σ
Output: a family-free DCJ similarity between A and B
1: $M := \emptyset$; $C := \emptyset$.
2: Build the gene similarity graph $GS_\sigma(A, B)$.
3: Build the capped weighted adjacency graph $AG_\sigma(A, B)$.
4: List all cycles C of $AG_\sigma(A, B)$ in increasing order of their lengths $|C|$.
5: Iterate over the list of cycles as follows. Select a set of consistent cycles of length 2 trying to maximize the sum of densities by a WMIS algorithm. Then, repeat for consistent cycles of length 4. And so on, until there are no more cycles. Let C' be the set of cycles selected at each iteration. Add C' to C, let $AG_\sigma(A, B) := AG_\sigma(A, B) \backslash C'$, update M by adding the new gene connections induced by C'.
6: Find and delete blocking genes, returning to Step 4 if there are genes in $GS_\sigma(A, B)$ unsaturated by M.
7: Return $s_\sigma(A, B) = \sum_{C \in C} \left(\frac{w(C)}{|C|} \right)$

---

was set to solve ILP instances with default parameters, time limit of 1800 s and 4 threads, and the heuristics were implemented in C++.

We generated datasets with 10 genome samples each, running pairwise comparisons between all genomes in the same dataset. Each dataset has genomes of sizes around 25, 50 or 1000 (the latter used only for running the heuristics), generated based on a sample from the tree of life with 10 leaf species and PAM distance of 100 from the root to the deepest leaf. Gamma distribution with parameters $k = 3$ and $\theta = 133$ was used for gene length distribution. For amino acid evolution we used the WAG substitution model with default parameters and the preset of Zipfian indels with rate 0.00005. Regarding genome level events, we allowed gene duplications and gene losses with rate 0.002, and reversals and translocations with rate 0.0025, with at most 3 genes involved in each event. To test different proportions of genome level events, we also generated simulated datasets with 2- and 5-fold increase for reversal and translocation rates.

Results are summarized in Table 1. Each dataset is composed of 10 genomes, totaling 45 comparisons of pairs per dataset. Rate $r = 1$ means the default parameter set for genome level events, while $r = 2$ and $r = 5$ mean the 2- and 5-fold increase of rates, respectively. For the ILP the table shows the average time for instances for which the optimal solution was found, the number of instances for which the optimizer did not find the optimal solution after time limit and, for the latter class of instances, the average relative gap between the best solution found and the upper bound found by the solver, given by $\left(\frac{\text{upper bound}}{\text{best solution}} - 1 \right) \times 100$. For heuristics, the running time for all instances of sizes 25 and 50 was negligible, therefore the table shows only the average relative gap between the solution found and the upper bound given by the ILP solver (if any).

Results clearly show the average relative gap of heuristics increases proportionally to the rate of reversals and translocations. This is expected, as higher mutation rates often result in higher normalized weights on longer cycles, thus the association of genes with greater gene similarity scores will be subject to the

Table 1. Results of experiments for simulated genomes

	ILP			Greedy-Density	Greedy-Length	Greedy-wmis
	Time (s)	Not finished	Gap (%)	Gap (%)	Gap (%)	Gap (%)
25 genes, $r = 1$	19.50	0	–	5.03	5.84	5.97
25 genes, $r = 2$	84.60	2	69.21	30.77	43.57	43.00
25 genes, $r = 5$	49.72	0	–	43.83	55.38	55.38
50 genes, $r = 1$	445.91	7	19.56	18.74	19.36	18.90
50 genes, $r = 2$	463.50	29	38.12	65.41	66.52	64.78
50 genes, $r = 5$	330.88	29	259.72	177.58	206.60	206.31

selection of longer cycles. Interestingly, for some larger instances the relative gap for heuristics is very close to the values obtained by the ILP solver, suggesting the use of heuristics may be a good alternative for some classes of instances or could help the solver finding lower bounds quickly. It is worth noting that the Greedy-Density heuristic found solutions with gap smaller than 1% for 38% of the instances with 25 genes.

In a single instance (25 genes, $r = 2$), the gap between the best solution found and the upper bound was much higher for the ILP solver and for the heuristics. This instance in particular is precisely the one with the largest number of edges in $GS_\sigma(A, B)$ in the dataset. This may indicate that a moderate increase in degree of vertices (1.3 on average to 1.8 in this case) may result in much harder instances for the solver and the heuristics, as after half of the time limit the solver attained no significant improvement on solutions found, and the heuristics returned solutions with a gap even higher.

Although we have no upper bounds for comparing the results of our heuristics for genome sizes around 1000, the algorithms are still very fast. The average running times are 0.30 s, 15.11 s and 12.16 s for Greedy-Density, Greedy-Length and Greedy-wmis, respectively, showing nevertheless little difference on results. However, in 25% of the instances with $r = 5$, the solutions provided by the heuristics varied between 10% and 24%, the best of which were given by Greedy-Density. That is probably because, instead of prioritizing shorter cycles, Greedy-Density attempts to balance both normalized weight and length of the selected cycles. The average running times for the instances with $r = 5$ are 2.35 s, 97.28 s and 102.67 s for Greedy-Density, Greedy-Length and Greedy-wmis, respectively.

To better understand how cycles scale, we generated 5-fold instances with 100, 500, 1000, 5000, and 10000 genes, running the Greedy-Density heuristic for these instances and counting different cycle lengths. The running times were 0.008 s, 0.667 s, 1.98 s, 508 s and 2896 s, respectively, on average. Results (Fig. 5) show that most of the cycles found are of short lengths compared to the genome sizes, providing some insight on why heuristics are fast despite having to enumerate a number of cycles that could be exponential. Besides, even the maximum number of longer cycles found for any instance is reasonably small.

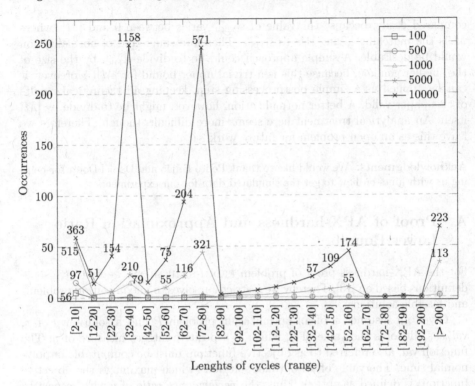

Fig. 5. Average count by lengths of cycles for the GREEDY-DENSITY heuristic for instances with $r = 5$ and genome sizes of 100, 500, 1000, 5000, and 10000 genes. Numbers above marks denote the maximum number of cycles for a pair of genomes in an instance (only for values greater than 50). As seen, the number of cycles may be exponential, therefore the heuristic implementation finds cycles of lengths up to 10, then up to 20, and so on. Moreover, when finding cycles of lengths up to 20, the algorithm does not try to find cycles composed by adjacencies in $AG_\sigma(A, B)$ already covered by shorter cycles chosen previously. The same occurs for longer lengths.

6   Conclusion

In this paper we studied the family-free DCJ similarity, which is a large-scale rearrangement measure for comparing two given genomes. We first presented formally the (NP-hard) problem of computing the family-free DCJ similarity. Then, we proposed an exact ILP algorithm to solve it. Following, we showed the APX-hardness of the family-free DCJ similarity problem and presented three combinatorial heuristics, with computational experiments comparing their results to the ILP. Results show that while the ILP program is fast and accurate for smaller instances, the GREEDY-DENSITY heuristic is probably the best choice for general use on larger instances.

One drawback of the function s_{FFDCJ} as defined in Eq. (3) is that distinct pairs of genomes might give family-free DCJ similarity values that cannot be compared easily, because the value of s_{FFDCJ} varies between 0 and $|M|$, where M is the matching giving rise to s_{FFDCJ}. Therefore some kind of normalization would be desirable. A simple approach could be to divide s_{FFDCJ} by the size of the smaller genome, because this is a trivial upper bound for $|M|$. Moreover, it can be applied as a simple postprocessing step, keeping all theoretical results of this paper valid. A better normalization, however, might be to divide by $|M|$ itself. An analytical treatment here seems more difficult, though. Therefore we leave this as an open problem for future work.

Acknowledgments. We would like to thank Pedro Feijão and Daniel Doerr for helping us with hints on how to get the simulated data for our experiments.

A   Proof of APX-hardness and Approximation Ratio Lower Bound

For the APX-hardness proof of problem FFDCJ-SIMILARITY, we first give some definitions based on [13]. Thereby we restrict ourselves to maximization problems and feasible solutions.

Given an instance x of an optimization problem P and a solution y of x, val(x, y) denotes the value of y, which is a positive integer measure of y. The function val, also referred to as objective function, must be computable in polynomial time. The value of an optimal solution (which maximizes the objective function) is defined as opt(x). Thus, the *performance ratio* of y with respect to x is defined as:

$$R_P(x, y) = \frac{\text{opt}(x)}{\text{val}(x, y)}. \tag{6}$$

Given two optimization problems P and P', let f be a polynomial-time computable function that maps an instance x of P into an instance $f(x)$ of P', and let g be a polynomial-time computable function that maps a solution y for the instance $f(x)$ of P' into a solution $g(x, y)$ of P. A *reduction* is a pair (f, g). A reduction from P to P' is frequently denoted by $P \leq P'$, and we say that P is *reduced* to P'. A reduction $P \leq P'$ *preserves membership* in a class $\mathcal{C}$ if $P' \in \mathcal{C}$ implies $P \in \mathcal{C}$. An *approximation-preserving* reduction preserves membership in either APX, PTAS, or both classes. The *strict reduction*, which is the simplest type of approximation-preserving reduction, preserves membership in both APX and PTAS classes and must satisfy the following condition:

$$R_P(x, g(x, y)) \leq R_{P'}(f(x), y). \tag{7}$$

We consider the following optimization problem, to be used within the proof of Theorem 1 below:

Problem MAX-2SAT3(ϕ): Given a 2-CNF formula (i.e., with at most 2 literals per clause) $\phi = \{C_1, \dots, C_m\}$ with n variables $X = \{x_1, \dots, x_n\}$, where each variable appears in at most 3 clauses, find an assignment that satisfies the largest number of clauses.

The formula ϕ as defined above is called a 2SAT3 formula. MAX-2SAT3 [4,8] is a special case of MAX-2SATB (also known as B-OCC-MAX-2SAT), where each variable occurs in at most B clauses for some B, which in turn is a restricted version of MAX-2SAT [21].

Theorem 1. FFDCJ-SIMILARITY *is APX-hard and cannot be approximated with approximation ratio better than* $22/21 = 1.0476\ldots$, *unless* $P = NP$.

Proof (Theorem 1, first part). We give a strict reduction (f, g) from MAX-2SAT3 to FFDCJ-SIMILARITY, showing that

$$R_{\text{MAX-2SAT3}}(\phi, g(f(\phi), \gamma)) \leq R_{\text{FFDCJ-SIMILARITY}}(f(\phi), \gamma),$$

for any instance ϕ of MAX-2SAT3 and solution γ of FFDCJ-SIMILARITY with instance $f(\phi)$. Since variables occurring only once imply their clauses and others to be trivially satisfied, we consider only clauses that are not trivially satisfied in their instance. Similar for clauses containing literals x_i and $\overline{x_i}$, for some variable x_i.

(Function f.) We show progressively how to build $GS_\sigma(A, B)$ and define genes and their sequences in chromosomes of A and B. For each variable x_i occurring three times, let Cx_i^1, Cx_i^2 and Cx_i^3 be *aliases* for the clauses where x_i occurs (notice that a clause composed of two literals has two aliases). We define a *variable component* $\mathcal{C}_i$ adding vertices (genes) x_i^1, x_i^2 and x_i^3 to $\mathcal{A}$, vertices (genes) Cx_i^1, Cx_i^2 and Cx_i^3 to $\mathcal{B}$, and edges $ex_i^j = (Cx_i^j, x_i^j)$ and $e\overline{x_i}^j = (Cx_i^j, x_i^k)$ for $j \in \{1, 2, 3\}$ and $k = (j + 1) \bmod 3 + 1$. An edge ex_i^j ($e\overline{x_i}^j$) has weight 1 (0) if the literal x_i ($\overline{x_i}$) belongs to the clause Cx_i^j. Edges in the variable component $\mathcal{C}_i$ form a cycle of length 6 (Fig. 6). Variable components for variables occurring two times are defined in a similar manner. Genomes are $A = \{(x_i^j)$ for each occurrence j of each variable $x_i \in X\}$ and $B = \{(Cx_i^j) : Cx_i^j$ is an alias to a clause in ϕ with only one literal$\} \cup \{(Cx_i^j \, Cx_{i'}^{j'}) : Cx_i^j$ and $Cx_{i'}^{j'}$ are aliases to the same clause in $\phi\}$.

The function f as defined here maps an instance ϕ of MAX-2SAT3 (a 2-CNF formula) to an instance $f(\phi)$ of FFDCJ-SIMILARITY (genomes A and B and $GS_\sigma(A, B)$) and is clearly polynomial. Besides, since all chromosomes are circular, the corresponding weighted adjacency graph $AG_\sigma(A, B)$ (or $AG_\sigma(A^M, B^M)$ for some matching M) is a collection of cycles only.

Now, notice that any maximal matching in $GS_\sigma(A, B)$ covers all genes in both A and B, inducing in $AG_\sigma(A, B)$ only cycles of length 2, composed by (genes in) chromosomes (x_i^j) and $(Cx_i^{j'})$, or cycles of length 4, composed by chromosomes (x_i^j), (x_k^l) and $(Cx_i^{j'} \, Cx_k^{l'})$.

Define the *normalized weight* of cycle C as $\mu(C) = w(C)/|C|$. In this transformation, each cycle C is such that $\mu(C) = 0, 0.5$ or 1. A cycle C such that $\mu(C) > 0$ is a *helpful cycle* and represents a clause satisfied by one or two literals ($\mu(C) = 0.5$ or $\mu(C) = 1$, respectively). See an example in Fig. 7.

In this scenario, however, a solution of FFDCJ-SIMILARITY with performance ratio r could lead to a solution of MAX-2SAT3 with ratio $2r$, since the total

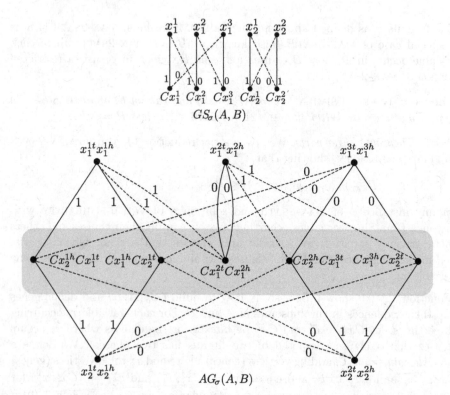

Fig. 6. $GS_\sigma(A, B)$ and $AG_\sigma(A, B)$ for genomes $A = \{(x_1^1), (x_1^2), (x_1^3), (x_2^1), (x_2^2)\}$ and $B = \{(Cx_1^1\ Cx_2^1), (Cx_1^2), (Cx_1^3\ Cx_2^2)\}$ given by function f (Theorem 1) applied to 2SAT(3) clauses $C_1 = (x_1 \vee x_2)$, $C_2 = (\overline{x_1})$ and $C_3 = (\overline{x_1} \vee x_2)$. In $GS_\sigma(A, B)$, solid edges correspond to ex_i^j and dashed edges correspond to $e\overline{x_i}^j$. In $AG_\sigma(A, B)$, shaded region corresponds to genes of genome B, and solid (dashed) edges correspond to solid (dashed) edges of $GS_\sigma(A, B)$.

normalized weight for two cycles C_1 and C_2 with $\mu(C_1) = \mu(C_2) = 0.5$ (two clauses satisfied by one literal each) is the same for one cycle C with $\mu(C) = 1.0$ (one clause satisfied by two literals). Therefore, achieving the desired ratio requires some modifications in f. It is not possible to make these two types of cycles have the same weight, but it suffices to get close enough.

We introduce special genes into the genomes called *extenders*. For some p even, for each edge $ex_i^j = (Cx_i^j, x_i^j)$ of weight 1 in $GS_\sigma(A, B)$ we introduce p extenders $\alpha_1, \ldots, \alpha_p$ into A (as a consequence, they are also introduced into $\mathcal{A}$) and p extenders $\alpha_{p+1}, \ldots, \alpha_{2p}$ into B (each ex_i^j of weight 1 has its own set of extenders). Edge ex_i^j is replaced by edges (Cx_i^j, α_1) with weight 1 (which we consider equivalent to ex_i^j) and (α_{p+1}, x_i^j) with weight 0, and edges (α_k, α_{p+k}) with weight 0 are added to $GS_\sigma(A, B)$ for each $1 \leq k \leq p$ (extenders α_1 and α_{p+1} are now part of the variable component $\mathcal{C}_i$). Regarding new chromosomes in genomes A and B, A is updated to $A \cup \{(\alpha_1 - \alpha_p)\} \cup \{(\alpha_k - \alpha_{k+1}) : k \in \{2, 4, \ldots,$

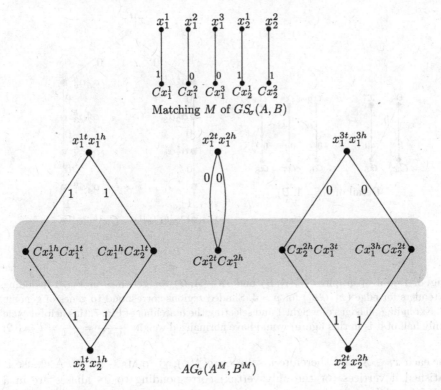

Fig. 7. A matching M of $GS_\sigma(A, B)$ and cycles induced by M in $AG_\sigma(A^M, B^M)$ for genomes of Fig. 6. This solution of FFDCJ-SIMILARITY represents clauses C_1 and C_3 of MAX-2SAT3 satisfied.

$p - 2\}\}$ and B to $B \cup \{(\alpha_k \ -\alpha_{k+1}) : k \in \{p + 1, p + 3, \ldots, 2p - 1\}\}$. By this construction, which is still polynomial, the path from x_i^{jt} to Cx_i^{jt} in $AG_\sigma(A, B)$ is extended from 1 to $1 + p$ edges, from $\{(x_i^{jt}, Cx_i^{jt})\}$ to $\{(x_i^{jt}, \alpha_p^t), (\alpha_{p+1}^t, \alpha_2^t), (\alpha_3^t, \alpha_{p+2}^t), (\alpha_{p+3}^t, \alpha_4^t), \ldots, (\alpha_1^t, Cx_i^{jt})\}$. The same occurs for the path from x_i^{jh} to Cx_i^{jh} (see Fig. 8). Now, cycles in $AG_\sigma(A, B)$ induced by edges of weight 0 in $GS_\sigma(A, B)$ have normalized weight 0, cycles previously with normalized weight 1 are extended and have normalized weight $\frac{1}{1+p}$, and cycles previously with normalized weight 0.5 are extended and have normalized weight $\frac{1}{2+p}$. Notice that, for a sufficiently large p, $\frac{1}{1+p}$ is quite close to $\frac{1}{2+p}$, hence the problem of finding the maximum similarity in this graph is very similar to finding the maximum number of helpful cycles.

(Function g.) By the structure of variable components in $GS_\sigma(A, B)$, and since solutions of FFDCJ-SIMILARITY are restricted to maximal matchings only, any solution γ for $f(\phi)$ is a matching that covers only edges ex_i^j or $e\overline{x_i}^j$ for each variable component $\mathcal{C}_i$. For a $\mathcal{C}_i$, if edges ex_i^j $(e\overline{x_i}^j)$ are in the solution then the variable x_i is assigned to true (false), inducing in polynomial time an assignment

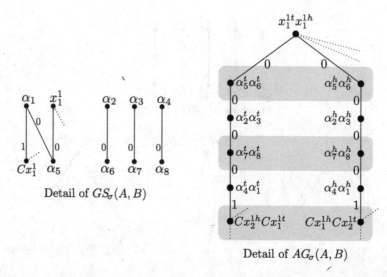

$$x_1^{1t}x_1^{1h}$$

Detail of $GS_\sigma(A, B)$

Detail of $AG_\sigma(A, B)$

Fig. 8. Detail of graphs $GS_\sigma(A, B)$ and $AG_\sigma(A, B)$ for genomes of Fig. 6 including extenders for edge (x_1^1, Cx_1^1) for $p = 4$. Shaded regions correspond to genes of genome B. Extending all edges of weight 1 and selecting the matching of Fig. 7, this helpful cycle (only half of it is in this figure) would have normalized weight $\frac{4}{4(p+1)} = \frac{1}{p+1} = \frac{1}{5} = 0.2$.

for each $x_i \in X$ and therefore a solution $g(f(\phi), \gamma)$ to MAX-2SAT3. A clause is satisfied if vertices (or the only vertex) corresponding to its aliases are in a helpful cycle.

(Approximation Ratio.) Given $f(\phi)$ and a feasible solution γ of FFDCJ-SIMILARITY with the maximum number of helpful cycles, denote by c' the number of helpful cycles in γ. Notice that c' is also the maximum number of satisfied clauses of MAX-2SAT3, that is, the value of an optimal solution for MAX-2SAT3 for any instance ϕ, denoted here by $\mathrm{opt}_{2SAT3}(\phi)$. Thus, $c' = \mathrm{opt}_{2SAT3}(\phi)$.

To achieve the desired ratio we must establish some properties and relations between the parameters of MAX-2SAT3 and FFDCJ-SIMILARITY and set some parameters to specific values.

Let $n := |A| = |B|$ before extenders are added. We choose for p (the number of extenders added for each edge of weight 1 in $GS_\sigma(A, B)$) the value $2n$ and define $\omega = \frac{1}{2+p} = \frac{1}{2+2n}$ and

$$\varepsilon = \frac{1}{1+p} - \frac{1}{2+p} = \frac{1}{4n^2 + 6n + 2},$$

which implies that $\omega + \varepsilon = \frac{1}{p+1}$. Thus, it is easy to see that $n\varepsilon < \omega$, i.e.,

$$\varepsilon < \frac{\omega}{n} < 1. \tag{8}$$

If $\mathrm{opt}_{SIM}(f(\phi))$ denotes the value of an optimal solution for FFDCJ-SIMILARITY with instance $f(\phi)$ and c^* denotes the number of helpful cycles in an optimal solution of FFDCJ-SIMILARITY, then we have immediately that

$$\frac{\text{opt}_{\text{SIM}}(f(\phi))}{\omega + \varepsilon} \le c^* \le \frac{\text{opt}_{\text{SIM}}(f(\phi))}{\omega}. \tag{9}$$

Besides that

$$0 \le c^* \le n, \tag{10}$$

and

$$c^* \omega \le \text{opt}_{\text{SIM}}(f(\phi)) \le c^*(\omega + \varepsilon). \tag{11}$$

Thus, we have

$$c^*(\omega + \varepsilon) = c^*\omega + c^*\varepsilon$$

$$< c^*\omega + \frac{c^*\omega}{n} \tag{12}$$

$$\le c^*\omega + 1 \cdot \omega \tag{13}$$

$$= c^*\omega + \omega, \tag{14}$$

where (12) comes from (8) and (13) is valid due to (10).

Now, let c^r be the number of helpful cycles given by an approximate solution for the FFDCJ-SIMILARITY with approximation ratio r. Then,

$$R_{\text{MAX-2SAT3}}(\phi, g(f(\phi), \gamma)) = \frac{\text{opt}_{\text{2SAT3}}(\phi)}{c^r} = \frac{c'}{c^r} \le r,$$

where the last inequality is given by Proposition 2 below. This concludes the first part of the proof.   □

Proposition 1. *Let c' be the number of helpful cycles in a feasible solution of* FFDCJ-SIMILARITY *with the greatest number of helpful cycles possible. Let c^* be the number of helpful cycles in an optimal solution of* FFDCJ-SIMILARITY. *Then,*

$$c' = c^*.$$

Proof. Since c' is the greatest number of helpful cycles possible, it is immediate that $c^* \le c'$.

Let us now show that $c^* \ge c'$. Suppose for a moment that $c^* < c'$. Since c^* and c' are integers, this implies that $c^* + 1 \le c'$, i.e.,

$$c^* \le c' - 1. \tag{15}$$

Let $\mathcal{C}'$ be the set of cycles with c' cycles, i.e., with the maximum number of helpful cycles possible. Let $\mu(\mathcal{C}') := \sum_{C \in \mathcal{C}'} \mu(C) = \sum_{C \in \mathcal{C}'} w(C)/|C|$. Then

$$\mu(\mathcal{C}') \ge c'\omega = (c' - 1)\omega + \omega$$

$$\ge c^*\omega + \omega \tag{16}$$

$$> c^*(\omega + \varepsilon) \tag{17}$$

$$\ge \text{opt}_{\text{SIM}}(f(\phi)), \tag{18}$$

where (16) follows from (15), (17) comes from (14), and (18) is valid due to (11). It means that $\mu(\mathcal{C}') > \text{opt}_{\text{SIM}}(f(\phi))$, which is a contradiction.

Therefore, $c' = c^*$.   □

Proposition 2. *Let c^r be the number of helpful cycles given by an approximate solution for* FFDCJ-SIMILARITY *with approximation ratio r. Let c' be the same as defined in Proposition 1. Then,*

$$c^r \geq \frac{c'}{r}.$$

Proof. Given an instance $f(\phi)$ of FFDCJ-SIMILARITY, let γ^r be an approximate solution of $f(\phi)$ with performance ratio r, i.e., $\mathrm{val}(f(\phi), \gamma^r) \geq \frac{\mathrm{opt_{SIM}}(f(\phi))}{r}$. Let c^r be the number of helpful cycles of γ^r. Then

$$c^r \geq \frac{\left(\frac{\mathrm{opt_{SIM}}(f(\phi))}{r}\right)}{\omega + \epsilon}$$

$$> \frac{\mathrm{opt_{SIM}}(f(\phi))}{r(\omega + \omega/n)} \tag{19}$$

$$= \frac{\mathrm{opt_{SIM}}(f(\phi))}{r\omega} \cdot \frac{n}{n+1}$$

$$\geq \frac{c'\omega}{r\omega} \cdot \frac{n}{n+1} \tag{20}$$

$$= \frac{c'}{r} \cdot \left(1 - \frac{1}{n+1}\right)$$

$$= \frac{c'}{r} - \frac{c'}{r(n+1)}$$

$$\geq \frac{c'}{r} - 1 \tag{21}$$

where (19) follows from (8), (20) is valid from (11) and Proposition 1. Then, from (21) we know that $c^r > \frac{c'}{r} - 1$ and, since c^r is an integer number, the result follows. $\qquad\square$

We now continue with the proof of Theorem 1.

Proof (Theorem 1, second part). First, notice that if a problem is APX-hard, the existence of a PTAS for it implies $P = NP$. Since a strict reduction preserves membership in the class PTAS, finding a PTAS for FFDCJ-SIMILARITY implies a PTAS for every APX-hard problem and $P = NP$. A PTAS for FFDCJ-SIMILARITY would also imply an approximation ratio better than $2012/2011 = 1.0005\ldots$, unless $P = NP$. This follows immediately from the reduction in Theorem 1 with $R_{\text{MAX-2SAT3}} = R_{\text{FFDCJ-SIMILARITY}}$ and the fact that MAX-2SAT3 is shown in [8] to be NP-hard to approximate within a factor of $2012/2011 - \varepsilon$ for any $\varepsilon > 0$.

However, our result is slightly stronger. Notice particularly that the reduction MAX-2SAT3 $\leq$ FFDCJ-SIMILARITY from the first part of the proof can be trivially extended to MAX-2SAT $\leq$ FFDCJ-SIMILARITY by extending variable components to arbitrary sizes. This increases the lower bound to $22/21 = 1.0476\ldots$ [17]. $\qquad\square$

References

1. Angibaud, S., Fertin, G., Rusu, I., Thévenin, A., Vialette, S.: Efficient tools for computing the number of breakpoints and the number of adjacencies between two genomes with duplicate genes. J. Comput. Biol. **15**(8), 1093–1115 (2008)

2. Angibaud, S., Fertin, G., Rusu, I., Thévenin, A., Vialette, S.: On the approximability of comparing genomes with duplicates. J. Graph Algorithms Appl. **13**(1), 19–53 (2009)

3. Angibaud, S., Fertin, G., Rusu, I., Vialette, S.: A pseudo-boolean framework for computing rearrangement distances between genomes with duplicates. J. Comput. Biol. **14**(4), 379–393 (2007)

4. Ausiello, G., Protasi, M., Marchetti-Spaccamela, A., Gambosi, G., Crescenzi, P., Kann, V.: Complexity and Approximation: Combinatorial Optimization Problems and Their Approximability Properties. Springer (1999)

5. Bafna, V., Pevzner, P.: Genome rearrangements and sorting by reversals. In: Proceedings of the FOCS 1993, pp. 148–157 (1993)

6. Bergeron, A., Mixtacki, J., Stoye, J.: A unifying view of genome rearrangements. In: Bücher, P., Moret, B.M.E. (eds.) WABI 2006. LNCS, vol. 4175, pp. 163–173. Springer, Heidelberg (2006). doi:10.1007/11851561_16

7. Berman, P.: A d/2 approximation for maximum weight independent set in d-claw free graphs. In: Halldórsson, M.M. (ed.) SWAT 2000. LNCS, vol. 1851, pp. 214–219. Springer, Heidelberg (2000). doi:10.1007/3-540-44985-X_19

8. Berman, P., Karpinski, M.: On some tighter inapproximability results (extended abstract). In: Wiedermann, J., van Emde Boas, P., Nielsen, M. (eds.) ICALP 1999. LNCS, vol. 1644, pp. 200–209. Springer, Heidelberg (1999). doi:10.1007/3-540-48523-6_17

9. Braga, M.D.V., Willing, E., Stoye, J.: Double cut and join with insertions and deletions. J. Comput. Biol. **18**(9), 1167–1184 (2011)

10. Braga, M.D.V., Chauve, C., Dörr, D., Jahn, K., Stoye, J., Thévenin, A., Wittler, R.: The potential of family-free genome comparison. In: Chauve, C., El-Mabrouk, N., Tannier, E. (eds.) Models and Algorithms for Genome Evolution, vol. 19, pp. 287–307. Springer, London (2013). doi:10.1007/978-1-4471-5298-9_13. Chap. 13

11. Bryant, D.: The complexity of calculating exemplar distances. In: Sankoff, D., Nadeau, J.H. (eds.) Comparative Genomics, pp. 207–211. Kluwer Academic Publishers, Dortrecht (2000)

12. Bulteau, L., Jiang, M.: Inapproximability of (1,2)-exemplar distance. IEEE/ACM Trans. Comput. Biol. Bioinf. **10**(6), 1384–1390 (2013)

13. Crescenzi, P.: A short guide to approximation preserving reductions. In: Twelfth Annual IEEE Conference on Proceedings of Computational Complexity, pp. 262–273 (1997). doi:10.1109/CCC.1997.612321

14. Dalquen, D.A., Anisimova, M., Gonnet, G.H., Dessimoz, C.: ALF - a simulation framework for genome evolution. Mol. Biol. Evol. **29**(4), 1115 (2012)

15. Dörr, D., Thévenin, A., Stoye, J.: Gene family assignment-free comparative genomics. BMC Bioinform. **13**(Suppl 19), S3 (2012)

16. Hannenhalli, S., Pevzner, P.: Transforming men into mice (polynomial algorithm for genomic distance problem). In: Proceedings of the FOCS 1995, pp. 581–592 (1995). doi:10.1109/SFCS.1995.492588

17. Håstad, J.: Some optimal inapproximability results. J. ACM (JACM) **48**(4), 798–859 (2001)

18. Hawick, K.A., James, H.A.: Enumerating circuits and loops in graphs with self-arcs and multiple-arcs. Technical report CSTN-013, Massey University (2008)
19. Johnson, D.: Finding all the elementary circuits of a directed graph. SIAM J. Comput. **4**(1), 77–84 (1975)
20. Martinez, F.V., Feijão, P., Braga, M.D.V., Stoye, J.: On the family-free DCJ distance and similarity. Algorithms Mol. Biol. **10**, 13 (2015)
21. Raman, V., Ravikumar, B., Rao, S.S.: A simplified NP-complete MAXSAT problem. Inf. Process. Lett. **65**(1), 1–6 (1998)
22. Rubert, D.P., Feijão, P., Braga, M.D.V., Stoye, J., Martinez, F.V.: Approximating the DCJ distance of balanced genomes in linear time. Algorithms Mol. Biol. **12**, 3 (2017)
23. Sankoff, D.: Edit distance for genome comparison based on non-local operations. In: Apostolico, A., Crochemore, M., Galil, Z., Manber, U. (eds.) CPM 1992. LNCS, vol. 644, pp. 121–135. Springer, Heidelberg (1992). doi:10.1007/3-540-56024-6_10
24. Sankoff, D.: Genome rearrangement with gene families. Bioinformatics **15**(11), 909–917 (1999)
25. Shao, M., Lin, Y.: Approximating the edit distance for genomes with duplicate genes under DCJ, insertion and deletion. BMC Bioinform. **13**(Suppl 19), S13 (2012)
26. Shao, M., Lin, Y., Moret, B.: An exact algorithm to compute the DCJ distance for genomes with duplicate genes. In: Sharan, R. (ed.) RECOMB 2014. LNCS, vol. 8394, pp. 280–292. Springer, Cham (2014). doi:10.1007/978-3-319-05269-4_22
27. Yancopoulos, S., Attie, O., Friedberg, R.: Efficient sorting of genomic permutations by translocation, inversion and block interchanges. Bioinformatics **21**(16), 3340–3346 (2005)

New Algorithms for the Genomic Duplication Problem

Jarosław Paszek[✉] and Paweł Górecki

Faculty of Mathematics, Informatics and Mechanics, University of Warsaw,
Warsaw, Poland
{jpaszek,gorecki}@mimuw.edu.pl

Abstract. One of evolutionary molecular biology fundamental issues is
to discover genomic duplication events and their correspondence to the
species tree. Such events can be reconstructed by clustering single gene
duplications that are inferred by reconciling a set of gene trees with a
species tree. Here we propose the first solution to the genomic duplication
problem in which every reconciliation with the minimal number of single
gene duplications is allowed and the method of clustering called minimum
episodes under the assumption that input gene trees are unrooted. We
also present an evaluation study of proposed algorithms on empirical
datasets.

Keywords: Genomic duplication · Duplication episode · Minimum
episodes problem · Reconciliation · Unrooted gene tree · Species tree

1 Introduction

The phenomenon of genomic duplication is fundamental to understand the evolution of life on Earth [1–5]. The research in phylogenetics focus on the way
how the gene families and genomes evolve by discovering the locations of gene
duplications. *Multiple gene duplications* occur when large parts of a genome are
duplicated. In particular, the *whole-genome duplication* occurred for numerous
species and had a crucial impact on the evolution of crops [6–9]. The studies
of this phenomenon focus on detecting its occurrences as well as its influence
on introgressing novel metabolic traits [10] or its association with periods of
increased environmental stress [11]. The methods of detecting whole-genome
duplications can be divided into three categories based on: synteny and colin-
earity comparison of genomes [1,12,13], the estimation of the age distribution of
paralogous gene pairs [3,14], and phylogenetic tree inference [15–17].

The reconstruction of the evolution of individual genes has been thoroughly
studied [18–22] also with the focus on gene trees [23–26], networks [27,28], from
the perspective of population genetics [29] or the evolution of entities (which can
be genes, gene domains, or parts of genes) [30].

The reconciliation model, introduced by Goodman [31] and formalized by
Page [18], interprets the differences between a gene tree and its species tree
[32–34]. In this model, each node from a rooted gene family tree is mapped into

© Springer International Publishing AG 2017
J. Meidanis and L. Nakhleh (Eds.): RECOMB CG 2017, LNBI 10562, pp. 101–115, 2017.
DOI: 10.1007/978-3-319-67979-2_6

the species tree and classified as a single gene duplication or related to a speciation event. In our work, we model a biologically consistent scenario as the embedding of a gene tree into a species tree which represents the location of evolutionary events in the species tree [35]. Identification of such a scenario is done by a function called duplication mapping that assigns a gene tree node, interpreted as a duplication event, to a node of a species tree [20,36–42]. Reconciliation becomes more complex when considering multiple gene duplications. The general formulation is as follows: *given a set of gene trees and a species tree find evolutionary scenarios for the collection of gene trees that yields the minimal number of multiple gene duplication events* [43]. Two fundamental issues arise when dealing with multiple gene duplications: a model of allowed evolutionary scenarios [20,43,44] and the rules of clustering gene duplications from gene trees into a multiple duplication event. We distinguish three variants of problems depending on the clustering: *episode clustering (EC)* [20,37,38], gene duplication clustering [44], and *minimum episodes (ME)*. EC is to find scenarios having the minimal number of locations of duplication episodes in a species tree. EC for rooted gene trees has a linear time solution [41], while for unrooted trees an FPT algorithm is known [36]. In ME a duplication and its ancestor duplication cannot be clustered together [20,38]. The first polynomial time algorithm for ME with rooted gene trees, called *RME*, under the model from [20] was proposed in [38], whereas the optimal linear time algorithm in [40,41]. The concept of assigning every duplication to an interval of allowed locations in a species tree was introduced in [45] in a more general framework without the requirement that the intervals induce a biologically consistent scenario. The naïve implementation of the iterative algorithm from [45] has cubic time complexity. The solution to RME for a variety of models was presented in [43]. In particular, the algorithm proposed in [43] solves RME in linear time.

Our Contribution. We propose the solution to the *unrooted minimum episodes problem, UME*, in which allowed scenarios have the minimal number of gene duplications [36]. According to our knowledge, UME is an open problem. We expanded the theory of unrooted reconciliation by presenting new properties of the *plateau* which is the subtree of an unrooted gene tree containing edges whose rootings have the minimal duplication cost. Next, we show that these properties lead to a decomposition of an unrooted gene tree that allows limiting significantly the possible search space. We show that every instance of UME can be transformed into at most 5^k "simpler" instances that can be solved in linear time, where k is bounded above by infrequent special cases of S2 stars [46] in input trees.

2   Results

Basic Notation

Let S denote a *species tree* which is a rooted binary tree with leaves uniquely labeled by the names of species. We assume that S is fixed throughout this work.

A *rooted gene tree* is a rooted binary tree with leaves labeled by the names of species. The rooted tree (T_1, T_2) has two subtrees T_1 and T_2 whose roots are the children of the tree root. Additionally, for nodes a and b, we write $a \preceq b$ when a and b are on the same path from the root, with b being closer to the root than a. Notation $a \prec b$ means that $a \preceq b$ and $a \neq b$. The root of a tree T we denote by root (T). By T_v we denote the subtree of T rooted at v. An cluster for a node v is the set of all species present in T_v.

Let $T = \langle V_T, E_T \rangle$ be a rooted gene tree such that the set of species present in T is a subset of the set of species present in S. The *least common ancestor (lca) mapping*, $\mathsf{M}_T : V_T \to V_S$, is defined as follows. If v is a leaf in T then $\mathsf{M}_T(v)$ is the leaf in S labeled by the label of v. For an internal node v in T having two children a and b, mapping $\mathsf{M}_T(v)$ is the least common ancestor of $\mathsf{M}_T(a)$ and $\mathsf{M}_T(b)$ in S. An internal node $g \in V_T$ is called a *duplication* if $\mathsf{M}_T(g) = \mathsf{M}_T(a)$ for a child a of g. The *duplication cost*, the total number of duplications in T, is denoted by $\mathsf{D}(T, S)$. Every non-duplication node of T we call a *speciation* (including leaves).

Evolutionary Scenarios

Here, we present the model of DLS trees [35] that will be used to represent evolutionary scenarios. A *DLS tree* is a binary tree having two types of internal nodes, denoting *gene duplications* and *speciations*, and two types of leaves denoting *gene losses* and *gene sequences*. DLS trees are defined as follows [43]:

1. a is a single-noded DLS tree denoting a *gene sequence* from the species a,
2. $A-$ is a single-noded DLS tree denoting a *lost gene* lineage, where A is a non-empty set of species,
3. $(R_1, R_2)+$ is a DLS tree whose root is a duplication node and its children are DLS trees R_1 and R_2 such that the set of species present in R_1 and the set of species present in R_2 are equal,
4. $(R_1, R_2)\sim$ is a DLS tree whose root represents a speciation and its children are DLS trees R_1 and R_2 such that the set of species present in R_1 and the set of species present in R_2 are disjoint.

Let T be a DLS-tree with at least one gene sequence. A gene tree can be extracted from T by contracting nodes of degree 2 from the smallest subgraph of T containing all gene sequences. Such an operation will be denoted by $\mathsf{gt}(T)$.

We say that a DLS-tree T is a *scenario* for a gene tree G and a species tree S if $\mathsf{gt}(T) = G$ and T is *compatible* with S, that is, every cluster of T is present in S. In such a case, every node g in G uniquely corresponds to a node in T denoted by $\xi(g)$. We can define mappings $\xi \colon G \to T$ and $F_T \colon G \to S$, such that $F_T(g)$ is the node in S whose cluster equals the cluster of $\xi(g)$. An example is depicted in Fig. 1.

Unrooted Reconciliation

The unrooted gene tree is an undirected acyclic connected graph in which each internal node has degree 3 and the leaves are labeled by the names of species.

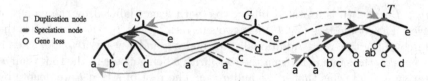

Fig. 1. An example of scenario T for a gene tree G and a species tree S and two corresponding mappings: $\xi\colon G \to T$ and $F_T\colon G \to S$ shown for internal nodes of G. Here, $T=(((((a,a)+,b\text{-})\sim,(c,d\text{-})\sim)\sim,(ab\text{-},(c\text{-},d)\sim)\sim)+,e)\sim$, note that $\mathsf{gt}(T) = G$.

The rooting of an unrooted gene tree $U = \langle V_U, E_U \rangle$ obtained from U by placing the root on an edge $e \in E_U$ is denoted by U_e. Such a rooting induces the duplication cost $\mathsf{D}(U_e, S)$. An edge e is called *optimal* if $\mathsf{D}(U_e, S)$ is minimal in the set of all rootings of U. It is known that the set of optimal edges, called the *plateau*, is a full subtree of U [46,47][1]. In this work, the subtree induced by the set of all optimal edges will be denoted by U^*. For X, the set of edges of unrooted tree U, by $U|_X$ we denote the smallest subgraph of U containing all edges from X.

Without loss of generality we assume that every root of a gene tree is mapped into the root of S, and both trees are non-trivial. An edge $e = \langle v, w \rangle$ of U can be classified as one of three following types: (a) *empty* if the root of U_e is a speciation, i.e., $\mathsf{M}_e(v) \neq \mathsf{root}(S) \neq \mathsf{M}_e(w)$, (b) *double* if $\mathsf{M}_e(v) = \mathsf{root}(S) = \mathsf{M}_e(w)$, and (c) *single* otherwise, where M_e is the lca-mapping between U_e and S. Let v be an internal node of U, then a *star* with the *center* v consists of three edges, sharing v. There are five possible types of stars present in unrooted gene trees [46,47], however, in this article we only use the star called $S2$ having one empty edge. In such a case the remaining edges are single, and by using the notation from Fig. 2, for $x \in \{a, b\}$ we have that $\mathsf{M}_{U_{\langle v,x \rangle}}(x) \neq \mathsf{root}(S) = \mathsf{M}_{U_{\langle v,x \rangle}}(v)$.

It follows from unrooted reconciliation that plateau has either exactly one empty edge or at least one double edge [46]. We say that a node is a *super-duplication* (respectively, a *super-speciation*) if it is a duplication (respectively, a speciation) in every rooting with the minimal duplication cost.

Lemma 1 (adapted from [36]). *Assume that an unrooted tree has a double edge. Then, every leaf of the plateau is a super-speciation, and every internal node of the plateau is a super-duplication.*

On the other hand, when there is an empty edge in an unrooted tree, we have:

Lemma 2. *Let U be an unrooted gene tree with an empty edge e. A node incident to e is a speciation in U_e if and only if it is a leaf of the plateau.*

---

[1] In this article, the notion of the plateau is used exclusively with the duplication cost. In literature, it is often called D-plateau in order to distinguish between plateaus for other costs, e.g. DL-plateau [47].

Proof. We use the notation from Fig. 2 where e is $\langle v, c \rangle$. We may assume that c is an internal node of U, otherwise, we have a trivial case where c is a leaf in the rooting of U which is a speciation. Thus, we have two S2 stars sharing the empty edge. ($\Leftarrow$) Without loss of generality, we may assume that v is a leaf of U^*. If v is not a speciation in $U_{\langle v,c \rangle}$ then it is a duplication. From the definition of the empty edge the root of $U_{\langle v,c \rangle}$ and node v in $U_{\langle v,a \rangle}$ are speciations. Moreover, the node v in $U_{\langle v,a \rangle}$ is mapped to $\mathrm{root}(S)$ thus the root of $U_{\langle v,a \rangle}$ is a duplication. Both rootings $U_{\langle v,c \rangle}$ and $U_{\langle v,a \rangle}$, have the same number of duplications having the same setting of duplications in subtrees T_a, T_b and T_c as indicated in Fig. 2. Hence, $\langle v, a \rangle$ is a U^* edge, a contradiction. ($\Rightarrow$) The proof is similar to the first case.

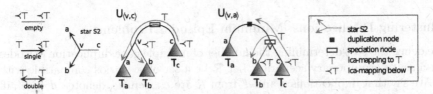

Fig. 2. Types of edges, star S2, and two rootings of an unrooted gene tree U: on the empty edge $\langle v, c \rangle$ and on the single edge $\langle v, a \rangle$. Here, $\top$ denotes the root of S.

The conclusion from the above Lemma 2 is that either only empty edge or the whole S2 star is included in the plateau. Moreover, we can describe the plateau having an empty edge by the following lemma:

Lemma 3. *If the unrooted gene tree has an empty edge then every leaf of the plateau is a super-speciation, and every internal node of the plateau not incident to an empty edge is a super-duplication.*

Proof. For the first part of the proof, let assume that v is a leaf of U^* which consists of $\langle v, c \rangle$ edge. Assume that v is a duplication in some plateau rooting. Then, the subtree T_v in this rooting is also a subtree in all plateau rootings because v is a leaf of U^*. Hence, v is a super-duplication. If $\langle v, c \rangle$ is an empty edge we have a contradiction from Lemma 2. Assume that $\langle v, c \rangle$ is non-empty. The edge $\langle v, a \rangle$ does not belong to U^*, therefore, the rooting $U_{\langle v,a \rangle}$ has more duplications than $U_{\langle v,c \rangle}$. Hence, $U_{\langle v,a \rangle}$ has two duplications in v and in the root. Therefore, the root of $U_{\langle v,c \rangle}$ is not a duplication. However, this is possible only when T_a and T_v are mapped below the $\mathrm{root}(S)$, thus the $\langle v, c \rangle$ is an empty edge, a contradiction. For the next part of the proof, if the U^* consists of exactly one empty edge then the property holds trivially. Let assume that the U^* has more than one edge. We show that every internal node v of U^*, that is, not incident to an empty edge is a super-duplication. Let us consider a path $p = v_1, v_2, \ldots, v_n$ ($n > 1$) consisting of nodes not incident with the empty edge connecting $v = v_1$ with a leaf v_n of U^*. Hence, when rooting on p, v is mapped to $\mathrm{root}(S)$ as it is the ancestor of nodes incident with the empty edge. Moreover, when rooting on $\langle v_{n-1}, v_n \rangle$ we have n gene duplications: for $v_1, v_2, \ldots, v_{n-1}$ and one for the root. All edges from p are elements of U^*, thus moving the root to other edges on p will preserve the total number of gene duplications. We showed that the first $n-1$ nodes on p are duplications for every rooting placed on this path. If v is incident to an empty edge it is a speciation mapped to the $\mathsf{root}(S)$ when rooting on p. When rooting on an empty edge the root is a speciation. Moreover, from Lemma 2 a child of the root is a duplication if it is an internal node of U^*. Hence, all plateau rootings have the same number of duplications equalling the number of internal nodes of U^*. When rooting on an empty edge, the root is a speciation and all internal nodes of U^* are duplications. Otherwise, if we place the root on the edge from U^*, the root is a duplication node and the only speciation is that node among nodes incident to an empty edge which is an ancestor to the other.

Clustering Duplications: Minimum Episodes Problems

We define the cost determining the number of multiple gene duplication episodes for a set of evolutionary scenarios. Let $\mathcal{R}$ be a set of scenarios compatible with S. We say that duplications d and d' from $\mathcal{R}$ are *clusterable*, denoted $d \sim_c d'$, iff (1) d and d' have the same cluster and (2) if d and d' are present in the same DLS-tree then either d and d' are incomparable or equal. Then, the minimum number of duplication episodes for $\mathcal{R}$, denoted $\mathsf{MES}(\mathcal{R}, S)$, is the size of the smallest partition of the set of all duplication nodes from $\mathcal{R}$ induced by an equivalence relation contained in $\sim_c$.

It can be shown that for a collection $\mathcal{R}$ of scenarios compatible with a species tree S,

$$\mathsf{MES}(\mathcal{R}, S) = \sum_{v \in V_S} \max_{T \in \mathcal{R}} \mathsf{duppath}(T, v), \tag{1}$$

where $\mathsf{duppath}(T, v)$ is the maximal (node) length of the path in T that consists of all comparable duplication nodes whose cluster equals the cluster of v [43].

Let $\mathcal{A}(G, S)$ be the set of all scenarios for a rooted gene tree G and a species tree S having the minimal number of gene duplications. Every element of $\mathcal{A}(G, S)$ will be referred to as an *allowed scenario*. Here, allowed scenarios are defined as in [36], for the comprehensive overview see [43]. Now, we formulate the general problem in which the input consists of mixed types of gene trees: rooted and unrooted.

Problem 1 (General Minimum Episodes, GME).
Given a collection of gene trees (rooted or not) $\mathcal{U} = \{U^1, U^2, \ldots, U^n\}$ and a species tree S.
Compute minimum episodes score $\mathsf{ME}(\mathcal{U}, S)$, or ME score, as the minimal value of $\mathsf{MES}(\{R_i\}_{i=1,2,\ldots,n}, S)$ in the sets of scenarios R_i such that $R_i \in \mathcal{A}(U^i, S)$ if U^i is rooted or $R_i \in \mathcal{A}(U^i_e, S)$ if U^i is unrooted, where e is an optimal edge.

Observe that we allow only scenarios that preserve the minimal number of gene duplications. We distinguish two variants of GME Problem: unrooted minimum episodes (UME) and rooted minimum episodes (RME) in which the instances consist entirely of unrooted and rooted gene trees, respectively. RME Problem has a linear time and space solution [43]. See also [38, 41] for more details on RME Problem.

2.1 Unrooted Tree Decomposition

In this section, we show that every unrooted gene tree can be decomposed into a set of trees having at most one unrooted tree with a simplified structure allowing to solve UME in a more efficient way. We start with the following observation.

Lemma 4. *Let U be an unrooted gene tree and T be a rooted subtree of U rooted at v. Let $X \subseteq U^*$ such that*

- *X is disjoint with $V_T \setminus \{v\}$,*
- *v is a speciation in every scenario from $\mathcal{A}(U_e, S)$ for all $e \in E_X$.*

Then, for any set of scenarios $\mathcal{X}$:

$$\min_{R \in \mathcal{A}(U_e, S), e \in E_X} \mathsf{MES}(\mathcal{X} \cup \{R\}, S) = \min_{\substack{R' \in \mathcal{A}(U'_e, S), e \in E_X, \\ R'' \in \mathcal{A}(T, S)}} \mathsf{MES}(\mathcal{X} \cup \{R', R''\}, S), \quad (2)$$

where U'_e is the unrooted tree obtained from U by replacing T with $S(M(v))$.

Proof. In every allowed scenario R from the left side, $F_{U_e}(v)$ is a speciation node. Thus, scenarios R' and R'' can be obtained from R as follows: R'' is the subtree rooted at $F_{U_e}(v)$ in R, while R' is obtained from R by replacing the subtree with the copy of $S(M(v))$, where every internal node is a speciation. Such a transformation is a bijection that preserves the clusterability of duplication nodes. We omit technical details.

Given a species tree S and a rooted tree G by $\widetilde{G}$ we denote the set of all $\preceq$-maximal elements in the set of all non-root speciation nodes from G. Lets $\sim$ be a relation on edges of U^* for an unrooted gene tree U such that $e \sim e'$ if $\widetilde{U}_e = \widetilde{U}_{e'}$. It should be clear that $\sim$ is an equivalence relation. The set of equivalence classes of this relation we denote by $U/_\sim$. An example is depicted in Fig. 3.

Lemma 5. *If an empty edge is present in an unrooted gene tree then every plateau edge present in S2 star uniquely defines one $\sim$-equivalence class. Otherwise, the tree has exactly one $\sim$-equivalence class.*

Proof. Let U be an unrooted gene tree. We have two cases: (a) either U has a double edge or (b) U has an empty edge. In the case (a), it follows from Lemma 1, that $\widetilde{U}_e$ consists of all U^* leaves for every e from U^*. Thus, we have one equivalence class consisting of all U^* edges. Let use the notation from Fig. 2. For the case (b), from the proof of Lemma 3 we conclude that for the empty edge $\langle v, c \rangle$ the set $\widetilde{U}_{\langle v,c \rangle}$ consists of all U^* leaves. Moreover, from the conclusion from the proof of Lemma 2, there are $0, 2$ or 4 single edges in U^* present in S2 stars. Let $\langle v, a \rangle$ be such an edge. The set $\widetilde{U}_{\langle v,a \rangle}$ consists of: (a) v which is the

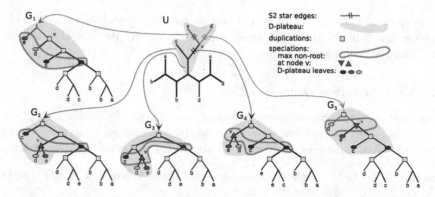

Fig. 3. Equivalence relation $\sim$. An example of an unrooted gene tree U with one S2 star and all plateau rootings reconciled with a species tree S=(((a,b),(c,d)),e). U^* contains five edges and induces three $\sim$-equivalence classes. The first consists of an empty edge $\langle e, v \rangle$, the second of $\langle d, v \rangle$ while the last class consists of the remaining three edges. These three classes induce rootings $\{G_1\}$, $\{G_5\}$ and $\{G_2, G_3, G_4\}$, respectively. Observe, that $\tilde{G}_2 = \tilde{G}_3 = \tilde{G}_4$ consist of a subset of U^* leaves and a speciation (different for each class) at node v which is a center of S2 star.

root of the subtree $T_v = (T_b, T_c)$ and thus it is a speciation (it maps to root(S) and both its children map below the root(S)) and (b) all leaves of U^* present in T_a. From Lemma 3 for every edge e of U^* present in T_a, we have $\tilde{U}_e = \tilde{U}_{\langle v,a \rangle}$. Summing up there can be 1,3 or 5 $\sim$-equivalence classes uniquely defined by every edge of U^* present in S2 star (see Fig. 3).

If an empty edge is an element of a class $X \in U^*/_\sim$, X will be called *plain*. Otherwise, we call X *complex*.

Lemma 6. *If $X \in U^*/_\sim$ is complex then the leaves from $U|_X$ are speciations in every tree U_e for every e in X.*

Proof. U has either an empty or a double edge. The leaves of U^* are super-speciations from Lemmas 1 and 3. If U has a double edge, then there is only one $\sim$-equivalence class (Lemma 5) and every leaf v of $U|_X$ is also a leaf in U^*. If U has an empty edge, say e, then there are 0, 2 or 4 classes X disjoint with $\{e\}$. For all of them the set of the leaves of $U|_X$ consists of a subset of the leaves of U^* (disjoint with subsets corresponding to other classes see Fig. 3) and a node v which is the center of a star S2 and a speciation when rooting on edges from X (see the proof of Lemma 5).

Definition 1 (Unrooted Decomposition). *Let U be an unrooted gene tree, and $X \in U^*/_\sim$, then:*

- *If X has an empty edge e then $\Delta(U, X) = \{U_e\}$.*
- *Otherwise, $\Delta(U, X)$ is the set of all maximal subtrees T_v of U such that v is a leaf of $U|_X$ and $T_v \cap U|_X = \{v\}$.*

For a complex class X, U^X denotes a tree obtained from $U|_X$ by replacing every leaf v with the subtree $S(M(\text{root}(T_v)))$. For example, for the largest class X from Fig. 3, we have: $\Delta(U, X) = \{c, \ (d, e), \ ((a, b), b), ((c, d), d))\}$ and $U^X = (((((a, b), (c, d)), e), ((a, b), (c, d)), c)$.

The intuition is that $\Delta(U, X)$ is the set of rooted trees T induced by X with the following properties: (a) the root of T is a speciation, and (b) T is a subtree present in all rootings induced by X. For example, when we consider an empty class there is only one possible rooting U_e. Hence, $\Delta(U, X) = \{U_e\}$. Lemma 6 describes the properties of $\Delta(U, X)$ for a complex class X. Finally, for an unrooted tree U we have the following formula:

Lemma 7 (Decomposition Lemma). *For a given set of input gene trees $\mathcal{G}$, an input unrooted gene tree U and a species tree S we have,* $\mathsf{ME}(\mathcal{G} \cup \{U\}, S) =$

$$\min_{X \in U/\sim} \begin{cases} \mathsf{ME}(\mathcal{G} \cup \{U_e\}) & \text{if } X = \{e\} \text{ and } e \text{ is empty,} \\ \min_{e \in X} \mathsf{ME}(\mathcal{G} \cup \{U_e^X\} \cup \Delta(U, X), S) & \text{otherwise.} \end{cases}$$

Proof. Let us consider the set of allowed DLS scenarios induced by rootings of edges from each $X \in U/\sim$. If X is plain, then the set is $\mathcal{A}(U_e, S)$. If X is complex, then by Lemma 6, X and every leaf v from $U|_X$, satisfies assumptions from Lemma 4. Thus, the subtree of U disjoint with $X \setminus \{v\}$ can be detached and replaced by $S(M(v))$ in U. By Lemma 4 the MES score is preserved. The rest follows by induction on the set of leaves v, where we show that the unrooted tree after all transformations is U^X and the set of detached subtrees is $\Delta(U, X)$.

3 Algorithms

Solution to RME

We start with the linear time algorithm for RME from [43] adapted to the model of allowed scenarios presented here.

Algorithm 1. Solution to RME (adapted from [43, 45])

1: **Input:** A species tree S, rooted gene trees $G_1, G_2, \ldots, G_n$ and interval $\mathsf{I}(d)$ defined for every duplication node. **Output:** $\mathsf{ME}(G_1, G_2, \ldots, G_n, S)$.

2: Let s be the lowest among top nodes of intervals, i.e., $s = \min_d \max \mathsf{I}(d)$.

3: Let $\lambda(s)$ be the maximal length of the s-chain, where s-chain is a path consisting of duplication nodes d such that $\max \mathsf{I}(d) = s$.

4: For every duplication d such that $\min \mathsf{I}(d) \preceq s \preceq \max \mathsf{I}(d)$, the level of d, denoted $\mathsf{level}_s(d)$, is the maximal number of duplications below d in an s'-chain containing d.

5: Assign every duplication d to s if $\mathsf{level}_s(d) \leq \lambda(s)$.

6: Remove all assigned duplication intervals, add $\lambda(s)$ to the score and repeat steps 2–6 until there is no interval left.

For the input consisting of rooted gene trees, every duplication d is associated with the interval consisting of all possible locations of d in the species tree.

Our model of allowed scenarios is equivalent to the model from [43], in which $I(d)$ is an interval defined by a pair $\langle M(d), s \rangle$, where $s \succeq M(d)$ is the child of $M(g)$ such that g is the lowest speciation satisfying $g \succ d$, or s is the root if such a speciation does not exist. Algorithm 1 is a greedy bottom-up algorithm that iteratively assigns duplications to the top-end of intervals. In every step, it finds the lowest top node s of available intervals and assigns to s all duplications d having max $I(d)$ equal to s. Additionally, the algorithm assigns other duplications to s but only if the ME score is not increased, which is controlled by $\lambda(s)$. For details please refer to [43]. In our case, Algorithm 1 will be used only with rooted gene trees in which the root is a speciation.

Solution to UME

A naïve solution to UME is to run RME algorithm for every combination of plateau rootings from input gene trees. In many cases the plateau can be large, hence, the time complexity of such a solution is $O(\prod_i |U_i|(\sum_i |U_i| + |S|))$. Here, we propose an algorithm based on Lemma 7 to limit the cases that have to be checked to the number of classes of $\sim$ relation.

Algorithm 2. Solution to UME

1: **Input:** Unrooted gene trees $U_1, U_2, \ldots, U_n$, a species tree S.
   Output: ME($\{U_i\}_{i=1,2,\ldots,n}, S$).
2: **For** every sequence $X_1, X_2, \ldots, X_n$ of classes
   from the product $U_1^*/_\sim \times U_2^*/_\sim \times \cdots \times U_n^*/_\sim$:
3:     $\mathcal{X}_r := \bigcup_i \Delta(U_i, X_i)$ and $\mathcal{X}_u := \bigcup_i \{U^{X_i} : X_i$ has no empty edge$\}$
4:     mex $:= \max_{U^X \in \mathcal{X}_u}$ gnaw($U^X, \mathcal{X}_r, S$)
5: **Return** the minimal value of mex computed in the above loop,
   where gnaw is defined below:
6: **Function** gnaw($U^X, \mathcal{X}_r, S$):
7:     Compute $r = \text{ME}(\mathcal{X}_r, S)$ and $\lambda(v)$ for every $v \in S$ by Algorithm 1.
       # Solve an instance of RME
8:     **Let** $\lambda(\text{root}(S)) = +\infty$ and
       $\lambda(v) = 0$ for every $v \neq \text{root}(S)$ not visited in Alg. 1 in line 2.
9:     **For** every $s \in S$, **Let** $\phi(s) = \begin{cases} \text{root}(S) & \text{if } s = \text{root}(S), \\ \text{par}(s), & \text{if } \lambda(\text{par}(s)) > 0, \\ \phi(\text{par}(s)) & \text{otherwise.} \end{cases}$
10:    **For** every ordered pair $\langle u, v \rangle$ of adjacent nodes in X:
11:    $\text{me}(u, v) = \begin{cases} \langle M_{\langle u,v \rangle}(v), 0 \rangle & u \text{ is a leaf in } X, \\ \text{next}(\max(\text{me}(x, u), \text{me}(y, u))) & u \text{ is internal in } X \text{ and } \{x, y, v\} \\ & \text{are all nodes adjacent to } u, \end{cases}$
       where $\text{next}(s, n) = \begin{cases} (s, n+1) & \text{if } n < \lambda(s), \\ (\phi(s), 1) & \text{otherwise.} \end{cases}$
12:    **For** $e = \{u, v\} \in X$,
       $m_e := \max\{n : \text{for } \langle s, n \rangle \in \{\text{me}(u, v), \text{me}(v, u)\} \text{ such that } s = \text{root}(S)\}$
13:    **Return** $r + 1 + \min_{e \in X} m_e$ *# End of gnaw*

Lemma 8 (Correctness of gnaw). *Let U be an unrooted gene tree and X be a complex class. Let $\mathcal{X}_r$ be a set of rooted gene trees T such that the root of every T is a speciation. Let $\mathsf{me}(u, v) = \langle s, n \rangle$, in a call of gnaw with U^X and $\mathcal{X}_r$, such that v is internal in X. Then,*

- *for every rooting U_e^X such that $e \in X$, and having v below the root, if Algorithm 1 (RME) is executed for $\mathcal{X}_r \cup \{U_e^X\}$, then v is assigned to a node s and $n = \mathsf{level}_s(d)$,*
- *the call of gnaw returns $\min_{e \in X} ME(\mathcal{X}_r \cup \{U_e^X\})$.*

Proof. First, observe that every call of gnaw satisfies the assumptions (see Definition 1). Assume that $e \in X$. Then, by the properties of a complex class X, we have in U_e^X that the root and all internal nodes of X, are duplications, while all leaves of X are speciations. Let X'_e be the set of duplication nodes from X including the root. Thus, for every $d \in X'_e$, we have $\mathsf{I}(d) = \langle M_e(v), \mathsf{root}(S) \rangle$, where M_e is the lca-mapping from U_e^X to S. Hence, all duplications from $\mathcal{X}_r$ have the top interval node below the root, therefore, if Algorithm 1 (RME) would be called with the input consisting of $\mathcal{X}_r \cup \{U_e^X\}$, then, for v being the root of S (in line 2 of Algorithm 1), all $\mathcal{X}_r$ duplications are already processed. Additionally, a duplication d from X'_e can be assigned earlier to a node $v \succeq M_e(d)$ only in step 5, if the condition is satisfied. Thus, we can separate the process of RME computation for $\mathcal{X}_r$ (line 3 of Algorithm 2) and the rootings of U^X. Furthermore, processing U^X can be done collectively for all rootings from X, by using a dynamic programming that jointly executes the assignment operation. Note, that in line 11 the first elements of $\mathsf{me}(x, u)$ and $\mathsf{me}(y, u)$ are comparable (i.e., u is a duplication), therefore, max is well defined by using lexicographical order. The proof of the first part follows by induction, in which a node in a rooted subtree of U^X is assigned to the first next free "slot" in a species node. Such a slot can be located by using next. When all slots of non-root nodes are occupied then duplications have to be assigned to the root. Such assignments create new episode events. Thus, the score of every rooting placed on $e = \{u, v\}$ can be easily computed by verifying if such additional episodes were created. This information is stored for the two subtrees of the root in $\mathsf{me}(u, v)$ and $\mathsf{me}(v, u)$, respectively, i.e., if $\mathsf{me}(u, v) = \langle \mathsf{root}(S), n \rangle$, then n additional episodes are required. This value for both subtrees is stored in m_e. Note that, max in line 12 is well defined, otherwise, X cannot be complex. Additionally, the root of U_e^X creates one more episode. Therefore, the score returned by gnaw consists of r (from rooted trees), the minimal value of m_e (the contribution of X) and 1 (the root duplication). An example is depicted in Fig. 4.

Lemma 9 (Correctness). *Given a collection of unrooted gene trees $\mathcal{U}$ and a species tree S, Algorithm 2 returns $ME(\mathcal{U}, S)$.*

Proof. The proof follows from Decomposition Lemma 4 and Lemma 8.

Lemma 10 (Complexity). *Algorithm 2 requires $O((|S| + \sum_i |U_i|)5^m)$ time and $O(\sum_i |U_i| + |S|)$ space, where m is the number of gene trees with S2 star having more than one class of $U/_\sim$.*

Proof. Time: The number of iterations of the main loop is bounded above by 5^m. Locating classes of $\sim$ and transforming trees can be done in linear time. Each call of gnaw requires $O(\sum_{T \in \mathcal{X}_r} |T| + |U^X|)$ time. *Space:* It follows from the complexity of Algorithm 1 and gnaw.

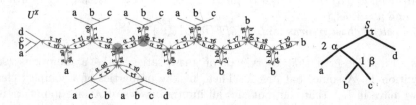

Fig. 4. Illustration of gnaw for U with a double edge. Here, τ denotes the root of S. Assume that S has two positive lambda's computed in line 3 by Algorithm 1: $\lambda(\alpha) = 2$ and $\lambda(\beta) = 1$. Every edge $e = \langle u, v \rangle$ of the plateau is split into two directed edges: $\langle u, v \rangle$ and $\langle v, u \rangle$. Each directed edge $\langle u, v \rangle$ is decorated with the lca-mapping $M_{\langle u,v \rangle}(v)$ and me(u, v). For example, $\tau 6$ denotes the lca-mapping to τ and me$(u, v) = \langle \tau, 6 \rangle$. Here, gnaw returns $3 + 1 + 3$ induced by the marked edge.

Experimental Evaluation

We evaluated the dataset from [20] consisting of 53 rooted gene trees from 16 Eucaryotes. Multiple gene duplication events were inferred for two species trees: S_1 from [48] and S_2 from [20]. The comparison of the results for RME [43] and UME algorithms is shown in Fig. 5, where the original rooting of every gene tree was ignored in UME.

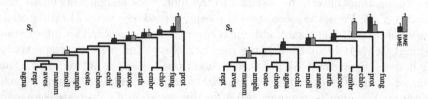

Fig. 5. Duplication episodes in Guig dataset [20] inferred by RME [43] and UME algorithms for species trees S_1 [48] and S_2 [20].

The result for S_1 indicates that UME algorithm found a better scenario than RME, i.e., 5 episodes vs. 6. Additionally, the duplication locations are generally in agreement with the solution to the unrooted variant of episode clustering (see more in [36]). Next, the result of RME for S_2 is consistent with [20,38]. However, in [37] authors suggested different evolutionary scenarios with a greater number of duplication episodes. The results differ, i.e., for the gene tree for

β-nerve growth factor precursor (NGF) of topology $(rept, (mamm, (amph, aves)))$ in the placement of two duplications inferred by that gene tree and S_2. In the optimal solution from UME algorithm the rooting of NGF gene tree is $(aves, ((mamm, rept), amph))$ and it infers one duplication with S_2.

4   Conclusions and Future Work

In this article, we proposed the first solution to the problem of minimum episodes clustering for the case when input gene trees are unrooted. We showed new properties of unrooted reconciliation for the duplication cost. Then, we proposed a decomposition of an unrooted gene tree and we showed that each part of the decomposition can be treated independently in the algorithm that solves UME. In theory, this is an FPT algorithm depending on the number of input trees having a special topology. The algorithm can be further improved by dealing with these special trees similarly to our solution to unrooted EC problem from [36].

Future work will focus on the open question of the complexity of UME (we conjecture that UME is intractable). Moreover, we plan to research on the applications of the developed theory to infer large genomic duplication events from simulated and empirical datasets of unrooted gene trees.

Acknowledgements. We would like to thank the reviewers for their detailed comments that allowed us to improve our paper. The support was provided by NCN grants #2015/19/N/ST6/01193 and #2015/19/B/ST6/00726.

References

1. Kellis, M., Birren, B.W., Lander, E.S.: Proof and evolutionary analysis of ancient genome duplication in the yeast Saccharomyces cerevisiae. Nature **428**, 617–624 (2004)
2. Guyot, R., Keller, B.: Ancestral genome duplication in rice. Genome **47**(3), 610–614 (2004)
3. Vision, T.J., Brown, D.G., Tanksley, S.D.: The origins of genomic duplications in Arabidopsis. Science **290**(5499), 2114–2117 (2000)
4. Costantino, L., Sotiriou, S.K., Rantala, J.K., Magin, S., et al.: Break-induced replication repair of damaged forks induces genomic duplications in human cells. Science **343**(6166), 88–91 (2014)
5. Cui, L., Wall, P.K., Leebens-Mack, J.H., Lindsay, B.G., et al.: Widespread genome duplications throughout the history of flowering plants. Genome Res. **16**(6), 738–749 (2006)
6. Aury, J.M., Jaillon, O., Duret, L., Noel, B., et al.: Global trends of whole-genome duplications revealed by the ciliate Paramecium tetraurelia. Nature **444**(7116), 171–178 (2006)
7. Van de Peer, Y., Maere, S., Meyer, A.: The evolutionary significance of ancient genome duplications. Nat. Rev. Genet. **10**(10), 725–732 (2009)
8. Vandepoele, K., Simillion, C., Van de Peer, Y.: Evidence that rice and other cereals are ancient aneuploids. Plant Cell. **15**(9), 2192–2202 (2003)

9. Sato, S., Tabata, S., Hirakawa, H., Asamizu, E., et al.: The tomato genome sequence provides insights into fleshy fruit evolution. Nature **485**(7400), 635–641 (2012)

10. Scossa, F., Brotman, Y., de Abreu e Lima, F., et al.: Genomics-based strategies for the use of natural variation in the improvement of crop metabolism. Plant Sci. **242**, 47–64 (2016)

11. Vanneste, K., Maere, S., Van de Peer, Y.: Tangled up in two: a burst of genome duplications at the end of the Cretaceous and the consequences for plant evolution. Philos. Trans. R. Soc. Lond. B Biol. Sci. **369**(1648), 20130353 (2014)

12. Tang, H., Bowers, J.E., Wang, X., Ming, R., et al.: Synteny and collinearity in plant genomes. Science **320**(5875), 486–488 (2008)

13. Holloway, P., Swenson, K., Ardell, D., El-Mabrouk, N.: Ancestral genome organization: an alignment approach. J. Comput. Biol. **20**(4), 280–295 (2013)

14. Blanc, G., Wolfe, K.H.: Widespread paleopolyploidy in model plant species inferred from age distributions of duplicate genes. Plant Cell **16**(7), 1667–78 (2004)

15. Bowers, J.E., Chapman, B.A., Rong, J., Paterson, A.H.: Unravelling angiosperm genome evolution by phylogenetic analysis of chromosomal duplication events. Nature **422**(6930), 433–8 (2003)

16. Jiao, Y., Wickett, N.J., Ayyampalayam, S., Chanderbali, A.S., et al.: Ancestral polyploidy in seed plants and angiosperms. Nature **473**(7345), 97–100 (2011)

17. Rabier, C.E., Ta, T., Ané, C.: Detecting and locating whole genome duplications on a phylogeny: a probabilistic approach. Mol. Biol. Evol. **31**(3), 750–62 (2014)

18. Page, R.D.M.: Maps between trees and cladistic analysis of historical associations among genes, organisms, and areas. Syst. Biol. **43**(1), 58–77 (1994)

19. Mirkin, B., Muchnik, I., Smith, T.F.: A biologically consistent model for comparing molecular phylogenies. J. Comput. Biol. **2**(4), 493–507 (1995)

20. Guigó, R., Muchnik, I.B., Smith, T.F.: Reconstruction of ancient molecular phylogeny. Mol. Phylogenet. Evol. **6**(2), 189–213 (1996)

21. Arvestad, L., Berglund, A.C., Lagergren, J., Sennblad, B.: Bayesian gene/species tree reconciliation and orthology analysis using MCMC. Bioinformatics **19**(Suppl 1), i7–15 (2003)

22. Bonizzoni, P., Della Vedova, G., Dondi, R.: Reconciling a gene tree to a species tree under the duplication cost model. Theor. Comput. Sci. **347**(1–2), 36–53 (2005)

23. Noutahi, E., Semeria, M., Lafond, M., Seguin, J., et al.: Efficient gene tree correction guided by genome evolution. PLoS ONE **11**(8), 1–22 (2016)

24. Schmidt-Böcking, H., Reich, K., Templeton, A., Trageser, W., Vill, V.: Reconstructing a supergenetree minimizing. BMC Bioinf. **16**(14), S4 (2015)

25. Dondi, R., Mauri, G., Zoppis, I.: Orthology correction for gene tree reconstruction: theoretical and experimental results. Proc. Comput. Sci. **108**, 1115–1124 (2017)

26. Scornavacca, C., Jacox, E., Szöllősi, G.J.: Joint amalgamation of most parsimonious reconciled gene trees. Bioinformatics **31**(6), 841–848 (2014)

27. Nakhleh, L.: Computational approaches to species phylogeny inference and gene tree reconciliation. Trends Ecol. Evol. **28**(12), 719–728 (2013)

28. Zhu, Y., Lin, Z., Nakhleh, L.: Evolution after whole-genome duplication: a network perspective. G3: Genes, Genomes. Genetics **3**(11), 2049–2057 (2013)

29. Zheng, Y., Zhang, L.: Effect of incomplete lineage sorting on tree-reconciliation-based inference of gene duplication. IEEE/ACM Trans. Comput. Biol. Bioinf. **11**(3), 477–485 (2014)

30. Duchemin, W., Anselmetti, Y., Patterson, M., Ponty, Y., et al.: DeCoSTAR: Reconstructing the ancestral organization of genes or genomes using reconciled phylogenies. Genome Biol. Evol. **9**(5), 1312–1319 (2017)

31. Goodman, M., Czelusniak, J., Moore, G.W., Romero-Herrera, A.E., et al.: Fitting the gene lineage into its species lineage, a parsimony strategy illustrated by cladograms constructed from globin sequences. Syst. Zool. **28**(2), 132–163 (1979)
32. Doyon, J.P., Chauve, C., Hamel, S.: Space of gene/species tree reconciliations and parsimonious models. J. Comput. Biol. **16**(10), 1399–1418 (2009)
33. Ma, B., Li, M., Zhang, L.: From gene trees to species trees. SIAM J. Comput. **30**(3), 729–752 (2000)
34. Stolzer, M., Lai, H., Xu, M., et al.: Inferring duplications, losses, transfers and incomplete lineage sorting with nonbinary species trees. Bioinformatics **28**(18), i409–i415 (2012)
35. Górecki, P., Tiuryn, J.: DLS-trees: a model of evolutionary scenarios. Theor. Comput. Sci. **359**(1–3), 378–399 (2006)
36. Paszek, J., Górecki, P.: Genomic duplication problems for unrooted gene trees. BMC Genom. **17**(1), 165–175 (2016)
37. Page, R.D.M., Cotton, J.A.: Vertebrate phylogenomics: reconciled trees and gene duplications. In: Pacific Symposium on Biocomputing, pp. 536–547 (2002)
38. Bansal, M.S., Eulenstein, O.: The multiple gene duplication problem revisited. Bioinformatics **24**(13), i132–8 (2008)
39. Burleigh, J.G., Bansal, M.S., Wehe, A., Eulenstein, O.: Locating multiple gene duplications through reconciled trees. In: Vingron, M., Wong, L. (eds.) RECOMB 2008. LNCS, vol. 4955, pp. 273–284. Springer, Heidelberg (2008). doi:10.1007/978-3-540-78839-3_24
40. Mettanant, V., Fakcharoenphol, J.: A linear-time algorithm for the multiple gene duplication problem. NCSEC, pp. 198–203 (2008)
41. Luo, C.W., Chen, M.C., Chen, Y.C., Yang, R.W.L., et al.: Linear-time algorithms for the multiple gene duplication problems. IEEE/ACM Trans. Comput. Biol. Bioinform. **8**(1), 260–265 (2011)
42. Burleigh, J.G., Bansal, M.S., Eulenstein, O., Vision, T.J.: Inferring species trees from gene duplication episodes. ACM BCB, pp. 198–203 (2010)
43. Paszek, J., Górecki, P.: Efficient algorithms for genomic duplicationmodels; APBC 2017. IEEE/ACM Trans. Comput. Biol. Bioinform. doi:10.1109/TCBB.2017.2706679
44. Fellows, M., Hallett, M., Stege, U.: On the multiple gene duplication problem. In: Chwa, K.-Y., Ibarra, O.H. (eds.) ISAAC 1998. LNCS, vol. 1533, pp. 348–357. Springer, Heidelberg (1998). doi:10.1007/3-540-49381-6_37
45. Czabarka, E., Szkely, L., Vision, T.: Minimizing the number of episodes and Gallai's theorem on intervals. arXiv:12095699;2012
46. Górecki, P., Tiuryn, J.: Inferring phylogeny from whole genomes. Bioinformatics **23**(2), e116–e122 (2007)
47. Górecki, P., Eulenstein, O., Tiuryn, J.: Unrooted tree reconciliation: a unified approach. IEEE/ACM Trans. Comput. Biol. Bioinform. **10**(2), 522–536 (2013)
48. Page, R.D.M., Charleston, M.A.: Reconciled trees and incongruent gene and species trees. Math. Hierarchies Biol. DIMACS 96 **37**, 57–70 (1997)

TreeShrink: Efficient Detection of Outlier Tree Leaves

Uyen Mai[1] and Siavash Mirarab[2(✉)]

[1] Computer Science and Engineering,
University of California at San Diego, San Diego, USA
umai@eng.ucsd.edu
[2] Electrical and Computer Engineering,
University of California at San Diego, San Diego, USA
smirarab@eng.ucsd.edu

Abstract. Phylogenetic trees include errors for a variety of reasons. We argue that one way to detect errors is to build a phylogeny with all the data then detect taxa that artificially inflate the tree diameter. We formulate an optimization problem that seeks to find k leaves that can be removed to reduce the tree diameter maximally. We present a polynomial time solution to this "k-shrink" problem. Given this solution, we then use non-parametric statistics to find an outlier set of taxa that have an unexpectedly high impact on the tree diameter. We test our method, TreeShrink, on five biological datasets, and show that it is more conservative than rogue taxon removal using RogueNaRok. When the amount of filtering is controlled, TreeShrink outperforms RogueNaRok in three out of the five datasets, and they tie in another dataset.

Keywords: Tree diameter · Rogue taxon removal · Gene tree discordance

1 Introduction

Modern phylogenetic datasets include a large number of genes and species. The sheer dataset size prevents analysts from carefully curating sequence alignments and trees manually. Moreover, manual curation is subject to biases of the curator and poses challenges in reproducibility. At the same time, phylogenetic analyses involve pipelines of many error-prone steps, including sequencing, contamination removal, homology and orthology detection, multiple sequence alignment, and tree inference. It has been long recognized that errors can propagate among these steps [1,2], but finding errors is difficult, especially when hundreds of genes are present [3]. Moreover, true gene trees can be discordant biologically or by estimation error; when these gene trees are fed to a species tree estimation method, the error may propagate [4–6]. This has motivated the development of

© Springer International Publishing AG 2017
J. Meidanis and L. Nakhleh (Eds.): RECOMB CG 2017, LNBI 10562, pp. 116–140, 2017.
DOI: 10.1007/978-3-319-67979-2_7

co-estimation methods that promise to more effectively break the chain of error propagation [7–9]. However, an end-to-end co-estimation of the entire pipeline has remained unachievable [9], leaving practitioner with other options.

More recently, promising methods of correcting gene trees based on a *given* species using reconciliation [10,11] have been developed [12–14]. When a species tree is not available, analysts often find creative and ad-hoc ways to find and eliminate erroneous data. Data filtering should be treated with care as it can remove useful signal in addition to error [15], and it also runs the risk of introducing bias. Thus, whether or how much filtering is justified remains unclear.

Data filtering by alignment masking has been extensively developed [16,17] despite some criticism [15]. Removing individual sites with a large impacts on the tree topology has also been suggested [18]. At the level of phylogenetic trees, two approaches are in wide use: rogue taxon removal (RTR) [19–23] and gene tree filtering. RTR methods look for taxa with unstable positions in a (gene) tree. They use bootstrapping [21,22] or jackknifing [19] to detect taxa that move often in replicate runs. A second approach is to remove genes that are believed to be problematic, perhaps due to missing data [24,25] or lack of signal [26]. The hope is that the inference of the species tree (i.e., by summarizing gene trees or concatenation) becomes more accurate if error prone genes are eliminated. Some analyses (e.g., [27]) take the middle road and filter individual taxa from individual gene trees based on some criteria (e.g., fragmentation). These aim to eliminate only the problematic data but nothing more. This raises a question: can we find minimal subsets of the data that are maximally responsible for errors?

One approach for finding problematics subset of the data is to examine branch lengths of an inferred phylogeny. Under a strict molecular clock assumption, one would expect all leaves to be equidistant from the root. Even when a strict clock cannot be assumed, one would not expect to see only a small minority of species that have *dramatically* different rates of evolution and hence root to tip distances. In other words, in a true phylogenetic tree, variations in root to tip distance are expected, but outliers in terms of root to tip distance have to be treated with suspicion. Many sources of error in a phylogenetic analysis, e.g., contamination, mistaken orthology, and mis-alignment, can produce sequences that will have to be put on very long branches compared to the other species. Thus, in cases of extremely long branches (e.g., Fig. 1a), it is likely that the sequence data of the outlier taxon or taxa contains errors, even if the cause of the error remains unclear. Even if the sequences of these extremely long branches are error-free, they still may be misplaced due to long branch attraction [28]. Thus, several studies have tried removing long branches [27,29], typically relying on rooted gene trees [27,29]. However, rooting is often challenging and prone to error.

Rooting is not necessary for finding distant taxa, as we can instead use the tree diameter (the maximum distance between any two leaves). In this paper, we introduce an optimization problem to find taxa that artificially inflate the diameter of a tree. We formulate the following k-shrink optimization problem.

Given: a tree on n leaves with branch lengths, and a number $1 \leq k \leq n$
Find: for each $1 \leq i \leq k$, the set of i leaves that should be removed to reduce
   the tree diameter maximally.

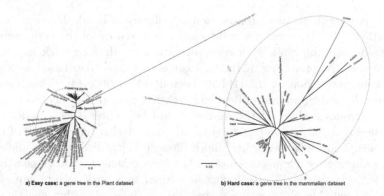

a) Easy case: a gene tree in the Plant dataset   **b) Hard case:** a gene tree in the mammalian dataset

Fig. 1. Example trees with suspicious long branches. (a) An unfiltered gene tree of a Plant dataset [27] with an obvious outlier leaf; (b) a gene tree in a mammalian dataset with a hard to detect outlier branch [30]. Outgroups are shown in blue and the suspicious long branches in the red. Dashed green line: the tree diameter after removal of red branches. Detecting the red branch is easy on the left but hard on the right. (Color figure online)

We develop an algorithm to solve the problem in $O(k^2h + n)$ time where h is the height of the tree. Given the results of the k-shrink problem, we face a non-trivial question: how to choose the number of leaves to remove?

Picking the number of taxa to remove requires trading off removing correct sequences on the one hand and keeping erroneous sequences on the other hand. This task is further complicated by long outgroups that greatly impact the diameter (Fig. 1b). We take a conservative approach and seek a minimal number of taxa that could be removed so that the reduction in the diameter of the tree is unexpectedly high. For $k = \Theta(\sqrt{n})$, we compute the reduction in the diameter when going from $i-1$ to i removals for $1 \leq i \leq k$ and then look for outlier values in the spectrum of these "delta diameter" values. We empirically show that the delta values have a tight and stable distribution after outliers are removed; we use statistical measures to find outliers in this distribution for a given level of false positive tolerance (α). On five biological datasets, we show that our method, TreeShrink, run with default settings $(\alpha = 0.05)$ improves the quality of gene trees effectively and reduces their discordance with putative species trees.

2   TreeShrink

We give main results in this section and present all proofs in Appendix A.

Notations and Definitions: For an unrooted tree t on the leaf-set $\mathcal{L}$, let $\delta(a, b)$ give the distance between a and b. The restriction of t to the leafset A is denoted by $t \lceil_A$, and we define $t \backslash_a = t \lceil_{\mathcal{L}-\{a\}}$. We refer to a pair of leaves in t with the highest pairwise distance as a *diameter pair* and call the two leaves *on-diameter*. Any tree has at least one diameter-pair. We define $\mathcal{P}(t)$ as a set of all diameter pairs of t: $\mathcal{P}(t) = \{(a,b)|(\forall x, y \in \mathcal{L})[\delta(a,b) \geq \delta(x,y)]\}$. The diameter set $\mathcal{D}(t)$ is defined as the set of all on-diameter leaves: $\mathcal{D}(t) = \{a|(\exists x)[(a,x) \in \mathcal{P}(t)]\}$.

We call a tree t *singly paired* if *all* the restricted trees of t (including t) have only one diameter pair; that is, $\forall A \subseteq \mathcal{L}, |\mathcal{P}(t \restriction_A)| = 1$. We refer to the process of removing *one* leaf from t as a *removal*. A removal of a leaf $a \in \mathcal{D}(t)$ is defined as a *reasonable removal*.

A *removing chain* of t is defined as an ordered list of removals. We refer to a removing chain of length k as a *k-removing chain*, and denote it by $\mathcal{H}_k(t)$. We refer to a removing chain that consists only of reasonable removals as a *reasonable removing chain*. An *optimal k-removing chain*, $\mathcal{H}_k^*(t)$, is a removing chain that results in a tree with the minimum diameter among all chains of length k. Any $\mathcal{R}_k(t) \subset \mathcal{L}$ with $|\mathcal{R}_k(t)| = k$ is called a *k-removing set* of t, and is called a *reasonable k-removing set* if there exists an ordering of $\mathcal{R}_k(t)$ that gives a reasonable k-removing chain. We refer to the set of all reasonable k-removing sets as the *k-removing space* of t, and denote it by $\mathcal{S}_k(t)$. We let $\mathcal{R}_k^*(t)$ denote an arbitrary *removing set* that gives the restricted tree with the minimum diameter. Finally, for a rooted version of t, we let $Cld(u)$ denote the set of leaves descended from u.

2.1 A Polynomial Time Solution to the k-shrink Problem

Only a reasonable removal can reduce the tree diameter. If t is singly paired, there are two reasonable removals to be considered, one of which may reduce the diameter more. If we are to remove more than one leaf, a simple greedy approach that chooses the optimal removal at each step is not guaranteed to produce the optimal solution (see Fig. 2a for a counter-example). Therefore, we need to consider a search space for this problem. However, a brute force search for all reasonable k-removing chains grows at least as $\Omega(2^k)$. The brute force method would first consider the initial diameter pair(s); then, to remove each of the two on-diameter leaves, it would consider the new diameter pair(s) after the removal and recurse on each diameter-pair. This recursive process produces all reasonable removing chains from 1 to k, and grows exponentially.

Three observations help us reduce the search space to grow linearly with k. First, if (a, b) is a diameter pair, then b remains on diameter after removing a.

Proposition 1. *If an on-diameter leaf is removed, the rest of the on-diameter set are on-diameter for the restricted tree:* $a \in \mathcal{D}(t) \Rightarrow \mathcal{D}(t) - \{a\} \subset \mathcal{D}(t \backslash_a)$.

All i-removing spaces for $1 \leq i \leq k$ can be represented as a DAG (Fig. 2c). In this graph, each node at row i represents an i-removing set $\mathcal{R}_i(t)$, and is also annotated by a diameter pair after the removal of $\mathcal{R}_i(t)$. Any path from the root ending at a node $\mathcal{R}_i(t)$ is a i-removing chain. Each row i in this graph corresponds to the $i-$removing space. Note that each node can be reached with multiple paths from the root; this leads to a second observation: any ordering of an i-removing chain gives the same restricted tree. Thus, we can reduce the search space from reasonable chains to reasonable sets. These two observations allow us to design a polynomial time algorithm for singly paired trees (described

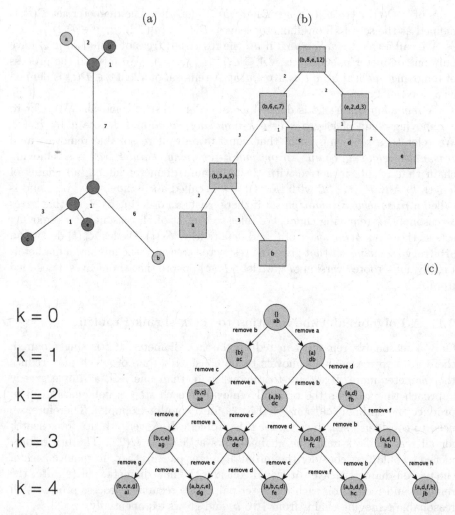

Fig. 2. Demonstration of the k-shrink algorithm (a) An example where a greedy method does not work. This tree only has one diameter pair (a, b) (colored in yellow). If $k = 2$, the greedy method removes b and c and gives a restricted tree with diameter 11, while the optimal solution is to remove a and d, which gives the restricted tree with diameter 10. (b) The preprocessing step. In a post-order traversal, we store $rec(u)$ for each internal node u. The record has four values: leaf x under u that has the longest distance to u, the distance $\delta(u, x)$, the leaf y in one of the other sides of u to x that has the largest $\delta(x, y)$, and the distance (c) Graphical representation of the reasonable search space. The root node represents the initial tree t; each node on row k represents a restricted tree with k leaves removed. Each node is annotated by the removing set (top) and a diameter pair of the induced tree (bottom). Each edge in the graph represents a reasonable removal. The path from the root to any node corresponds to a reasonable removing chain. Each row k in the graph gives the k-removing space of t ($\mathcal{S}_k(t)$). (Color figure online)

next). Our third observation (formalized later) is that when a tree is not singly paired, breaking ties arbitrarily still guarantees optimality.

Singly-Paired Trees: We now state our main results.

Theorem 1. *The k-removing space of a singly paired tree t includes all the optimal k-removing sets of t; that is:* $\forall k > 0 : \mathcal{R}_k^*(t) \in \mathcal{S}_k(t)$.

Theorem 2. *The size of the k-removing space for a singly paired tree t is $k+1$.*

Corollary 1. *The size of the reasonable search space up to level k is $O(k^2)$.*

The ith row of the reasonable search space graph (Fig. 2c) contains $i+1$ nodes and one of them gives an optimal i-removing set. Therefore, traversing all $O(k^2)$ nodes in this graph gives the optimal solution to our k-shrink problem.

Algorithm 2 describes how we build and traverse the reasonable search space graph. We start with the root node, corresponding to the initial tree t and build the rows iteratively. For any node at step i with (x, y) as its diameter pair, two nodes have to be added to the next row, one for removing x and another for removing y. As the proof of Theorem 2 indicates, two sister nodes in step i have to share one descendant in step $i+1$ (Fig. 2c). Thus, to construct each row from the previous row we simply need to find the diameter pair of the tree restricted to the removal-set of each node; as we do this, we also keep track of the length of the diameter at each node and the optimal i-removing set. At the end, we report an optimal-removing set for each i from 1 to k.

Computing the pair set for all nodes can be done in $O(n)$ with some pre-processing. First, we do a bottom-up traversal of t (rooted arbitrarily) and for each internal node u store four values: (i) the leaf $x \in Cld(u)$ with the longest distance to u, (ii) the distance $\delta(u, x)$, (iii) the leaf $y \in Cld(u) - Cld(c_1)$ with the longest distance to u (where c_1 is a child of u such that $x \in Cld(c_1)$), and (iv) the distance $\delta(x, y)$ (see Fig. 2b). We store these values for each node u as a tuple $rec(u) = (rec_1(u), rec_2(u), rec_3(u), rec_4(u))$. These values can be computed in a post-order traversal of the tree in the natural way. Let (a, b) be a diameter pair; note that regardless of the arbitrary rooting of the tree, at the LCA of a and b, the record includes a, b as the first and third fields and the tree diameter as the last. Thus, the tree diameter corresponds to the record with the largest fourth value. Given the results of this preprocessing, we have:

Proposition 2. *Given a rooted tree t of height h, $(a, b) \in \mathcal{P}(t)$, and $rec(u)$ for all nodes $u \in \mathcal{L}$, Algorithm 1 correctly finds one diameter pair of $t \backslash_a$ in $O(h)$.*

According to Proposition 2, finding the new diameter pair after removing any node can be done in $O(h)$ using Algorithm 1. From Corollary 1 and Proposition 2, we have:

Corollary 2. *Algorithm 2 solves the k-shrink problem in $O(k^2h + n)$.*

Algorithm 1. Assuming $(a, b) \in \mathcal{P}(t)$, find a leaf x such that $(x, b) \in \mathcal{P}(t_{\backslash a})$; assume t has $rec(u)$ computed for all of its nodes.

Function FindPair(t, a, b)
    $diameter \leftarrow 0$
    For each node u in the path from the parent of a to the root
        Update $rec(u)$ from its children records, ignore a if it is one of the children
        If $rec_4(u) > diameter$
            $diameter \leftarrow rec_4(u)$
            $diamPair \leftarrow (rec_1(u), rec_3(u))$
    For each node u in the path from the parent of b to $LCA(a, b)$
        If $rec_4(u) > diameter$
            $diameter \leftarrow rec_4(u)$
            $diamPair \leftarrow (rec_1(u), rec_3(u))$
    return $x \in diamPair$ if $x \neq a$

Algorithm 2. Polynomial time k-shrink algorithm

Function Shrink(t, k)
    Compute $rec(u)$ for all nodes of t using a postorder tree traversal
    $minD \leftarrow$ an array of k elements initialized to ∞
    $(a, b) \leftarrow (rec_1(u), rec_3(u))$ where u is the node with the maximum $rec_4(u)$
    add tuple $(0, a, b, \{\}, \delta(a, b))$ to an empty queue Q
    while Q not empty
        $(i, a, b, R, d) \leftarrow$ remove from Q
        $minD[i] = min(minD[i], d)$
        if $i < k$
            add $(i + 1, FindPair(t \restriction_{\mathcal{L} - R}, a, b), b, R \cup \{a\})$ to Q
            if index i is removed from Q for the first time
                add $(i + 1, a, FindPair(t \restriction_{\mathcal{L} - R}, b, a), R \cup \{b\})$ to Q
    return $minD$

Extension to Trees that Are Not Singly Paired: If the tree t is not singly paired, nodes in the search graph could have more than two children. Allowing this would increase the size of the search space. However, we prove that whenever there are ties, we can break them arbitrarily and still guarantee to find *an* optimal solution. Thus, Algorithm 2 works for trees that are not singly paired.

For any diameter pair $(a, b) \in \mathcal{P}(t)$, we define a *pair-restricted k-removing space* as a subset of $\mathcal{S}_k(t)$ such that each of its elements includes either a or b.

Theorem 3. *For any k, any arbitrary pair-restricted k-removing space includes at least one optimal k-removing set.*

The full proof, given in Appendix A.5, is involved and requires several steps. First, it is not hard to prove that any tree t has a single midpoint which partitions its diameter set into disjoint subsets. We call each of those subsets a *diameter group* of t (refer to Lemmas 2 and 3 in Appendix A.5). It is not hard to see that the tree cannot shrink in diameter unless all but one of its diameter groups are removed. We refer to the restricted tree of the removing set consisting of all but one of the diameter groups as a *minimum shrunk tree* of t. We can prove that any arbitrary pair-restricted removing space can produce *all* minimum shrunk tree (see the full proof in Appendix A.5). Besides, unless k is small enough so that there is no k-removing set can reduce the tree diameter (in such a case any solution is optimal and the result of Theorem 3 trivially follows), any optimal solution of k-shrink can be induced from one of those minimum shrunk trees (Lemma 4 in Appendix A.5). Thus, to find an optimal tree t^*, we can start from *any* pair-restricted removing space and concatenate the two removing chains: the chain that induce the minimum shrunk tree t_i^* from any arbitrary diameter pair, and the chain that starts from t_i^* to induce t^*.

Theorem 3 tells us that searching any pair-restricted k-removing space guarantees to give us at least one optimal solution. Therefore, for a tree that is not singly paired, we can simply restrict the search space arbitrarily to one of its diameter pairs *at any step* of the algorithm. This ensures that the search space size grows with $O(k^2)$, and that Algorithm 2 still correctly finds an optimal solution in $O(k^2 h + n)$.

2.2   Statistical Selection of the Filtering Level

Let's assume we have selected $k \leq n$, the maximum number of leaves that we allow to be removed. Having a solution to the k-shrink problem, we now want to find the smallest value r for $1 \leq r \leq k$ such that removing r leaves reduces the diameter substantially more than expected. Defining what is an expected reduction in the diameter as a result of each removal is not trivial and depends on many factors such as the rate of speciation, taxon sampling, and the tree topology. To avoid modeling such processes, we use empirical statistics.

Let Δ_i be the difference in the minimum diameter with $i - 1$ leaves removed and the minimum diameter with i leaves removed (i.e., the difference between the minimum at rows $i - 1$ and i of the reasonable search space of Fig. 2c). For a tree with no outlier branches, we would expect Δ_i values to be randomly distributed around some stable value (T1 in Fig. 3). For a tree with one outlier taxon on a very long branch, we would expect that Δ_1 is much larger than other Δ_i values (T2 in Fig. 3). If two taxa are on a very long branch, we would expect a small Δ_1, a large Δ_2, and small values again for $i > 2$ (T3 in Fig. 3). If there are two exceptionally long branches, one with three taxa and another with five taxa, we expect Δ_3 and Δ_8 to be large and other values to be small (T4 in Fig. 3). Taking this idea one step further, let $\mathcal{E} \subset \{1 \ldots k\}$ be the set of values i where Δ_i is "exceptionally large" (e.g., $\mathcal{E} = \{3, 8\}$ for T4). Then, it would be reasonable to take the maximum value in $\mathcal{E}$ as the cutoff (i.e., $r = \max(\mathcal{E})$). A crucial question remains to be unanswered: how do we define "exceptionally large"? While various methods could be used, we take a non-parameteric approach.

For any tree and a large enough k, we have a distribution over Δ values. We log-transform the Δ values and compute a kernel density function [31] over the empirical distribution of $\log(\Delta)$ (results without log transformation were not appreciably different; see Fig. 7). To estimate the density, we use Gaussian kernels with Silverman's rule of thumb smoothing bandwidth [31] (as implemented in

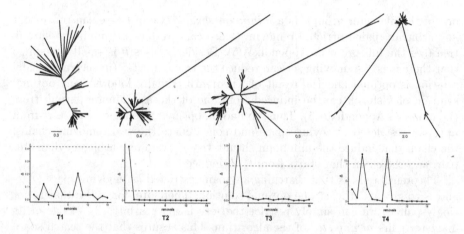

Fig. 3. Patterns of Δ_i as a function of i. Four unfiltered gene trees from a Plant dataset [27] are shown (*top*). For each tree, we also show Δ_i for $1 \leq i \leq 2\sqrt{n} = 16$ (*bottom*). The selected threshold for defining outliers is shown with a dashed red line. r will be set to the largest value above the threshold (0, 1, 2, and 8, respectively). (Color figure online)

the R package [32]). Given the density function and a false positive tolerance rate α, we define values with a CDF above $1 - \alpha$ as outliers. The maximum i that results in an outlier Δ_i is chosen as r, the threshold of removal. For datasets that include several gene trees, putting all the Δ values from different genes together in estimating the kernel increases the number of data points and may increase the accuracy (see an example in Fig. 7). When doing so, the branch lengths in trees from different genes have to be normalized; we chose to normalize by the median branch length of each gene tree.

Finally, we need to decide k, the maximum number of leaves that *could* be removed. Large values of k do not fit our goal of finding *outlier* taxa, and thus, we seek to limit k. Moreover, using a value of k that grows sublinearly with n would give us an algorithm that is fast enough for large n. For example, $k = \Theta(\sqrt{n})$ would result in $O(nh)$ running time, which on average is close to $O(n \log n)$ and is $O(n^2)$ in the worst case. While the choice must be ultimately made by the user, as a default, we set $k = 2\sqrt{n}$ where n is the number of leaves. This ensures that our running time does not grow worse than quadratically with n for any tree. This choice also serves the goal of limiting the number of leaves that *could* be removed. The only remaining parameter that needs to be specified for TreeShrink is α (we use 0.05 by default).

3   Results

Datasets: Since erroneous sequences in real data can have intricate patterns and varied causes, we use five real biological datasets instead of simulations to test the effectiveness of filtering methods. See Table 1 for summary of these datasets.

Table 1. Summary of the 5 biological datasets

Dataset	Species	Genes	Download
Plants [27]	104	852	doi:10.1186/2047-217X-3-17
Mammals [30]	37	420	doi:10.13012/C5BG2KWG
Insects [34]	144	1478	https://esayyari.github.io/InsectsData/
Cannon [35]	78	213	doi:10.5061/dryad.493b7
Rouse [36]	26	393	doi:10.5061/dryad.79dq1

Plants [27]: This dataset of 104 plant species and 852 genes was used to establish the sister to land plants and early diversification patterns within land plants. The transcriptomic nature of the data presented the authors with challenges in terms of gene identification, annotation, and missing data. To obtain reliable species trees using ASTRAL [33], the authors had to use various filtering, including removal of low occupancy genes and genes with fragmentary sequences. The ASTRAL tree obtained on these filtered gene trees was mostly congruent with results from concatenation, though some interesting clades (e.g., the Bryophytes) were differently resolved. In our analyses, we start with all gene trees estimated from nucleotide data with the third codon position removed.

Insects [34]: This phylotranscriptomic dataset includes 144 species and 1,478 genes. This large scale dataset was used to resolve controversial relationships among major insect orders, but only concatenation analyses were reported. We performed a species tree analysis of the same dataset using ASTRAL, obtained on RAxML gene trees that we estimated on all 1,478 gene trees.

Metazoa-Cannon [35] *and Rouse* [36]: Whether Xenacoelomorpha (a group of bilaterally symmetrical marine worms) are sister to all the remaining Bilateria (animals with bilateral symmetry) has been the subject of much recent debate [35–37]. This dataset of 213 genes from 78 species sampled from across the animal tree-of-life was used to confidently place Xenacoelomorpha as sister to Bilateria. Among other analyses, ASTRAL-II [38] was used on a collection of gene trees that the authors published, and we use. The dataset by Rouse *et al.* addresses the same question as Cannon *et al.* using 393 genes and 26 species.

Mammals [30]: This mammalian dataset consists of 37 taxa (36 mammals and Chicken) and 420 gene trees. Since the original gene trees had several issues (including insufficient ML searches and mislabeled taxa [39]), here we use RAxML gene trees that were inferred and used in a re-analysis of this dataset [5]. Several reanalyses of this dataset using various methodologies have largely been consistent, except, for the position of the tree shrews that often changes [5,33].

Methods: We implemented TreeShrink (https://github.com/uym2/TreeShrink) using the Dendropy package [40]. We compare TreeShrink with two alternative methods, and a control where we remove taxa randomly from the tree.

The main alternative to TreeShrink, used in some previous analyses [27,29] is to first root the gene trees using outgroups or midpoint rooting and to then remove taxa with outlier root-to-tip distances. We use this "RootedFiltering" approach where we define outliers as values that lie several standard deviations (we vary this threshold) above the average. For the Plant dataset, 681 genes included the outgroups; for the remaining, we used the midpoint rooting. In other datasets, each gene tree included at least one of the outgroups. We also compare our method with a RTR method called RogueNaRok [22]. Similar to other RTR methods, RogueNaRok defines a rogue taxon as one that has an unstable position in replicate bootstrap runs on the same alignment. The level of filtering of RogueNaRok can be adjusted by setting a "dropset" size parameter.

Evaluation Procedure: Judging the effectiveness of filtering methods on real data is challenging, as we do not know if a removed sequence is in fact erroneous. We use a simple intuition to overcome this difficulty. While true gene trees may have real discordances with the species tree, erroneous sequences with incorrect placement in gene trees will further increase the observed discordance. Thus, the amount of gene tree discordance with the species tree or among genes should reduce as a result of effective filtering. Note that none of the methods that we test take the species tree as input, and none is trying to directly reduce the discordance. When gene tree discordance is reduced as a result of an optimization problem that does not seek to reduce it, the reduction provides some independent evidence that removed sequences are problematic. To compute gene tree discordance, we use an ASTRAL tree on unfiltered gene trees as the reference species tree and use the MS (Matching Splits) metric [41], implemented in the TreeCmp [42] to measure distance. We also compare all pairs of gene trees to each other to eliminate the need for a reference species tree. Finally, to facilitate the interpretation of MS, we include random removal as a control.

A second difficulty in comparing methods is the uneven amount of filtering and the potential to remove true signal. To test this, we investigate the impact of filtering on the taxon occupancy for individual species, defined as the number of gene trees that include that species. Lowered occupancy may negatively impact downstream analyses such as species tree inference and functional analyses. Ideally, a filtering method would not reduce taxon occupancy dramatically. Moreover, removing the same taxon repeatedly from many genes could be even more problematic for downstream analyses such as species tree estimation.

Filtering methods typically have a knob that can control the amount of filtering. To avoid impacts of arbitrary choices, we vary the thresholds of the methods. For TreeShrink, we set the α parameter to several values: 0.01, 0.025, 0.05, 0.075, 0.1. For RogueNaRok, we change the weight factor to control the dropset size by setting it to 0.0 (default), 0.25, 0.5, 0.75, 1.0. For RootedFiltering, we change the number of standard deviations above the average that would constitute long branches: 2.0, 2.5, 3.0, 3.5, 4.0. For random filtering, for each threshold of TreeShrink on each gene tree, we remove the same exact number of leaves as suggested by TreeShrink, but we choose the taxa randomly.

Data Availability: All data used in this paper and the python implementation of TreeShrink are available at https://uym2.github.io/TreeShrink/.

3.1   The Impact of Filtering on Taxon Occupancy

First, we study methods in terms of their impact on the occupancy when run in the default mode ($\alpha = 0.05$ for TreeShrink, 3 std for RootedFiltering, and default settings for RogueNaRok). Overall, RogueNaRok reduces the occupancy much more than both methods that rely on branch length (Fig. 4).

Plants: Here, RogueNaRok removes three taxa from at least half of the gene trees where they are present, and removes twelve taxa from one third of the genes (Fig. 4a). For example, it removes *Kochia scoparia, Acorus americanus*, and *Larrea tridentata*, respectively, from 343 out of 654, 251 out of 693, and 221 out of 590 genes that include them. Several extensively removed species are single taxa at the base of large clades. For example, *Kochia scoparia* is sister to a group of 7 Eudicots, and *Acorus americanus* is basal to 10 Monocots [27]. Surprisingly, *Arabidopsis thaliana* is removed in 200 genes, even though its sequences are from a genome, not a transcriptome, and are arguably less prone to error compared to other species. Moreover, a focal point of this study is placing *Chara vulgaris* (basal to a large group of all land plants, Zygnematophyceae, and Coleochaetales in [27]). RogueNaRok would have removed Chara from 160 genes out of 302; such aggressive filtering could limit the scope of the biological questions that could be confidently answered. Thus, it appears that the default version of RogueNaRok removes many leaves and also tends to remove taxa that are basal to large and diverse groups.

In contrast, both RootedFiltering and TreeShrink remove only about ≈2% of the data. However, three outgroup species, *Pyramimonas parkeae, Monomastix opisthostigma*, and *Nephroselmis pyriformis*, are removed by TreeShrink in 196, 160, and 136 genes, respectively. As pointed out earlier, TreeShrink has a tendency to remove outgroups aggressively and these removals may not be warranted (we revisit this point in the discussion section). Interestingly, RogueNaRok removes these three taxa in nearly as many genes as TreeShrink did.

Insects: RogueNaRok again aggressively removes 17% of all the data also and tends to remove taxa basal to large diverse groups (Fig. 4b). For example, *Conwentzia psociformis*, which is basal among 8 Neuropterida [34]) is removed from 684 out of 1,412 genes that included it. *Zorotypus caudelli*, an enigmatic and hard to place taxon that is sister to a large clade in the ASTRAL species tree is also removed from 52% of genes that include it. Many of the outgroups, including *Speleonectes tulumensis* and *Cypridininae sp* are also removed with high probabilities (56% and 57%). TreeShrink and RootedFiltering remove around 1% of the data and do not impact taxon occupancy dramatically for any taxon.

Metazoa Datasets: Similar patterns are observed on both Metazoa datasets, where the default RogueNaRok removes more than 20% of the leaves overall, and

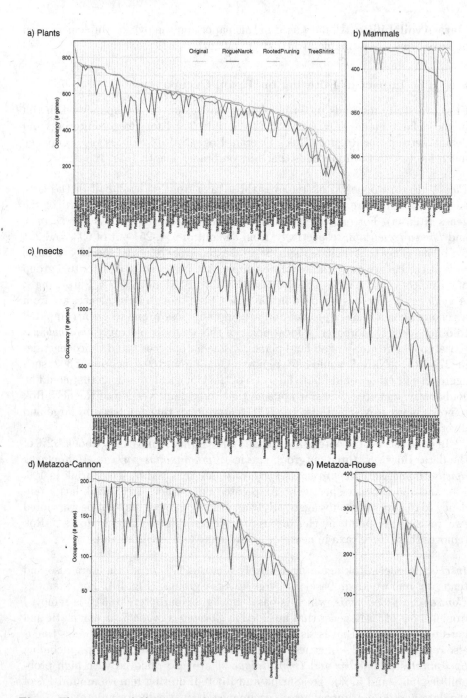

Fig. 4. The impact of filtering methods on taxon occupancy for the 5 datasets. For each taxon (x-axis, ordered by occupancy), we show the number of genes that include it before and after filtering (default settings).

many of the species are extensively removed from many genes (Fig. 4). One of the extensively removed taxa from the Cannon dataset is *Xenoturbella bocki*, which is removed in 93 out of the 208 genes that included it. Xenoturbella is the basal branch of the Xenacoelomorpha and for the question addressed in this study, it is arguably the most important taxon; removing it in 45% of genes would leave a long branch and could negatively impact the placement of Xenacoelomorpha. On these data, too, TreeShrink and RootedFiltering remove up to 2% of the data and no single taxon is extensively removed.

Mammals: Patterns of change in taxon occupancy are somewhat different for the mammalian dataset, which has full occupancy before filtering (Fig. 4e). Default RogueNaRok removes only 4% of the data overall, only marginally more than TreeShrink, which removes 2%. TreeShrink removes the outgroup, Chicken, in 147 of the 410 genes. Platypus is removed often by TreeShrink, RootedFiltering, and RogueNaRok. Several issues in the Platypus sequences have been identified [39], and perhaps, its extensive filtering by all methods is justified. However, Platypus is also close to the outgroup and the basal mammalian taxa, conditions that make both RogueNaRok and TreeShrink less reliable. Rogue-NaRok also removes the shrew, hedgehog, and the tree shrew from at least 60 genes; the shrew and the hedgehog are both basal branches to a larger clade of Laurasiatheria. The tree shrew is the most unstable taxon in the dataset with a very uncertain position [5,30,33,39], a fact that RogueNaRok's results also corroborate.

3.2   The Impact of Filtering on Gene Tree Discordance

We now compare methods run with different thresholds in terms of both the overall portion of taxa retained and the reduction in the gene tree discordance (Figs. 5 and 8). Methods that maximally reduce the discordance with a minimal removal of the data are preferred (top right corner). On all five datasets, all the three filtering methods are on average better than the control random filtering. Since extensive filtering is neither intended nor desired in our analyses, we focus on cases that remove at most 5% of the data (but also see Fig. 8).

Comparing TreeShrink and the two alternatives, three different patterns are observed on various datasets. On Plants (Fig. 5a) and Insects (Fig. 5c), TreeShrink provides the best reduction in the discordance between species trees and gene trees and among pairs of gene trees. On Metazoa-Cannon (Fig. 5d), TreeShrink reduces the discordance the most when judged with respect to the species tree, but TreeShrink and RogueNaRok are tied when discordance is measured among pairs of genes. On the Metazoa-Rouse dataset, all methods are comparable and barely outperform random pruning (Fig. 5e). On the mammalian dataset, RogueNaRok is the best (Fig. 5b).

On the insect dataset, RogueNarok is completely dominated by both TreeShrink and RootedFiltering, and TreeShrink is better than RootedFiltering, albeit marginally. For example, the MS discord with respect to the species tree is reduced by 20 units using TreeShrink but only 12 units using RogueNaRok

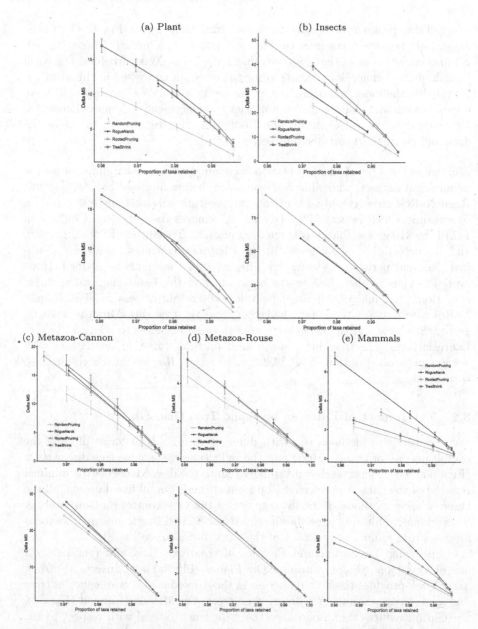

Fig. 5. The impact of filtering on gene tree discordance on 5 datasets. The reduction in the MS distance (*y-axis*) is shown versus the total proportion of the taxa retained in the gene trees after filtering (x-axis). A line is drawn between all points corresponding to different thresholds of the same method. Points below the 95% retainment are not shown here (but are shown in Fig. 8). (*Top*) MS is computed for the species tree and each gene tree. Average and standard error are shown over all genes of each dataset. (*Bottom*): MS is computed for each pair of gene trees (*y-axis*). Average and standard error are shown over all pairs of genes of each dataset.

when removing around 1% of the data (2346 and 2165 taxa in total, respectively). Allowing RogueNaRok to remove closer to 2% (3370 taxa) still reduces MS by 18 units and does not match TreeShrink. On the plant dataset, Rooted-Filtering and RogueNaRok are essentially tied and TreeShrink is narrowly but consistently better than both. For example, TreeShrink with a 0.02 threshold removes 603 taxa in total from all genes and reduces the MS discordance with the species tree by 7 (as opposed to 3 for the control), whereas RogueNaRok and RootedFiltering need to remove 770 and 689 taxa to achieve the same reduction in the MS discord. The Metazoa-Cannon dataset reveals a consistent but narrow advantage for TreeShrink compared to RogueNaRok and RootedFiltering when judged by discordance with the species tree, but the two methods are tied when judged by pairwise gene tree discordance. On the Rouse dataset, all methods seem essentially tied, and they barely outperform random removal.

On the mammalian dataset, RogueNaRok is clearly the best method, followed by RootedFiltering and then by TreeShrink. It appears that on this dataset, TreeShrink is not successful in reducing discordance, perhaps because most removals are concentrated on Chicken and Platypus, which are the outgroup taxa.

4 Discussion

In this paper, we introduced TreeShrink, a method for finding leaves of a phylogenetic tree that disproportionately impact its diameter. TreeShrink is also very scalable; we tested its running time on the GreenGenes tree [43] with $203,452$ tips ($k = 902$), and it was able to finish in under 6 min.

The potential errors in long branches is noted by many others. Gatesy and Springer used the presence of long branches in gene trees estimated in two mammalian datasets, to argue against specific coalescent-based analyses [6] (cf., Figs. 9 and 10). Typically, these branches are dealt with by eliminating long root to tip distances [27,29]. TreeShrink gives a principled automated way to achieve this *without* rooting. Outgroups, however, remain problematic.

An outgroup can look like an outlier in terms of it impact on the tree diameter. As shown in Fig. 1b, in some cases, an outgroup can contribute to the tree diameter as much as a potentially erroneous taxon. To address this challenge, we learn statistics about delta diameter from not one gene tree but a collection of gene trees (when available). This enables us to better distinguish outgroups, which are a natural part of the delta diameter distribution, from true outliers. Using single gene trees to define the diameter does in fact reduce the effectiveness of our method in terms of reducing discordance (Fig. 7). Another solution is using multiple outgroups, when possible, to break long branches leading to the outgroup. Consistent with this observation, the only dataset where TreeShrink did not perform well was the mammalian dataset, which is also the only dataset that included only a single distant outgroup. Finally, we concede that with all these cautious steps taken, distant outgroups are still more prone to deletion in our approach. However, users of the method can simply choose to overrule the suggestions of TreeShrink and retain the outgroup(s). Alternatively, they can remove the outgroups from the gene trees before running TreeShrink, and then remove the suggested set of leaves from a tree that does include the outgroup.

In this study, we simply removed taxa from existing trees. When used in practice, a better approach is to reestimate alignments and gene trees after removing the problematic sequences. The inclusion of these problematic taxa could impact gene alignments and trees negatively, even for the remaining leaves.

Finally, while we compared our method to RogueNaRok as an alternative, we point out that the two methods have very different objectives and can therefore be used in a complementary fashion. RogueNaRok is based on topological stability while TreeShrink is based on branch length. Future studies should examine the impact of combining the results of the two methods.

Acknowledgments. This work was supported by the NSF grant IIS-1565862 to SM and UM. Computations were performed on the San Diego Supercomputer Center (SDSC) through XSEDE allocations, which is supported by the NSF grant ACI-1053575.

A   Proofs

A.1   Proof of Proposition 1

We start with a Lemma:

Lemma 1. *If (a, b) is a diameter pair of t, then for any $c, d \in L(t) - \{a\}$,*

$$max(\delta(c, b), \delta(d, b)) \geq \delta(c, d).$$

Proof. Consider the quartet formed by the 4 leaves a, b, c, d in t.

Case 1: (Fig. 6a) a and b are on the same side of the quartet:

$$\delta(a, b) \geq \delta(a, d) \implies \delta(m, b) \geq \delta(m, d) \implies \delta(c, m) + \delta(m, b) \geq \delta(c, m) + \delta(m, d)$$

$$\implies \delta(c, b) \geq \delta(c, d)$$

(a)   (b)

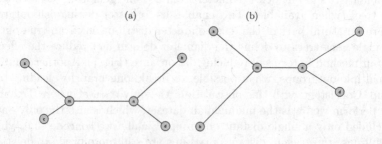

Fig. 6. Example quartet trees

Case 2: (Fig. 6b) Without loss of generality, we assume $\delta(n,c) \geq \delta(n,d)$. We will prove that $\delta(b,c) \geq \delta(c,d)$.

We have:

$$\delta(a,b) \geq \delta(a,c) \implies \delta(b,m) \geq \delta(c,m)$$

$$\implies \delta(b,m) + \delta(n,c) \geq \delta(c,m) + \delta(n,c)$$

$$\implies \delta(b,m) + \delta(n,c) \geq 2\delta(n,c)$$

$$\implies \delta(b,c) = \delta(b,m) + \delta(n,c) + \delta(m,n) \geq \delta(n,c) + \delta(n,d) = \delta(c,d)$$

$\square$

We now provide the proof of Proposition 1.

Proof. Consider an arbitrary leaf $b \in \mathcal{D}(t) - \{a\}$. We prove that $b \in \mathcal{D}(t\backslash_a)$.

Case 1: $(a,b) \notin \mathcal{P}(t)$. Because $b \in \mathcal{D}(t)$, there exists $c \in \mathcal{D}(t) - \{a\}$ such that (c,b) is a diameter pair of $t\backslash_a$. Therefore, $b \in \mathcal{D}(t\backslash_a)$.

Case 2: $(a,b) \in \mathcal{P}(t)$. Let (c,d) be a diameter-pair of $t\backslash_a$. According to Lemma 1, $max(\delta(c,b), \delta(d,b)) \geq \delta(c,d)$. Therefore, either (c,b) or (d,b) is a diameter-pair of $t\backslash_a$. Thus, $b \in \mathcal{D}(t\backslash_a)$. $\square$

A.2   Proof of Theorem 1

Proof. We need to prove that a $\mathcal{R}_k(t)$ is either a reasonable removing set or it is not an optimal removing set. We proceed by contradiction. Assume $\mathcal{R}_k(t)$ is optimal but not a reasonable removing set. Let $\mathcal{R}_m(t)$ be the largest reasonable removing set that is a subset of $\mathcal{R}_k(t)$ (note $0 \leq m \leq k$). If $m = k$, then $\mathcal{R}_k(t)$ is a reasonable set, contradicting the assumption. For $m < k$, consider the tree $t\lceil_{\mathcal{R}_m(t)}$ and let a_m, b_m be its diameter pair. if $a_m \in \mathcal{R}_k(t)$ or $b_m \in \mathcal{R}_k(t)$, adding them to $\mathcal{R}_m(t)$ would generate a reasonable chain of size $m+1$, contradicting our assumption. If $a_m \notin \mathcal{R}_k(t)$ and $b_m \notin \mathcal{R}_k(t)$, all removals after m in $\mathcal{R}_k(t)$ fail to reduce the diameter, but removing either a_m or b_m would reduce the diameter. Thus, $\mathcal{R}_k(t)$ cannot be optimal, contradicting our assumption. $\square$

A.3   Proof of Theorem 2

Proof. To remove k leaves from a singly paired tree t that has (a,b) as a diameter pair, at least one of a or b has to be removed (or else the diameter never decreases). Thus, three types of reasonable chains exist: those that contain only a, those that contain only b, and those that contain both a and b. Note that after removing a, by Proposition 1, removing b is a reasonable removal (and vice versa), and thus, removing both a and b is always reasonable (in either order).

Case 1: $a \in \mathcal{R}_{k-2}, b \notin \mathcal{R}_{k-2}$: If a reasonable chain has a but not b, by Proposition 1, b is in the diameter set at each step of the chain. Since b by definition is never removed and recalling that the tree is singly paired, at each step, there is only one reasonable removal (whatever taxon is on-diameter in addition to b). Therefore, only one reasonable chain does not include b.

Case 2: $a \notin \mathcal{R}_{k-2}, b \in \mathcal{R}_{k-2}$: Similar to Case 1, one such chain exists.

Case 3: $a \in \mathcal{R}_{k-2}, b \in \mathcal{R}_{k-2}$: In this case, the reasonable removing chain must start with a, b or b, a. In either ordering, we are left with the same induced tree, and need to remove $k-2$ more leaves. Therefore, the set of all reasonable removing sets in this case is: $\{(\{a, b\} \cup R) \text{ for } R \text{ in } \mathcal{S}_{k-2}(t \restriction_{L(t)-\{a,b\}})\}$.

Combining the three cases together, we have:

$$|\mathcal{S}_k(t)| = |\mathcal{S}_{k-2}(t \restriction_{L(t)-\{a,b\}})| + 2$$

Let $s_k = |\mathcal{S}_k(t)|$. We have the following recursion:

$$s_k = \begin{cases} 1, & k = 0 \\ 2, & k = 1 \\ s_{k-2} + 2 & k \geq 2 \end{cases} \tag{1}$$

Thus, $s_k = k + 1$ $\qquad\qquad\qquad\qquad\qquad\qquad\qquad\qquad\qquad\qquad\square$

A.4   Proof of Proposition 2

Proof. Recall that the record of each internal node keeps track of the most distant leaves below two children of the node. When we remove a, only those nodes on the path from a to the root can have a change in the their record. The first traversal of Algorithm 1 updates the records for those nodes, using simple recursive functions that can be computed in $O(1)$ per node.

According to Proposition 1, $b \in \mathcal{D}(t \backslash_a)$. Therefore, one of the longest paths in $t \backslash_a$ must include b; let c be the other taxon. The record of the LCA of c and b, after the update in the first round, will have the value of this longest value. Thus, by checking the updated record for all nodes in the path from b to the root we will find the maximum value. Moreover, when updating the records in the first traversal from a to the root, we have already checked all the nodes from the $LCA(a, b)$ to the root. In the second traversal, we check the nodes from b to $LCA(a, b)$, completing the search. Each of the two traversals of Algorithm 1 visits at most h nodes and only need constant time operations in each visit. Therefore, the overall time complexity of Algorithm 1 is $O(h)$. $\qquad\square$

A.5   Proof of Theorem 3

First, we prove the following lemmas:

Lemma 2. *All the longest paths in any tree have the same midpoint.*

Proof. If t has only one diameter pair, then Lemma 2 is trivially correct.

If t has more than one diameter pair, let (a, b) and (c, d) be two distinct diameter pairs of t and let m be the midpoint of the path between a and b. We prove that m is also the midpoint of the path between c and d, that is m lies on that path between c and d and $\delta(m, c) = \delta(m, d)$. w.l.o.g, we suppose $\delta(m, c) \geq \delta(m, d)$.

- We prove that the path between c and d must pass m; that is, c and d belong to two different subsets in the partition defined by m on $\mathcal{L}$ (we call elements of the partition a "side"). We prove by contradiction, assuming c and d belong to the same side of m. Then $\delta(m, c) + \delta(m, d) > \delta(c, d)$. Also, either a or b must be on a different side from c and d to m (by definition, a and b cannot be on the same side to m). Suppose a is in a different side from c and d to m. Then: $\delta(a, b) \geq \delta(a, c) \implies \delta(a, m) + \delta(m, b) \geq \delta(a, m) + \delta(m, c) \implies \delta(m, b) \geq \delta(m, c) \implies \delta(m, a) \geq \delta(m, c)$. So we have, $\delta(a, c) = \delta(m, a) + \delta(m, c) \geq 2\delta(m, c) \geq \delta(m, c) + \delta(m, d) > \delta(c, d)$; this leads to a contradiction because (c, d) is a diameter pair.
- Prove that $mc = md$. Suppose $\delta(m, c) > \delta(m, d)$.
  We have: $2\delta(m, a) = 2\delta(m, b) = \delta(a, b) = \delta(c, d) \leq \delta(m, c) + \delta(m, d) < 2\delta(m, c)$. Therefore: $\delta(m, a) < \delta(m, c)$ and $\delta(m, b) < \delta(m, c)$.
  Case 1: c belongs to a different side of a to m. Then, $\delta(m, a) + \delta(m, c) = \delta(a, c) \implies \delta(m, a) + \delta(m, b) < \delta(a, c) \implies \delta(a, b) < \delta(a, c)$. This is a contradiction because (a, b) is a diameter pair of t.
  Case 2: c belongs to the same side of a to m. Then c belongs to a different side of b to m. Similar to case 1, in this case we can prove that $\delta(a, b) < \delta(b, c)$ which also leads to a contradiction.
  Thus, m is the midpoint of the path between c and d.   □

This lemma allows us to define some new concepts that are useful in the rest of the proof.

New Definitions: The single midpoint of any tree t partitions the diameter set into disjoint subsets; we call each of those subsets a *diameter group* of t (if the midpoint is in the middle of the branch, we have two diameter groups; a midpoint coinciding on an internal node would give three or more groups). We call any restriction of t with k leaves removed a *k-optimal restricted tree* if no other restriction removing k leaves has a lower diameter. We call a tree t *k-shrinkable* if there exists a k-removing set that *strictly* reduces its diameter. We call any induced tree on t that has a smaller diameter than t a *shrunk tree* of t. Note that unless all but one of the diameter groups of a tree t are removed, the tree cannot shrink in diameter. When all but one of the diameter groups of a tree t is removed, we refer to the resulting tree as a *minimum shrunk tree* of t.

It is easy to see the following lemma.

Lemma 3. *For all a and b, $(a, b) \in \mathcal{P}(t)$ if and only if a and b belong to two distinct diameter groups.*   □

Now we prove a less obvious Lemma.

Lemma 4. *If tree t is k-shrinkable, any k-optimal restricted tree t^* can be induced from one of the minimum shrunk trees of t.*

Proof. Because t is k-shrinkable, the diameter of t^* must be strictly smaller than the diameter of t. Suppose t^* is not an induced tree of any minimum shrink tree of t; then, t* has at least two leaves from two different diameter groups of t. Based on Lemma 3, t^* shares with t at least one diameter pair and therefore, has the same diameter as t, which is a contradiction.   □

We now turn to the proof of Theorem 3. Recall:

Theorem 3. *For any k, any arbitrary pair-restricted k-removing space includes at least one optimal k-removing set.*

Proof. If t is not k-shrinkable, any k-removing set is optimal and the result trivially follows. We now focus on a case where t is k-shrinkable.

Suppose t has m diameter groups:

$$D^1 = \{d_1^1, d_2^1, \ldots, d_{p_1}^1\}, D^2 = \{d_1^2, d_2^2, \ldots, d_{p_2}^2\}, \ldots, D^m = \{d_1^m, d_2^m, \ldots, d_{p_m}^m\}.$$

For $i = 1 \ldots m$, let $k^i = |\bigcup_{j \neq i} D_j|$ be the size of all groups except group i, and let t^i denote the minimum shrunk tree of t that excludes all groups $D^j, j \neq i$. Let $k^p = \max_i(k^i)$. For the tree t to be k-shrinkable, we need that $k^i \leq k$; thus, $k \geq k^p$.

To produce any minimum shrunk tree t^i with $k^i \leq k^p$, we can start from any removal (a, b) such that $a \in D^x$ and $b \in D^y$ (for $x \neq y$), and continue to produce t^i. To see this, note that if $x \neq y \neq i$, any chain that starts with either a or b and continues to select from any groups other than D^i will produce the minimum shrunk tree t^i after k^i removals. Now, w.l.o.g, consider $x = i$ and $y \neq i$. Then, consider the chain that starts by removing y and continues by removals from any group other than D^i. This chain will also produce t^i after k^i removals. In other words, each pair-restricted k-removing space of t can produce all the minimum shrunk trees t^i that have $k^i \leq k$.

Based on Lemma 4, when t is k-shrinkable, at least one of the minimum shrunk trees (say t_i^*) can induce any k-optimal restricted tree t^*. We also just proved that any pair-restricted space can produce *all* minimum shrunk trees. Therefore, any arbitrary pair-restricted removing space will include a chain that induces t_i^* from t and another chain that produces t^* starting from t_i^*. Thus, the union of the removing sets corresponding to these two chains will produce t^* and will be part of any arbitrary pair-restricted k-removing space.

B   Supplementary Figures

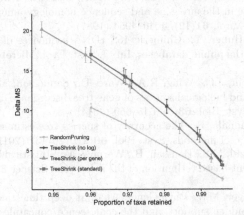

Fig. 7. Alternative implementations of TreeShrink. Using Δ values instead of $\log \Delta$ values had no discernible impact on the trajectory, though it did slightly change the results. Fitting kernels on individual genes instead of fitting it on the full set of genes reduced the accuracy, but still was much better than the control random.

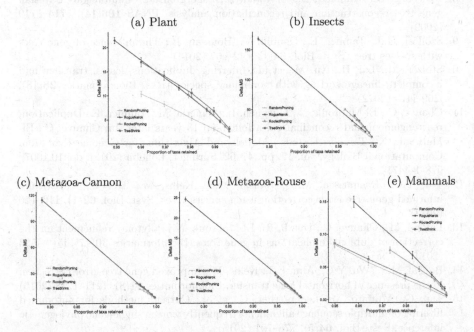

Fig. 8. Impact of taxon removal on gene tree discordance. For each of the four biological dataset, we show the reduction in the distance between the species tree and gene trees measured by the MS metric (y-axis) versus the total proportion of the taxa retained in the gene trees after filtering (x-axis). Average delta MS values and standard error bars are shown over all genes for each dataset. A line is drawn between all five points corresponding to each method.

References

1. Braun, M.J., Clements, J.E., Gonda, M.A.: The visna virus genome: evidence for a hypervariable site in the env gene and sequence homology among lentivirus envelope proteins. J. Virol. **61**(12), 4046–4054 (1987)
2. Hugenholtz, P., Huber, T.: Chimeric 16S rDNA sequences of diverse origin are accumulating in the public databases. Int. J. Syst. Evol. Microbio. **53**(1), 289–293 (2003)
3. Zwickl, D.J., Stein, J.C., Wing, R.A., Ware, D., Sanderson, M.J.: Disentangling methodological and biological sources of gene tree discordance on Oryza (Poaceae) chromosome 3. Syst. Biol. **63**(5), 645–659 (2014)
4. Leaché, A.D., Rannala, B.: The accuracy of species tree estimation under simulation: a comparison of methods. Syst. Biol. **60**(2), 126–137 (2011)
5. Mirarab, S., Bayzid, M.S., Boussau, B., Warnow, T.: Statistical binning enables an accurate coalescent-based estimation of the avian tree. Science **346**(6215), 1250463 (2014)
6. Gatesy, J., Springer, M.S.: PhyloGenet. Anal. at deep timescales: unreliable gene trees, bypassed hidden support, and the coalescence/concatalescence conundrum. Mol. Phylogenet. Evol. **80**, 231–266 (2014)
7. Arvestad, L., Berglund, A.C., Lagergren, J., Sennblad, B.: Gene tree reconstruction and orthology analysis based on an integrated model for duplications and sequence evolution. In: RECOMB, pp. 326–335. ACM Press, New York (2004)
8. Akerborg, O., Sennblad, B., Arvestad, L., Lagergren, J.: Simultaneous Bayesian gene tree reconstruction and reconciliation analysis. PNAS **106**(14), 5714–5719 (2009)
9. Szöllősi, G.J., Tannier, E., Daubin, V., Boussau, B.: The inference of gene trees with species trees. Syst. Biol. **64**(1), e42–e62 (2014)
10. Stolzer, M., Lai, H., Xu, M., et al.: Inferring duplications, losses, transfers and incomplete lineage sorting with nonbinary species trees. Bioinformatics **28**(18), i409–i415 (2012)
11. Chauve, C., El-Mabrouk, N., Guéguen, L., Semeria, M., Tannier, E.: Duplication, rearrangement and reconciliation: a follow-up 13 years later. In: Chauve, C., El-Mabrouk, N., Tannier, E. (eds.) Models and Algorithms for Genome Evolution. Computational Biology, vol. 19, pp. 47–62. Springer, London (2013). doi:10.1007/978-1-4471-5298-9_4
12. Wu, Y.C., Rasmussen, M.D., Bansal, M.S., Kellis, M.: TreeFix: statistically informed gene tree error correction using species trees. Syst. Biol. **62**(1), 110–120 (2013)
13. Lafond, M., Chauve, C., Dondi, R., El-Mabrouk, N.: Polytomy refinement for the correction of dubious duplications in gene trees. Bioinformatics **30**(17), i519–i526 (2014)
14. Bansal, M.S., Wu, Y.C., Alm, E.J., Kellis, M.: Improved gene tree error correction in the presence of horizontal gene transfer. Bioinformatics **31**(8), 1211–1218 (2015)
15. Tan, G., Muffato, M., Ledergerber, C., et al.: Current methods for automated filtering of multiple sequence alignments frequently worsen single-gene phylogenetic inference. Syst. Biol. **64**(5), 778–791 (2015)
16. Castresana, J.: Selection of conserved blocks from multiple alignments for their use in PhyloGenet. Anal. Mol. Biol. Evol. **17**(4), 540–552 (2000)
17. Capella-Gutiérrez, S., Silla-Martínez, J.M., Gabaldón, T.: trimAl: a tool for automated alignment trimming in large-scale phylogenetic analyses. Bioinformatics **25**(15), 1972–1973 (2009)

18. Shen, X.X., Hittinger, C.T., Rokas, A.: Studies can be driven by a handful of genes. Nature 1(April), 1–10 (2017)
19. Krüger, D., Gargas, A.: New measures of topological stability in phylogenetic trees - taking taxon composition into account. Bioinformation 1(8), 327–330 (2006)
20. Westover, K.M., Rusinko, J.P., Hoin, J., Neal, M.: Rogue taxa phenomenon: a biological companion to simulation analysis. Mol. Phylogenet. Evol. 69(1), 1–3 (2013)
21. Pattengale, N.D., Swenson, K.M., Moret, B.M.E.: Uncovering hidden phylogenetic consensus. In: Borodovsky, M., Gogarten, J.P., Przytycka, T.M., Rajasekaran, S. (eds.) ISBRA 2010. LNCS, vol. 6053, pp. 128–139. Springer, Heidelberg (2010). doi:10.1007/978-3-642-13078-6_16
22. Aberer, A.J., Krompass, D., Stamatakis, A.: Pruning rogue taxa improves phylogenetic accuracy: an efficient algorithm and webservice. Syst. Biol. 62(1), 162–166 (2013)
23. Goloboff, P.A., Szumik, C.A.: Identifying unstable taxa: efficient implementation of triplet-based measures of stability, and comparison with Phyutility and Rogue-NaRok. Mol. Phylogenet. Evol. 88, 93–104 (2015)
24. Hosner, P.A., Braun, E.L., Kimball, R.T.: Land connectivity changes and global cooling shaped the colonization history and diversification of New World quail (Aves: Galliformes: Odontophoridae). J. Biogeogr. 42, 1883–1895 (2015)
25. Streicher, J.W., Schulte, J.A., Wiens, J.J.: How should genes and taxa be sampled for phylogenomic analyses with missing data? An empirical study in iguanian lizards. Syst. Biol. 65(1), 128–145 (2016)
26. Salichos, L., Rokas, A.: Inferring ancient divergences requires genes with strong phylogenetic signals. Nature 497(7449), 327–331 (2013)
27. Wickett, N.J., Mirarab, S., Nguyen, N., et al.: Phylotranscriptomic analysis of the origin and early diversification of land plants. PNAS 111(45), 4859–4868 (2014)
28. Bergsten, J.: A review of long-branch attraction. Cladistics 21(2), 163–193 (2005)
29. Hampl, V., Hug, L., Leigh, J.W., et al.: Phylogenomic analyses support the monophyly of Excavata and resolve relationships among eukaryotic "supergroups". PNAS 106(10), 3859–3864 (2009)
30. Song, S., Liu, L., Edwards, S.V., Wu, S.: Resolving conflict in eutherian mammal phylogeny using phylogenomics and the multispecies coalescent model. PNAS 109(37), 14942–14947 (2012)
31. Silverman, B.: Density estimation for statistics and data analysis. In: Monographs on Statistics and Applied Probability. Chapman & Hall (1986)
32. R Core Team: R: A Language and Environment for Statistical Computing. R Foundation for Statistical Computing, Vienna, Austria (2016)
33. Mirarab, S., Reaz, R., Bayzid, M.S., et al.: ASTRAL: genome-scale coalescent-based species tree estimation. Bioinformatics 30(17), i541–i548 (2014)
34. Misof, B., Liu, S., Meusemann, K., et al.: Phylogenomics resolves the timing and pattern of insect evolution. Science 346(6210), 763–767 (2014)
35. Cannon, J.T., Vellutini, B.C., Smith, J., et al.: Xenacoelomorpha is the sister group to Nephrozoa. Nature 530(7588), 89–93 (2016)
36. Rouse, G.W., Wilson, N.G., Carvajal, J.I., Vrijenhoek, R.C.: New deep-sea species of Xenoturbella and the position of Xenacoelomorpha. Nature 530(7588), 94–97 (2016)
37. Philippe, H., Brinkmann, H., Copley, R.R., et al.: Acoelomorph flatworms are deuterostomes related to Xenoturbella. Nature 470(7333), 255–258 (2011)

38. Mirarab, S., Warnow, T.: ASTRAL-II: coalescent-based species tree estimation with many hundreds of taxa and thousands of genes. Bioinformatics **31**(12), i44–i52 (2015)
39. Springer, M.S., Gatesy, J.: The gene tree delusion. Mol. Phylogenet. Evol. **94**(Part A), 1–33 (2016)
40. Sukumaran, J., Holder, M.T.: DendroPy: a Python library for phylogenetic computing. Bioinformatics **26**(12), 1569–1571 (2010)
41. Bogdanowicz, D., Giaro, K.: Matching split distance for unrooted binary phylogenetic trees. IEEE/ACM Trans. Comput. Biol. Bioinform. **9**(1), 150–160 (2012)
42. Bogdanowicz, D., Giaro, K., Wróbel, B.: TreeCmp: comparison of trees in polynomial time. Evol. Bioinform. **2012**(8), 475–487 (2012)
43. DeSantis, T.Z., Hugenholtz, P., Larsen, N., et al.: Greengenes, a chimera-checked 16S rRNA gene database and workbench compatible with ARB. Appl. Environ. Microbiol. **72**(7), 5069–5072 (2006)

Rearrangement Scenarios Guided by Chromatin Structure

Sylvain Pulicani[1,4], Pijus Simonaitis[2], Eric Rivals[1,3],
and Krister M. Swenson[1,3(✉)]

[1] LIRMM, CNRS – Université Montpellier, 161 Rue Ada, 34392 Montpellier, France
{pulicani,swenson}@lirmm.fr
[2] ENS Lyon, 46 Allée D'Italie, 69364 Lyon, France
[3] Institut de Biologie Computationnelle (IBC), Montpellier, France
[4] Institut de Génétique Humaine (IGH), UMR9002 CNRS-UM, Montpellier, France

Abstract. Genome architecture can be drastically modified through a succession of large-scale rearrangements. In the quest to infer these rearrangement scenarios, it is often the case that the parsimony principal alone does not impose enough constraints. In this paper we make an initial effort towards computing scenarios that respect chromosome conformation, by using Hi-C data to guide our computations. We confirm the validity of a model – along with optimization problems MINIMUM LOCAL SCENARIO and MINIMUM LOCAL PARSIMONIOUS SCENARIO – developed in previous work that is based on a partition into equivalence classes of the adjacencies between syntenic blocks. To accomplish this we show that the quality of a clustering of the adjacencies based on Hi-C data is directly correlated to the quality of a rearrangement scenario that we compute between *Drosophila melanogaster* and *Drosophila yakuba*. We evaluate a simple greedy strategy to choose the next rearrangement based on Hi-C, and motivate the study of the solution space of MINIMUM LOCAL PARSIMONIOUS SCENARIO.

Keywords: Genome rearrangement · Double cut and join · DCJ · Hi-C · Chromatin conformation

1 Introduction

Genome architecture is modified on a large scale by *rearrangements*. Even somewhat distantly related species, such as human and mouse, can share almost all of the same genes yet have drastically different gene orders [6]. These differences are a result of a succession of rearrangements from an ancestral architecture, each rearrangement acting on *breakpoints* between conserved stretches of DNA. In the quest to infer accurate rearrangement *scenarios* that transform one modern-day architecture into another, it is often the case that the parsimony principal alone does not impose enough constraints [5].

One potentially useful biological factor that drives chromosome evolution is the 3-dimensional conformation of chromosomes in the nucleus; a current

© Springer International Publishing AG 2017
J. Meidanis and L. Nakhleh (Eds.): RECOMB CG 2017, LNBI 10562, pp. 141–155, 2017.
DOI: 10.1007/978-3-319-67979-2_8

hypothesis is that breakpoints which are close in 3D space are more likely to take part in a rearrangement than those which are distant. Evidence supporting this hypothesis has been reported for inter-species rearrangements [13], as well as for somatic rearrangements [3,17]. We call this the *locality hypothesis*.

Although the 3D conformation is not static, loci tend to group together into topologically associating domains (TADs). Existing techniques to get an accurate estimate of the distance between pairs of loci (like for example FISH), are not scalable to the whole genome, therefore the Hi-C method was developed as a surrogate [8,10]. Hi-C operates on several cells at once, binning genomic coordinates into windows and producing a matrix where position i, j contains the number of loci within window i that come in close physical contact with a loci from window j. Thus the *weight* of a rearrangement that acts on breakpoints in window i and j can be obtained by looking up that value in the matrix.

In previous work, we developed algorithms to use this 3D structure to guide the computation of rearrangement scenarios. We sought out *local* scenarios, relying on a partition of the adjacencies such that each class of the partition contains mutually proximal adjacencies. Rearrangements that act on breakpoints from the same class then have no cost, and are referred to as *local*, whereas those that act on breakpoints from different classes incur a cost of one, and are referred to as *non-local* [11,12]. We called these problems MINIMUM LOCAL SCENARIO (MLS) and MINIMUM LOCAL PARSIMONIOUS SCENARIO (MLPS). The former asks for the minimum number of non-local moves necessary when transforming one genome into the other, while the latter asks the same question for parsimonious scenarios. A partition of the adjacencies is part of the input to MLS and MLPS. It remains an open question how to create such a partition given genome adjacencies and Hi-C data.

In this paper, we explore practical aspects of finding rearrangement scenarios using *Drosophila melanogaster* and *Drosophila yakuba*. To this end, we compute a double cut and join (DCJ) scenario [2,15] that greedily chooses a next move based on the weight (Hi-C value) for its pair of breakpoints. In Sect. 2 we show that this scenario has exceptionally high weight with respect to scenarios where DCJs are sampled at random.

In Sect. 3 we address practical aspects of computing globally optimal local scenarios using MLS and MLPS. In particular, we show that MLS can be computed efficiently despite being NP-Hard (proved in [11]). We also show that a simple k-medoid clustering can be used as an informative partition of the adjacencies.

In Sect. 4.1 we evaluate how the greedy scenario from Sect. 2 performs with respect to the clusterings. Our results show that the scenario does not minimize the number of non-local moves, indicating that there is room for improvement. On the other hand, we show that a hybrid method which computes an MLPS before sampling random parsimonious DCJ moves finds higher weight scenarios than sampling alone. A hybrid method that attempts to greedily choose parsimonious local moves after computing non-local moves with MLPS does not, however, outperform the generic greedy algorithm. This indicates that further work needs to be done to choose a set of non-local moves from the many possible MLPSs that maximize the weight with respect to the Hi-C data.

Finally, Sect. 6.1 shows that the difference between the MLPS and the MLS is generally very low. As shown in Simonaitis and Swenson [11], this indicates that an algorithm which allows for an arbitrary difference of costs between local and non-local moves is possible.

1.1 Definitions and Experimental Setup

When comparing large-scale genome architecture, *syntenic blocks* of similar sequences of genes between a group of species are first inferred using sequence similarity [7]. The *adjacencies* between these blocks are the stretches of nucleotides with no homology between the genomes, and are the potential locations for *breakpoints* that rearrangements act on. These breakpoints could be close or far in three dimensional space, as indicated by Hi-C data. We refer to this spacial proximity as the *weight* of a pair of breakpoints, and by extension as the weight of a rearrangement. The *weight of a scenario* is the mean of the weights of its rearrangements.

Our general experimental setup is the following. Genomes of *Drosophila melanogaster* and *Drosophila yakuba* are first partitioned into 64 syntenic blocks (see Sect. 5.1). The adjacencies between the blocks are the potential breakpoints. Normalized Hi-C data for *melanogaster* are used as a locality measurement (see Sect. 5.2). We construct a partition of the adjacencies by clustering around medoids [9] using Hi-C data as the similarity function (see Sect. 3.2).

To weight a rearrangement we take the intervals $(i, j), (k, l)$ corresponding to the coordinates of the breakpoints of the DCJ. Say the interval (i, j) $((k, l)$, respectively) spans windows i' through j' (k' through l', respectively) of the Hi-C matrix. The weight of the move then is the average matrix value over all combinations of indices $\{i', i' + 1, \ldots, j'\} \times \{k', k' + 1, \ldots, l'\}$. Intervals are updated through DCJs as described in [12].

Figure 1 shows a toy example describing our model of genome rearrangement with respect to Hi-C data. Consider a DCJ that operates on $s_1 x s_2$ and $s_3 y s_4$ where s_i is a syntenic block or telomere, and x and y are adjacencies. Then there are two different ways to make s_1 adjacent to s_3: we can have either $s_1 x s_3$ (and $s_2 y s_4$) or $s_1 y s_3$ (and $s_2 x s_4$). The two possibilities are illustrated by the first DCJ in panels (a) and (b); they invert the same blocks, yet act on the adjacencies in the two different ways.

The distinction between *cost* and *weight* is an important one in our presentation. Weight always refers to a value or average value taken directly from a Hi-C matrix. Two examples are: (1) the weight of a scenario, which is a function of the values in the Hi-C matrix, and (2) the weight of a clustering, which is a function of the Hi-C values between pairs of adjacencies in the same cluster. Cost always refers to an assessment of a scenario of rearrangements with respect to a clustering of the adjacencies. For example, the cost of a scenario can be computed with respect to a clustering (of any weight) by assigning a cost of zero to adjacencies that are in the same cluster, and a cost of one to the others.

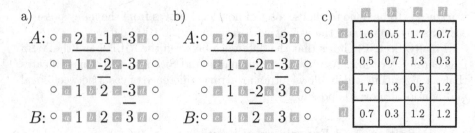

Fig. 1. Parsimonious DCJ scenarios transforming genome A into B. Syntenic blocks are represented by numbers, adjacencies by gray boxes, and telomeres (chromosome ends) by ∘. Each DCJ changes the blocks or telomeres surrounding two adjacencies. Panel (c) shows a matrix such that position i, j is the average value in the Hi-C matrix for the genomic intervals corresponding to adjacencies i and j. Panels (a) and (b) show two possible greedy parsimonious scenarios computed of the first having average weight 1.4 and the second with average weight 0.97 respectively.

2   Greedy Computation of Rearrangement Scenarios

2.1   Greedy Scenarios

In order to maximize the weight of a scenario, we designed Algorithm 1. This greedy algorithm computes a parsimonious rearrangement scenario, performing at each step the DCJ with the highest weight from all the available parsimonious DCJs.

> **Data:**
>   – The genomes A and B,
>   – M_A, the corresponding Hi-C maps for A.
>
> **Result:** L, a parsimonious scenario of DCJs transforming A into B.
> $C := A$;
> **while** C *is not equal to* B **do**
> > $d :=$ a DCJ that transforms C into C' such that
> > $dist(C', B) = dist(C, B) - 1$ and the weight of d in M_A is maximum;
> > Apply d to C;
> > Append d to L;
>
> **end**

Algorithm 1. Greedy parsimonious scenario

2.2   Weight and Greedy Scenarios

We compared the weight of a greedy scenario with those of 1000 parsimonious scenarios where each DCJ is sampled uniformly at random. The results are shown in Fig. 2.

The greedy scenario has a significantly higher weight than the sampled scenarios. This is expected as the greedy scenario is a local maximum. We also

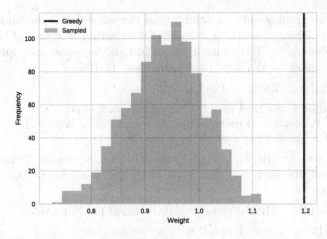

Fig. 2. Weight of the greedy scenario as a vertical bar along with the distribution of 1000 sampled parsimonious scenarios. The greedy scenario has a low probability of being sampled due to the exponential number of parsimonious scenarios (with respect to scenario length).

note that the weight along the greedy scenario decreases linearly ($r^2 = 0.92$), and that only 10 DCJs out of 51 are performed with inter-chromosomal adjacencies. Moreover, the last 8 DCJs (those with lowest weights) are done using inter-chromosomal adjacencies.

3   Global Computation of Locality

We saw in the last section that a scenario with exceptionally high weight can be obtained by greedily computing a parsimonious scenario according to Hi-C. In this section we discuss practical aspects behind the computation of scenarios that globally minimize non-local moves. First we define the optimization problems used for this task in previous work [11,12]. In Sect. 3.3 we show that the first of the two problems is tractable when computing scenarios between *melanogaster* and *yakuba*, despite being NP-hard.

These optimization problems rely on a partition of the adjacencies as input. In the final two subsections we show how we compute clusters of adjacencies to serve as these partitions, and how the quality of these clusters show a strong correlation to the number of non-local moves that must be done (the MINIMUM LOCAL SCENARIO).

3.1   Optimization Problems

The idea behind the clustering of adjacencies is that pairs of adjacencies in the same cluster are more likely to take part as breakpoints in a rearrangement than pairs between different clusters. Thus we evaluate rearrangements by giving the inter-cluster rearrangements a cost of 1, and the intra-cluster rearrangements a cost of 0. The *locality cost* (or just cost) of a scenario is the sum of the cost of the constituent moves. With such a model in hand we posed two optimization problems for which we developed algorithms for [11,12].

Problem 1 (MLS). MINIMUM LOCAL SCENARIO.

INPUT: Adjacency sets A and B with a clustering of A.
OUTPUT: A scenario of rearrangements transforming A into B.
MEASURE: The locality cost of the scenario.

The problem MINIMUM LOCAL PARSIMONIOUS SCENARIO introduces the constraint that the output is also a parsimonious scenario, *i.e.* a scenario of minimum length.

Problem 2 (MLPS). MINIMUM LOCAL PARSIMONIOUS SCENARIO.

INPUT: Adjacency sets A and B with a clustering of A.
OUTPUT: A *parsimonious* scenario of rearrangements transforming A into B.
MEASURE: The locality cost of the scenario.

We will use the term MLS and MLPS to denote the locality cost of the scenario for an optimal solution of the problem.

3.2 Clustering

Both algorithms of the previous section require a clustering of the adjacencies for genome A. To that end, we use Hi-C data as a similarity measure for the clustering; the higher the weight for a pair of adjacencies, the more likely we are to have them in the same cluster. Clustering around medoids [9] was chosen for its simplicity and speed. A *medoid* of a cluster is an element that maximizes the sum of the similarities to the rest of a cluster. This sum is the cluster's *weight*, and when summed over all the clusters it provides us with a *weight function* for a clustering. An important property of this clustering method is that it provides many local optima that we can compare to the solutions obtained for our optimization problems (MLS and MLPS). We use three clustering algorithms: K-MEDOIDS, RANDOM, and MIXED that generate k clusters for a fixed k, which in our case ranges from 1 to 70.

The K-MEDOIDS algorithm starts with randomly initialized centroids. The rest of the elements are then associated to the centroids that are most similar to them. The medoids of the obtained clusters are then computed and they become the new centroids around which the elements will be clustered. We continue this procedure until the weight of a clustering stops increasing.

The RANDOM algorithm partitions the elements at random into k non-empty clusters. On our data we observe that the weights of the random clusters are always smaller than those provided by the clustering around medoids. In order to bridge the gap between the obtained weights we mix K-MEDOIDS and RANDOM algorithms to obtain a MIXED algorithm that initializes the centroids randomly and then chooses at random the number of resulting elements to be assigned to the clusters based on the similarity function, and how many of them will be assigned at random (without performing further iterations).

3.3   Feasibility of Computing MLS

In general, finding the MLS is computationally costly, as we have proven it to be NP-hard [11]. In other words, we cannot expect to be able to compute a scenario that minimizes the number of non-local moves on any pair of genomes. We established an exact algorithm, however, that runs efficiently if a certain parameter called the number of *simple cycles* is "small enough". We expect "small enough" to roughly be in the hundreds of thousands.

Although the number of simple cycles between two genomes is hard to predict since it depends on the entire problem input (*i.e.* the syntenic blocks and the clustering), we find that between *melanogaster* and *yakuba* the number of simple cycles is always very small using the clusters computed by the K-MEDOIDS algorithm. Particularly encouraging is the fact the even completely random clustering yield a practical number of simple cycles.

In particular, we ran 100 instances of K-MEDOIDS using Hi-C data from *melanogaster* for every k ranging from 2 to 70, and computed the number of simple cycles for those clusterings. The values for all 6900 runs are presented in Fig. 3a. The average number of simple cycles is 16.9 for $k = 42$, the maximum that we observed. The average number of simple cycles over all k is 8.5. Figure 3b presents a similar histogram for the runs of RANDOM for *melanogaster*. As expected, RANDOM clusterings produce larger numbers of simple cycles.

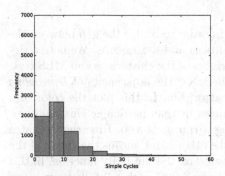

(a) Using K-MEDOIDS clusterings. Average is 8.5, standard deviation is 8.4. The highest average is 16.9 for $k = 42$.

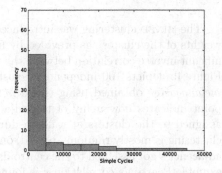

(b) Using RANDOM clusterings. Average is 9,164, standard deviation is 18,227. Four values higher than 50,000 were detected with a maximum of 86,319.

Fig. 3. The frequency of the number of simple cycles computed for all possible values of k (ranging from 2 to 70).

3.4   MLS and the Weight of the Clustering

Figure 4a presents 200 independent clusterings of the adjacencies of *Drosophila melanogaster* into $k = 15$ clusters. Half of them are generated using the RANDOM algorithm, the others using the K-MEDOIDS algorithm. There is a clear separation on both axes between these two sets of clusterings. The MLS and clustering weight are always significantly better on a K-MEDOIDS clustering than a RANDOM clustering. Similar results are found for values of k ranging from 5 to 50.

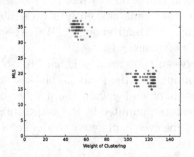

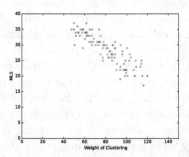

(a) 100 RANDOM clusterings (blue dots) and 100 K-MEDOIDS clusterings (red dots).

(b) 100 MIXED clusterings with varying amounts of randomly assigned adjacencies. Point colors go from blue (random) to red. Pearson's correlation: $r = -0.87$ (p-value $= 1 \times 10^{-31}$)

Fig. 4. The number of non-local moves computed for MLS compared to the weight of the clustering for $k = 15$ clusters on *D. melanogaster* Hi-C data. (Color figure online)

The MIXED clustering was introduced in order to bridge the gap between the weights of the clusterings provided by RANDOM and K-MEDOIDS. We note a significant inverse correlation between the weights of the clusterings and MLS cost. Figure 4b depicts 100 independent clusterings for the adjacencies of *Drosophila melanogaster* obtained using the MIXED algorithm. In this plot the color of a point indicates how many of the adjacencies in that particular clustering got assigned to the clusters at random during a run of MIXED. Blue shows that a clustering is mostly random and red, on the other hand, means that most of the adjacencies got assigned to the centroids based on the similarity function. In this example Pearson's correlation r is found to be -0.87 with a 2-tailed p-value of 1×10^{-31}. Similar results were found for values of k ranging from 5 to 50, where the correlation slightly increases up to $k = 25$ before slowly decreasing.

4   Evaluation

As we have seen, the MLS and MLPS are the results of a binary cost function. On the other side, the greedy scenario is a mean of Hi-C values. Therefore we cannot compare them directly. In this section, we compare the scenarios in two complementary ways. First we compute the locality cost of the greedy scenario using the clustering. Second we compute the weight of hybrid algorithms that first compute non-local moves with MLPS, and then complete the scenario using sampling or greedy methods.

4.1   Evaluating Greedy on Clusters

Using the same clusters as below, we computed the cost for the sampled scenarios and for the greedy one. The results are shown in Fig. 5, along with the costs of MLS and MLPS.

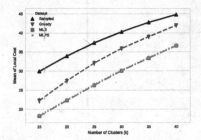 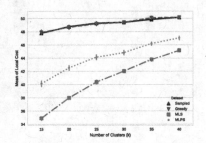

(a) Locality costs using K-MEDOIDS cluster-ings. The MLS and MLPS curves almost perfectly fit each other.

(b) Locality costs using RANDOM cluster-ings. The greedy and sampled curves almost perfectly fit each other.

Fig. 5. Locality costs for the greedy, sampled, MLS and MLPS scenarios. Since the DCJ distance is 51, the maximum possible cost is 51.

As a sanity check, note that for the same values of k, the cost using RANDOM clusterings is significantly higher than when using K-MEDOIDS clusterings. For K-MEDOIDS clustering, the cost of the greedy scenario is always lower than that of a sampled scenario. By greedily preferring the adjacencies that maximize the Hi-C weight, we give preference to the intra-cluster adjacencies. Moreover, the difference in cost between the greedy scenario and the sampled ones increases for small numbers of clusters. The costs of the MLS and MLPS are always lower than for greedy. This shows that room for improvement remains over the purely greedy algorithm.

4.2   The Weight of MLPS

A solution to MLPS does not directly give a Hi-C weight. In order to calculate it, we arbitrarily choose one of the many optimal solutions to MLPS, each giving a set of non-local moves. We then compute the local moves to build the rest of the scenario by either sampling, or by choosing moves greedily. Call these algorithms MLPS-*sampled* and MLPS-*greedy*, respectively. We compare the distributions computed using the sampling methods to the values obtained using the greedy methods in Fig. 6.

As expected, the MLPS-sampled distribution is significantly higher (*i.e.* more weight) than the sampled distribution. Moreover, there is no MLPS-sampled scenario with a particularly low weight. This indicates that the clustering captures the Hi-C information, allowing us to build biologically meaningful scenarios with MLPS. This further buttresses the link between clustering and MLPS observed in Sect. 6.1.

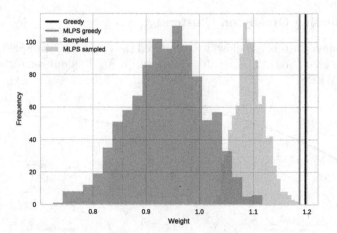

Fig. 6. Weights of the greedy and MLPS-greedy scenarios as vertical bars along with the distribution of 1000 parsimonious sampled and MLPS-sampled scenarios. The number of cluster is $k = 20$.

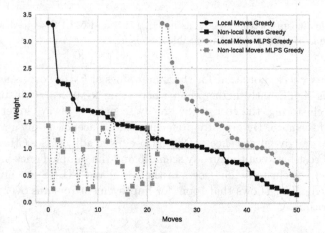

Fig. 7. Weight per move of the greedy and the MLPS-greedy scenarios.

The greedy scenario remains better than the MLPS-greedy scenario. We conjecture that this is due to the fact that we choose an arbitrary set of non-local moves that correspond to an optimal MLPS. Moreover, we plotted the weight of each move of the greedy and the MLPS-greedy scenarios in Fig. 7. The effect of choosing an arbitrary MLPS is apparent: the largest non-local moves chosen by the greedy scenarios are not chosen by the MLPS-greedy method.

We conjecture that choosing – out of all possible MLPS solutions – the solution that optimizes the weight of the non-local moves would yield a heavier MLPS-greedy solution than the purely greedy solution. Indeed, by replacing in the MLPS-greedy solution only the two lowest weighted non-local moves, with the two highest weighted non-local moves from the pure greedy scenario, the score of the new scenario would already be as good as the purely greedy algorithm.

5 Data Treatment

5.1 Creation of Syntenic Blocks

We computed syntenic blocks in two steps. First, we took the orthologs for
D. melanogaster and *D. yakuba*. This was done using the OMA groups database [1]. We removed each gene that overlaps or intersects another, along with
its ortholog in the other species. Then, the blocks were constructed using the
Orthocluster tool [16]. The basic idea of Orthocluster is to aggregate orthologs to
make the biggest possible blocks without breaking certain constraints that define
the synteny; these constraints are the maximal and minimal number of genes per
blocks, the absolute gene order between the genomes, the genes strandedness,
the quantity of non-ortholog genes and the possibility to make nested blocks.
We forbid the creation of nested blocks as we wanted a 1-1 block mapping. All
other parameters were default.

Orthocluster outputs clusters of genes. We interpret the clusters as syntenic
blocks by taking the smallest gene position in a cluster as the start position of
the block from this cluster, and the largest position as the end for that block.

5.2 Hi-C as a Measure of the Locality

The Hi-C experiment provides a rough estimate of how many times a pair of
genomic loci are in close proximity [8]. Roughly speaking, formaldehyde is introduced in a population of cells so that parts of the genomes that are in spacial
proximity are linked together. Each side of the link contains segments of DNA
that are then sequenced. The sequences are mapped to a reference genome so
that the pair of spatially proximal loci are determined. Finally, the raw data are
corrected for experimental biases (see [14] for the method applied to the matrices
that we use, published in [10]).

The chromosomes in the nucleus have a dynamic structure. Thus, cells from
the same cell type will yield different sets of pairs of loci. Therefore, since the
experiment is done on a cell population, we observe the average of all these sets
of loci among all the cells of the population.

Due to the nature of contact probability, which decreases dramatically with
respect to chromosomal distance (it roughly follows a power law), we applied the
normalization done by Lieberman-Aiden *et al.* (see the appendix of [8]) to the
matrices published in [10]. This gives the long rearrangements (in the genetic
coordinate sense) to have the same importance as the short ones. Specifically, a
normalized intrachromosomal heatmap entry $INTRA_{ij}$ gets the value

$$INTRA_{ij} = H_{ij}/averageAtDist(|i - j|),$$

where $averageAtDist(d)$ is the expected value of an entry d off of the diagonal of
any intrachromosomal matrix. A normalized inter-chromosomal heatmap entry
$INTER_{ij}$ gets the value

$$INTER_{ij} = H_{ij}/\left(\frac{interaction_i * interaction_j}{interaction_{all}}\right),$$

where $interaction_x$ is the sum of all interactions for position x, and $interaction_{all}$ is the total sum of all entries in all matrices (intra *and* inter-chromosomal).

6   Future Work and Conclusions

We wrap things up by introducing future directions. The first subsection discusses the feasibility of using more general weight functions when computing globally optimal scenarios. We showed in [11] that this question is directly related to the difference MLPS − MLS. The second subsection briefly discusses the possibility of using Hi-C data from two different species.

6.1   MLPS − MLS and the Weight of the Clustering

We study the value of the difference MLPS − MLS due to its potential role in computing a more general cost function. When this difference is low the number of non-parsimonious rearrangements in the MLS is few. In this case a non-binary cost function that has an arbitrary, but fixed, difference between local and non-local move cost is easier to compute and approximate [11]. In this section we show that the difference is usually very low for *Drosophila*, and that higher quality clusters reduce this difference.

A significant correlation between the difference MLPS − MLS and the weights of the clusterings is found, as depicted in Fig. 8a and b for $k = 10$. Similar results hold for values of k from 5 to 50.

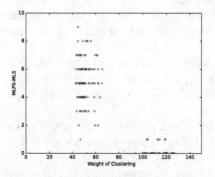 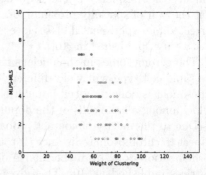

(a) 100 RANDOM (blue dots) and 100 K-MEDOIDS clusterings (red dots)

(b) 100 MIXED clusterings with varying amounts of randomly assigned adjacencies. Dots color goes from blue (random) to red. Pearson's correlation:
$r = -0.69$ (p-value $= 2 \times 10^{-15}$)

Fig. 8. The difference in the number of non-local moves computed for MLPS and MLS compared to the weight of the clustering for $k = 10$ clusters on *D. melanogaster* Hi-C data. (Color figure online)

Further, for the clusterings provided by K-MEDOIDS this value of MLPS −
MLS is low in general, as displayed in Fig. 9a. As for Fig. 3a we did 100 runs of
K-MEDOIDS for every k, and found the average for MLPS − MLS to be highest
at $k = 24$ with a value 0.29. The average over all k was 0.12. A similar histogram
for the runs of RANDOM for *melanogaster* can be found in Fig. 9b.

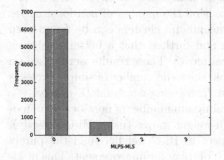

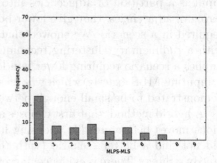

(a) Using K-MEDOIDS clusterings. The aver-   (b) Using RANDOM clusterings. The average
age is 0.12, the standard deviation is 0.34.   is 2.19, the standard deviation is 2.2. The
The highest average is 0.29 for $k = 24$.   highest average is 7.

Fig. 9. The frequency of the difference MLPS − MLS over all possible values of k
(ranging from 2 to 70).

6.2   Using Hi-C from *D. yakuba*

No data for *D. yakuba* are publicly available at this time. However, we gained
access to unpublished preliminary data. All results for greedy scenarios and clusterings on *melanogaster* also apply to *yakuba*, and tend to be more pronounced.

When computing the parsimonious scenario from species A to B we use the
Hi-C data of A. In doing so, we ignore the differences in chromatin conformation
between A and B that are specific to species B. Between human and mouse,
for example, the conformation is in general very similar with a small number
of important differences [4]. Thus, using the Hi-C data from both species is a
current challenge.

As a preliminary investigation, we *reverse* the greedy scenario from A to B
and use the Hi-C data of species B. In that case, the weight is closer to that
of the sampled scenarios. This could be a reflection of the 3D spacial features
specific to each species; by selecting adjacencies of very high weight we could
select some special features in the chromatin conformation of one species over
the other. If these characteristics have no correspondence in the other genome,
they lead to low weight in the reversed scenario. Assuming that the differences
in 3D are linked to post-speciation evolutionary events, there is the potential
to use these adjacencies to place rearrangements on phylogenetic branches. We
plan to further investigate this direction.

6.3   Conclusions

We have demonstrated the existence of scenarios corresponding exceptionally well to Hi-C data. They can be computed using a simple greedy strategy. In an effort to find a global optimum we had developed the MLS and MLPS methods in previous work, but it remained unclear how to use them since part of the input is a partition of adjacencies into equivalence classes. Given such a clustering of adjacencies, they give lower bounds on the number of non-local moves required in a scenario. We showed that meaningful clusters can be found even with a rudimentary clustering technique, and further, that a better clustering implies a scenario requiring fewer non-local moves. These results were based on computing MLS exactly, which was feasible since the number of simple cycles is demonstrated to be small enough between *D. melanogaster* and *D. yakuba*.

A hybrid method, that first chooses a minimum number of non-local parsimonious moves before greedily choosing high-weight moves (using Hi-C directly), did not find a better scenario (with respect to Hi-C weight) than the purely greedy strategy. There is evidence, however, of room for improvement. This motivates the study of the solution space of MLPS so that one could pick, among all optimal solutions, the one that also has the best Hi-C values. In general, the best way to balance weight and cost remains an open question.

From a practical perspective several improvements to this work are in order. First, development of the same methods on the inversion model would be appropriate for certain branches of the tree of life. In *Drosophila*, extra care would have to be taken since inversions and transpositions are the two main drivers behind architecture transformations. Second, a world of more sophisticated clustering methods exist. Application of the right method may partition the adjacencies in a more relevant way when choosing local rearrangement scenarios.

Acknowledgements. The authors would like to thank the reviewers for their helpful comments. Sylvain PULICANI is funded by NUMEV grant AAP 2014-2-028 and EPIGENMED grant ANR-10-LABX-12-01. This work is partially supported by the IBC (Institut de Biologie Computationnelle) (ANR-11-BINF-0002) and by the Labex NUMEV flaship project GEM.

References

1. Altenhoff, A.M., Škunca, N., Glover, N., Train, C.-M., Sueki, A., Piližota, I., Gori, K., Tomiczek, B., Müller, S., Redestig, H., et al.: The OMA orthology database in 2015: function predictions, better plant support, synteny view and other improvements. Nucleic acids research, p. gku1158 (2014)
2. Bergeron, A., Mixtacki, J., Stoye, J.: A unifying view of genome rearrangements. In: Bücher, P., Moret, B.M.E. (eds.) WABI 2006. LNCS, vol. 4175, pp. 163–173. Springer, Heidelberg (2006). doi:10.1007/11851561_16
3. Campbell, P.J., Stephens, P.J., Pleasance, E.D., O'Meara, S., Li, H., Santarius, T., Stebbings, L.A., Leroy, C., Edkins, S., Hardy, C., et al.: Identification of somatically acquired rearrangements in cancer using genome-wide massively parallel paired-end sequencing. Nat. Genet. **40**(6), 722–729 (2008)

4. Chambers, E.V., Bickmore, W.A., Semple, C.A.: Divergence of mammalian higher order chromatin structure is associated with developmental Loci. PLoS Comput. Biol. **9**(4), e1003017 (2013)
5. Chauve, C., Gavranovic, H., Ouangraoua, A., Tannier, E.: Yeast ancestral genome reconstructions: the possibilities of computational methods II. J. Comput. Biol. **17**(9), 1097–1112 (2010)
6. Chinwalla, A.T., Cook, L.L., Delehaunty, K.D., Fewell, G.A., Fulton, L.A., Fulton, R.S., Graves, T.A., Hillier, L.D.W., Mardis, E.R., McPherson, J.D., et al.: Initial sequencing and comparative analysis of the mouse genome. Nature **420**(6915), 520–562 (2002)
7. Ghiurcuta, C.G., Moret, B.M.E.: Evaluating synteny for improved comparative studies. Bioinformatics **30**(12), i9–i18 (2014)
8. Lieberman-Aiden, E., van Berkum, N.L., Williams, L., Imakaev, M., Ragoczy, T., Telling, A., Amit, I., Lajoie, B.R., Sabo, P.J., Dorschner, M.O., Sandstrom, R., Bernstein, B., Bender, M.A., Groudine, M., Gnirke, A., Stamatoyannopoulos, J., Mirny, L.A., Lander, E.S., Dekker, J.: Comprehensive mapping of long-range interactions reveals folding principles of the human genome. Science **326**(5950), 289–293 (2009)
9. Park, H.-S., Jun, C.-H.: A simple and fast algorithm for k-medoids clustering. Expert Syst. Appl. **36**(2, Part 2), 3336–3341 (2009)
10. Sexton, T., Yaffe, E., Kenigsberg, E., Bantignies, F., Leblanc, B., Hoichman, M., Parrinello, H., Tanay, A., Cavalli, G.: Three-dimensional folding and functional organization principles of the drosophila genome. Cell **148**(3), 458–472 (2012)
11. Simonaitis, P., Swenson, K.M.: Finding Local Genome Rearrangements. Springer, Heidelberg (2017)
12. Swenson, K.M., Simonaitis, P., Blanchette, M.: Models and algorithms for genome rearrangement with positional constraints. Algorithms Mol. Biol. **11**(1), 13 (2016)
13. Véron, A.S., Lemaitre, C., Gautier, C., Lacroix, V., Sagot, M.-F.: Close 3D proximity of evolutionary breakpoints argues for the notion of spatial synteny. BMC Genom. **12**(1), 303 (2011)
14. Yaffe, E., Tanay, A.: Probabilistic modeling of Hi-C contact maps eliminates systematic biases to characterize global chromosomal architecture. Nat. Genet. **43**(11), 1059–1065 (2011)
15. Yancopoulos, S., Attie, O., Friedberg, R.: Efficient sorting of genomic permutations by translocation, inversion and block interchange. Bioinformatics **21**(16), 3340–3346 (2005)
16. Zeng, X., Nesbitt, M.J., Pei, J., Wang, K., Vergara, I.A., Chen, N.: Orthocluster: a new tool for mining synteny blocks and applications in comparative genomics. In: Proceedings of the 11th International Conference on Extending database technology: Advances in Database Technology, pp. 656–667. ACM (2008)
17. Zhang, Y., McCord, R.P., Ho, Y.-J., Lajoie, B.R., Hildebrand, D.G., Simon, A.C., Becker, M.S., Alt, F.W., Dekker, J.: Spatial organization of the mouse genome and its role in recurrent chromosomal translocations. Cell **148**(5), 908–921 (2012)

A Unified ILP Framework for Genome Median, Halving, and Aliquoting Problems Under DCJ

Pavel Avdeyev[✉], Nikita Alexeev, Yongwu Rong, and Max A. Alekseyev

The George Washington University, Washington, D.C., USA
avdeyev@gwu.edu

Abstract. One of the key computational problems in comparative genomics is the reconstruction of genomes of ancestral species based on genomes of extant species. Since most dramatic changes in genomic architectures are caused by genome rearrangements, this problem is often posed as minimization of the number of genome rearrangements between extant and ancestral genomes. The basic case of three given genomes is known as the *genome median problem*. Whole genome duplications (WGDs) represent yet another type of dramatic evolutionary events and inspire the reconstruction of pre-duplicated ancestral genomes, referred to as the *genome halving problem*. Generalization of WGDs to whole genome multiplication events leads to the *genome aliquoting problem*.

In the present study, we provide polynomial-size integer linear programming formulations for the aforementioned problems. We further obtain such formulations for the restricted versions of the median and halving problems, which have been recently introduced for improving biological relevance.

1 Introduction

Genome rearrangements are large-scale evolutionary events that shuffle genomic architectures. Since genome rearrangements are rare, the number of such events between two genomes is used in phylogenomic studies to measure the evolutionary distance between them. Such measurement is often based on the maximum parsimony assumption, implying that the evolutionary distance can be estimated as the minimum number of rearrangements (known as *the rearrangement distance*) between genomes. This assumption further enables addressing the problem of ancestral genomes reconstruction by minimizing the total distance between genomes along the branches of the evolutionary tree. The basic case of this problem with just three given genomes is called the *genome median problem* (GMP), which asks to reconstruct a single ancestral genome (*median genome*) at the minimum total distance from the given genomes. Since solutions to the GMP are not always biologically adequate [12,19,32], there was recently proposed a version of the GMP, called the *intermediate genome median*

P. Avdeyevand and N. Alexeev are contributed equally to this work.

J. Meidanis and L. Nakhleh (Eds.): RECOMB CG 2017, LNBI 10562, pp. 156–178, 2017.
DOI: 10.1007/978-3-319-67979-2_9

problem (IGMP), which restricts solutions to the *intermediate genomes*, i.e., genomes appearing in a shortest rearrangement scenario between some pair of the given genomes [12].

Since genome rearrangements preserve the gene content, the aforementioned problems are considered for genomes with a uniform gene content. To account for genes present in varying number of copies in different genomes, one need to consider other types of evolutionary events. One of the important sources of duplicated genes in genomes is the *whole genome duplication* (WGD) events, which simultaneously duplicate each chromosome of a genome. WGD events are known to happen in the evolution of yeasts [21], fishes [26], plants [18], and even mammals [9]. A typical example of a WGD happening in the evolution is illustrated by the following phylogenetic tree:

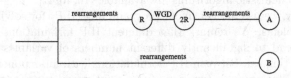

where B and R are *ordinary* genomes (i.e., genomes containing each gene in a single copy), $2R$ is a *perfect 2-duplicated* genome (i.e., a genome resulted from the WGD of R), and A is an *2-duplicated genome* (i.e., a genome containing each gene in two copies). Depending on availability of genomes, there exists a number of problems related to reconstruction of the genome R (and $2R$). The basic problem in this context is the *genome halving problem* (GHP) that relies only on the genome A and asks to find an ordinary genome R such that the rearrangement distance between A and $2R$ is minimized. The GHP solution space is typically huge [4], which makes it impractical to obtain biologically adequate solutions. The *guided genome halving problem* (GGHP) improves the biological relevance by using an additional ordinary genome B and asking for an ordinary genome R that minimizes the total rearrangement distance between the ordinary genomes B and R and between the 2-duplicated genome A and the perfect 2-duplicated genome $2R$. While the former distance is easy to compute (for a known genome R), computing the latter distance (called the *double distance* [33]) represents a much harder problem. Similarly to the GMP and IGMP, there was recently proposed a version of the GGHP, which restricts the constructed perfect 2-duplicated genomes to the GHP solution space, called the *restricted guided genome halving problem* (RGGHP) [4].

Besides the IGMP and RGGHP, we consider other (apparently new) restricted versions of the genome median and guided genome halving problems, where each gene adjacency in the solutions is conserved in at least one of the given genomes. We refer to these versions as *conserved* GMP and GGHP, respectively.

A WGD can be viewed as a particular case of a *whole genome multiplication* (WGM), which simultaneously creates $m \geq 2$ copies of each chromosome, forming a *perfect m-duplicated* genome. Correspondingly, the aforementioned family of halving problems can be generalized to a family of *aliquoting* problems [35].

A convenient model for the most common genome rearrangements is given by the *Double-Cut-and-Join* (DCJ) operations [38], also known as *2-breaks* [3], which make two "cuts" in a genome and "glues" the resulting genomic fragments in a new order. Namely, DCJs mimic *reversals* (that inverse contiguous segments of chromosomes), *translocations* (that exchange tails of two chromosomes), and *fissions/fusions* (that split/join chromosomes). Under the DCJ model, the GHP and rearrangement distance admit polynomial-time solutions [1,6,25,36]. On the other hand, the GMP, IGMP, double distance, GGHP, and RGGHP are known to be NP-hard [7,13,33]. The GMP is the most studied problem among these NP-hard problems, for which there exists a number of exact [7,37,39] and heuristic [15,27] algorithms. For the IGMP, only a heuristic algorithm is available [12]. For the double distance, there exists an efficient greedy heuristics [14,28]. For the GGHP, two heuristic algorithms are available [16,40,41].

In this study, we use integer linear programming (ILP) for solving the aforementioned problems. We remark that different ILP formulations of the same problem may lead to significantly different numbers of variables and/or constraints. We distinguish between ILP formulations with an exponential number of variables/constraints (as a function of the input genomes size) and those with a polynomial number of variables/constraints. Exponential-size ILP formulations are known for the following problems:

- the rearrangement distance under models, where the strands[1] of genes are ignored [8,22,23];
- the GMP under the reversals-only model [7].

Polynomial-size ILP formulations are known for the following problems:

- the reversal distance and the transposition distance, where the strands of genes are ignored [10];
- the rearrangement distance under models, where different types of rearrangements are weighted differently [20];
- the rearrangement distance between genomes with duplicated genes under the reversals-only model [24,31];
- the rearrangement distance between genomes with duplicated genes under the DCJ model [29,30].

In the present study, we restrict our analysis to genomes with circular chromosomes (extension to linear genomes will be described elsewhere) and present polynomial-size ILP formulations for the GMP, GGHP, double distance, and the family of aliquoting problems under the DCJ model. As a by-product, we also obtain ILP formulations for the IGMP and RGGHP. We remark that our ILP formulations are based on similar ideas as the ILP formulation for computing DCJ distance between two genomes with duplicated genes [29]. However, our approach is based on breakpoint graphs rather than adjacency graphs. This

---

[1] The strand of a gene is typically encoded by a sign. When the strands are ignored, the genomes are represented as permutations of (unsigned) genes.

further enables us to extend it to a large class of ancestral genomes reconstruction problems that can be reduced to finding a maximal cycle decomposition of breakpoint graphs under various constraints. To the best of our knowledge, the proposed approach gives first polynomial-size ILP formulations for these problems.

2   Background

Breakpoint Graphs and Median Problems. We represent a circular chromosome consisting of n genes as a graph cycle with n directed edges (encoding genes and their strands) alternating with n undirected edges (connecting the extremities of adjacent genes) (Fig. 2c). We label each directed edge with the corresponding gene x, and further label its tail and head endpoints with x^t and x^h, respectively. For an ordinary genome P with m chromosomes, the *genome graph* $\mathfrak{G}(P)$ is formed by m such cycles representing the chromosomes of P (Fig. 2c). The undirected edges in $\mathfrak{G}(P)$ (called *P-edges*) form a matching (called *P-matching*). A DCJ in the genome P corresponds in $\mathfrak{G}(P)$ to the replacement of a pair of P-edges with a different pair of P-edges[2] on the same set of four vertices.

For ordinary genomes P and Q composed of the same genes, the *breakpoint graph* $\mathfrak{G}(P, Q)$ is defined as the superposition of the genome graphs $\mathfrak{G}(P)$ and $\mathfrak{G}(Q)$ (Fig. 1). In other words, $\mathfrak{G}(P, Q)$ can be constructed by gluing the identically labeled directed edges in $\mathfrak{G}(P)$ and $\mathfrak{G}(Q)$. From now on, we will ignore directed edges and assume that the breakpoint graph $\mathfrak{G}(P, Q)$ consists only of (undirected) P-edges and Q-edges, forming P-matching and Q-matching, respectively. Then $\mathfrak{G}(P, Q)$ represents a collection of cycles formed by alternating P-edges and Q-edges, called *PQ-cycles* (or *QP-cycles*). Similarly, the breakpoint graph can be defined for three or more genomes [5].

A DCJ *scenario* between genomes P and Q is a sequence of DCJs transforming P into Q (Fig. 1). The *DCJ distance* (i.e., the length of a shortest DCJ scenario) between two genomes can be computed by the following formula [3,38]:

$$d_{\mathrm{DCJ}}(P, Q) = n - c(P, Q), \tag{1}$$

where $c(P, Q)$ is the number of PQ-cycles in $\mathfrak{G}(P, Q)$.

Let us formulate the genome median problem under the DCJ model, which we generalize to any number of genomes.

Genome Median Problem (GMP). *For a given ordinary genomes $P_1, \ldots, P_q$ on the same genes, find an ordinary genome M minimizing the total distance $\sum_{i=1}^{q} d_{\mathrm{DCJ}}(P_i, M)$.*

The classic formulation of the GMP corresponds to $q = 3$.

To formulate the intermediate genome median problem, we recall some equivalent definitions of an intermediate genome (Fig. 1):

---

[2] Here we view genome P as evolving and P-edges as changing.

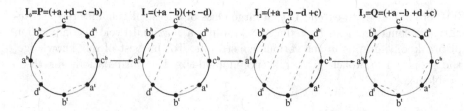

Fig. 1. A shortest DCJ scenario transforming an ordinary genome $P = (+a+d-c-b)$ (*grey solid edges*) into an ordinary genome $Q = (+a-b+d+c)$ (*black solid edges*). The intermediate genomes are shown in *grey dashed edges*.

Theorem 1 *[12]. An ordinary genome I is* intermediate *between ordinary genomes P_1 and P_2 composed of the same n genes if and only if any of the following conditions hold:*

1. *the genome I appears in a shortest DCJ scenario between P_1 and P_2; or*
2. $d_{\text{DCJ}}(P_1, I) + d_{\text{DCJ}}(I, P_2) = d_{\text{DCJ}}(P_1, P_2)$; *or*
3. *the total number of IP_1-cycles and IP_2-cycles in $\mathfrak{G}(P_1, P_2, I)$ equals $n + c(P_1, P_2)$.*

The intermediate genome median problem is formulated as follows.

***Intermediate Genome Median Problem (IGMP** [12]). For given ordinary genomes P_1, P_2, P_3 on the same genes, find an intermediate genome M between genomes P_1 and P_2 that minimizes $d_{\text{DCJ}}(P_3, M)$.*

We also propose to consider a *conserved* version of the GMP based on the assumption that each gene adjacency in a median genome is conserved in at least one of the given genomes.

***Conserved Genome Median Problem (CGMP).** For given ordinary genomes $P_1, \ldots, P_q$ on the same genes, find an ordinary genome M such that each gene adjacency in M is present in at least one of the genomes $P_1, \ldots, P_q$, and the total distance $\sum_{i=1}^{q} d_{\text{DCJ}}(P_i, M)$ is minimized.*

We remark that the CGMP is consistent with the empirical observation that solutions to the GMP tend to be close to one of the given genomes [19]. It is therefore expected that solutions to the CGMP well approximate solutions to the GMP. At the same time, the CGMP allows more "freedom" for solutions to have parts close to different given genomes.

Contracted Breakpoint Graphs and Aliquoting Problems. Let $m \geq 2$ be an integer and A be an *m-duplicated genome* (i.e., a genome containing each gene in m copies). We construct the genome graph $\mathfrak{G}(A)$, where the directed edges appear in m identically labeled copies (Fig. 2a). By gluing all copies of each edge into a single edge, we obtain the *contracted genome graph* $\hat{\mathfrak{G}}(A)$ (Fig. 2b). We define *A-components* in $\hat{\mathfrak{G}}(A)$ as the connected components formed by A-edges. We remark that each A-component is a connected m-regular graph (since each vertex is incident to m A-edges). If $m = 2$, A-components form cycles (called *A-cycles*).

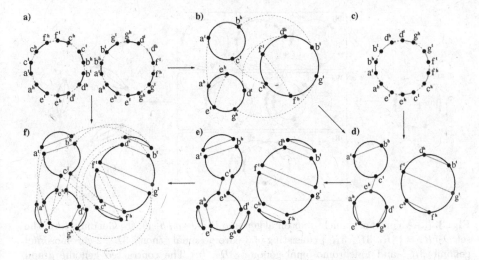

Fig. 2. For an 2-duplicated genome $A = (-a-b+g+d+f+g+e)(-a+c-f-c-b-d-e)$ and an ordinary genome $R = (-a-b-d-g+f-c-e)$, (a) the genome graph $\mathfrak{G}(A)$; (b) the contracted genome graph $\hat{\mathfrak{G}}(A)$; (c) the genome graph $\mathfrak{G}(R)$; (d) the contracted breakpoint graph $\hat{\mathfrak{G}}(A,R)$; (e) a maximal AR-cycle decomposition C of $\hat{\mathfrak{G}}(A,2R)$; and (f) the breakpoint graph $\mathfrak{G}(A,X)$ having the same cycle structure as C, for some genome $X \in D_2(R)$ and some labeling of gene copies of A and X.

Now let mR be a *perfect m-duplicated* genome resulted from a WGM of an ordinary genome R. Similarly to a WGD, which can simultaneously duplicate each chromosome of an ordinary genome into either a single 2-duplicated chromosome or two identical ordinary chromosomes [2], we assume that a WGM can create m identical copies of a chromosome or possibly combine some of these copies into single chromosomes (Fig. 3a, b). Hence, a WGM can multiplicate each chromosome of R in $p(m)$ ways, where $p(m)$ is the number of integer partitions of m ($p(2) = 2$, $p(3) = 3$, $p(4) = 5$, ... form the sequence A000041 in the Online Encyclopedia of Integer Sequences [34]). So, a perfect m-duplicated genome mR may contain m-duplicated chromosomes of different types, and we denote the set of all possible such genomes mR as $D_m(R)$ (Fig. 3b). It follows that $|D_m(R)| = p(m)^{\text{chr}(R)}$, where $\text{chr}(R)$ is the number of chromosomes in R. It can be easily seen that the contracted genome graph $\hat{\mathfrak{G}}(mR)$ does not depend on a particular choice of $mR \in D_m(R)$, and its R-edges form multi-edges composed of m parallel edges each (Fig. 3b,c). We refer to such multi-edges as mR-*edges*. It is clear that the mR-edges form a matching in $\hat{\mathfrak{G}}(mR)$. Replacing each mR-edge with an R-edge in $\hat{\mathfrak{G}}(mR)$ transforms it into the breakpoint graph $\mathfrak{G}(R)$.

For an m-duplicated genome A and an ordinary genome R composed of the same genes (present in m and 1 copies, respectively), the *contracted breakpoint graph* $\hat{\mathfrak{G}}(A,R)$ is defined as the superposition of $\hat{\mathfrak{G}}(A)$ and $\mathfrak{G}(R)$, and can be constructed in the same way as breakpoint graphs (Fig. 2b, c, d). The A-edges and the R-edges in $\hat{\mathfrak{G}}(A,R)$ form A-components (A-cycles if $m = 2$) and R-matching, respectively.

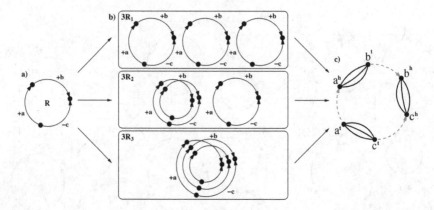

Fig. 3. (a) An ordinary unichromosomal genome $R = (+a+b-c)$. (b) Genomes form the set $D_3(R) = \{3R_1, 3R_2, 3R_3\}$ consisting of 3-chromosomal genome $3R_1$, 2-chromosomal genome $3R_2$, and unichromosomal genome $3R_3$. (c) The contracted genome graph $\hat{\mathfrak{G}}(3R)$ for any $3R \in D_3(R)$.

Similarly, the *contracted breakpoint graph* $\hat{\mathfrak{G}}(A, mR)$ of a m-duplicated genome A and a perfect m-duplicated genome mR is defined as the superposition of the contracted genome graphs $\hat{\mathfrak{G}}(A)$ and $\hat{\mathfrak{G}}(mR)$. The graph $\hat{\mathfrak{G}}(A, mR)$ can also be obtained from $\hat{\mathfrak{G}}(A, R)$ by replacing each R-edge with an mR-edge. Since each vertex in $\hat{\mathfrak{G}}(A, mR)$ is incident to m A-edges and m R-edges, $\hat{\mathfrak{G}}(A, mR)$ can be decomposed into a collection of AR-cycles (Fig. 2e). We remark that there may exist an exponential number of such decompositions, among which we distinguish *maximal decompositions* with the maximum possible number of AR-cycles, denoted $c_{\max}(A, R)$ (Fig. 2e). The ambiguity in chromosome m-duplication inspires the following problem, which generalizes the double distance:

Aliquoting Distance Problem (ADP). *Given an m-duplicated genome A and an ordinary genome R on the same genes, find*

$$d_{\mathrm{DCJ}}^m(A, R) = \min_{mR \in D_m(R)} d_{\mathrm{DCJ}}(A, mR).$$

The double distance represents a partial case of the ADP with $m = 2$. We establish a connection between solutions to the ADP and maximal AR-cycle decompositions in the following theorem, which generalizes Theorem 2 in [1].

Theorem 2. *Let A be an m-duplicated genome and R be an ordinary genome composed of the same n genes. For any AR-cycle decomposition C of $\hat{\mathfrak{G}}(A, mR)$, there exists a genome $X \in D_m(R)$ and a labeling of gene copies in A and X (turning A and X into ordinary genomes with mn genes) such that $\mathfrak{G}(A, X)$ has the same cycle structure as C.*

Proof. Let C be any AR-cycle decomposition of $\hat{\mathfrak{G}}(A, mR)$. We view C as a graph formed by A-edges and R-edges on labeled vertices such that each label is shared by m vertices. We notice that C and $\mathfrak{G}(A)$ have the same multiset of A-edges (viewed as 2-element sets formed by endpoint labels). Hence, there exists a one-to-one correspondence between the A-edges of C and the A-edges of $\mathfrak{G}(A)$ that preserves the endpoint labels. Gluing the corresponding A-edges in C and $\mathfrak{G}(A)$ into single A-edges (respecting the endpoint labels), we obtain a graph G that contains both C and $\mathfrak{G}(A)$ as subgraphs. Then G represents the breakpoint graph of A and some genome X, i.e., $G = \mathfrak{G}(A, X)$, such that $\hat{\mathfrak{G}}(A, X) = \hat{\mathfrak{G}}(A, mR)$, implying that $X \in D_m(R)$.   □

Using Theorem 2, one can restate the ADP as finding a maximal AR-cycle decomposition of $\hat{\mathfrak{G}}(A, mR)$, and thus

$$d_{\mathrm{DCJ}}^m(A, R) = m|R| - c_{\max}(A, R). \tag{2}$$

Now, let us formulate the guided aliquoting problem, which generalizes the guided halving problem:

Genome Aliquoting Problem (GAP [35]**).** *For a given m-duplicated genome A, find an ordinary genome R minimizing $d_{\mathrm{DCJ}}^m(A, R)$.*

From (2), it follows that the GAP is equivalent to finding an ordinary genome R that maximizes $c_{\max}(A, R)$. For $m \geq 3$, the computational complexity of the GAP is unknown. If $m = 2$, such ordinary genome R satisfies the following property:

Theorem 3. ([2,11]). *For any 2-duplicated genome A composed of n distinct genes,*

$$\max_R c_{\max}(A, R) = n + k,$$

where the maximum is taken over all ordinary genomes R on the same genes as A, and k is the number of even A-cycles in $\mathfrak{G}(A)$.

The solution space of the GHP is typically huge [4], which makes it infeasible to obtain a biologically adequate solution. To address this issue, one may restrict the GAP by taking into account an additional ordinary genome and posing the following problem:

Guided Genome Aliquoting Problem (GGAP). *Given an m-duplicated genome A and an ordinary genome B on the same genes, find an ordinary genome R minimizing the total distance to genomes A and B, i.e., $d_{\mathrm{DCJ}}^m(A, R) + d_{\mathrm{DCJ}}(R, B)$.*

The guided genome halving problem [33] represents a partial case of the GGAP with $m = 2$. Similarly to the IGMP representing a restricted version of the GMP, there exists a restricted version of the GGAP:

Restricted Guided Genome Aliquoting Problem (RGGAP). *Given an m-duplicated genome A and an ordinary genome B on the same genes, find an ordinary genome R that is a solution to the GAP for A and minimizes $d_{\mathrm{DCJ}}(B, R)$.*

The restricted guided genome halving problem [4,41] represents a partial case of the RGGAP with $m = 2$.

We also propose the following *conserved* version of the GGAP based on the assumption that each gene adjacency in the pre-multiplicated genome is conserved in at least one of the given genomes.

Conserved Guided Genome Aliquoting Problem (CGGAP). *Given an m-duplicated genome A and an ordinary genome B on the same genes, find an ordinary genome R minimizing the total distance $d_{\mathrm{DCJ}}^m(A, R) + d_{\mathrm{DCJ}}(R, B)$ such that each gene adjacency in R is present in at least one of the genomes A and B.*

We refer to the partial case $m = 2$ of the CGGAP as the *conserved guided genome halving problem* (CGGHP).

3   ILP Formulations

Genome Median Problem. Let us consider the GMP for given ordinary genomes $P_1, P_2, \ldots, P_q$ on the same n genes. By the formula (1), for any ordinary genome M we have $\sum_{l=1}^q d_{\mathrm{DCJ}}(M, P_l) = qn - \sum_{l=1}^q c(M, P_l)$. So, we need to construct an ordinary genome M that maximizes $\sum_{l=1}^q c(M, P_l)$. For each $l \in [q] = \{1, 2, \ldots, q\}$, the breakpoint graph $\mathfrak{G}(M, P_l)$ forms a subgraph of the breakpoint graph $\bar{G} = \mathfrak{G}(P_1, \ldots, P_q, M)$, and thus $c(M, P_l)$ equals the number of MP_l-cycles in $\bar{G}$.

Let $V = \{v_1, \ldots, v_{2n}\}$ be the set of vertices (indexed arbitrarily) in $\bar{G}$.[3] We represent a genome M as the collection of binary variables $r_{\{i,j\}}$ that indicate whether an unordered pair of vertices $\{v_i, v_j\}$ forms an M-edge in $\bar{G}$. Since M-edges form a matching on V, the following constraints hold

$$\sum_{\substack{i=1 \\ i \neq j}}^{2n} r_{\{i,j\}} = 1, \quad r_{\{i,j\}} \in \{0,1\}, \forall j \in [2n]. \tag{3}$$

For each $l \in [q]$, let $e_1^l, e_2^l, \ldots, e_n^l$ be the P_l-edges (indexed arbitrarily) in $\bar{G}$. For $j \in [2n]$, let $\mathrm{inc}_l(j) \in [n]$ be the index of the P_l-edge incident to the vertex v_j in $\bar{G}$. For each $k \in [n]$, we assign an integer variable p_k^l equal the smallest index of a P_l-edge from the MP_l-cycle containing e_k^l. Then we have

$$p_k^l \leq k, \quad \forall k \in [n], l \in [q]. \tag{4}$$

Clearly, for any $k, s \in [n]$, $p_k^l = p_s^l$ if and only if e_k^l and e_s^l belong to the same MP_l-cycle in $\bar{G}$. In particular, if $\{v_i, v_j\}$ is an M-edge in $\bar{G}$, then $p_{\mathrm{inc}_l(i)}^l = p_{\mathrm{inc}_l(j)}^l$, implying that[4]

$$\left| p_{\mathrm{inc}_l(i)}^l - p_{\mathrm{inc}_l(j)}^l \right| \leq 2n \left(1 - r_{\{i,j\}}\right), \quad \forall \{i, j\} \subset [2n], l \in [q]. \tag{5}$$

---

[3] Note that V is determined by the genes present in the genomes $P_1, P_2, \ldots, P_q$, and thus V does not depend on the choice of M.

[4] Under the inequality $|a - b| \leq c$, we understand a pair of linear inequalities $a - b \leq c$ and $b - a \leq c$.

For each $k \in [n]$ and $l \in [q]$, we define a binary variable $\tilde{p}_k^l$ that indicates whether e_k^l has the smallest index in its MP_l-cycle in $\bar{G}$ (in other words, $\tilde{p}_k^l$ indicates whether $p_k^l = k$). These variables satisfy the following constraints:

$$k\,\tilde{p}_k^l \le p_k^l, \quad \tilde{p}_k^l \in \{0,1\}, \forall k \in [n], l \in [q]. \tag{6}$$

Since in each MP_l-cycle exactly one edge has the smallest index, we have $c(M, P_l) = \sum_{k=1}^n \tilde{p}_k^l$.

Theorem 4. *Let $P_1, P_2, \ldots, P_q$ be ordinary genomes composed of the same n genes. A solution to the ILP with the objective function*

$$maximize \quad \sum_{l=1}^q \sum_{k=1}^n \tilde{p}_k^l, \tag{7}$$

under the constraints (3), (4), (5), (6) defines an ordinary genome M that is a solution to the GMP for the genomes $P_1, \ldots, P_q$.

Proof. First of all, the constraints (3) guarantee that the collection of binary variables $r_{\{i,j\}}$ defines a matching on the vertex set V. So, these variables define some ordinary genome M and the graph $\bar{G}$.

Second, for any $l \in [q]$, the constraints (5) guarantee that for any two edges e_k^l and e_s^l from the same MP_l-cycle in $\bar{G}$, $p_k^l = p_s^l$. Indeed, if e_k^l and e_s^l belong to the same MP_l-cycle, then there exists a sequence of P_l-edges $(e_k^l, e_{j_1}^l, e_{j_2}^l, \ldots, e_{j_t}^l, e_s^l)$, where every pair of adjacent P_l-edges are connected by an M-edge. From (5), it follows $p_k^l = p_{j_1}^l = \cdots = p_{j_t}^l = p_s^l$.

The constraints (5) and (4) together guarantee that p_k^l does not exceed the smallest index of a P_l-edge from the MP_l-cycle containing e_k^l. From (4) and (6), it follows that $k\tilde{p}_k^l \le p_k^l \le k$. Then we may have $\tilde{p}_k^l = 1$ only if $p_k^l = k$, i.e., e_k^l has the smallest index in its MP_l-cycle in $\bar{G}$. Hence, $\sum_{k=1}^n \tilde{p}_k^l \le c(M, P_l)$. Note that the equality is attained if and only if for all $k \in [n]$, $\tilde{p}_k^l$ equals the indicator of $p_k^l = k$. Thus, if the objective function in (7) is maximized, then $\sum_{l=1}^q \sum_{k=1}^n \tilde{p}_k^l = \sum_{l=1}^q c(M, P_l)$ and the corresponding genome M is a median for the genomes $P_1, P_2, \ldots, P_q$. $\qquad\square$

There are $O(n^2 + qn)$ variables and $O(qn^2)$ constraints in the GMP ILP formulation given by Theorem 4. Note that in the case $q = 3$ we obtain an ILP formulation for the classic GMP, which has quadratic size in the number of genes.

The ILP formulation in Theorem 4 can further be adjusted for solving the CGMP. Namely, in this case we have the following additional constraints:

$$r_{\{i,j\}} = 0 \text{ if } \{v_i, v_j\} \text{ is not a } P_k\text{-edge for any } k \in [q]. \tag{8}$$

Theorem 5. *Let $P_1, \ldots, P_q$ be ordinary genomes composed of the same n genes. A solution to the ILP with the objective function (7) under the constraints (3), (4), (5), (6), (8) defines an ordinary genome M that is a solution to the CGMP for the genomes $P_1, \ldots, P_q$.*

Thus, we obtain a linear-size ILP formulation for the CGMP with $O(qn)$ variables and $O(qn)$ constraints.

Intermediate Genome Median Problem. We can easily adjust the constructed ILP for the GMP with $q = 3$ to solve the IGMP. For given genomes P_1, P_2, P_3, it asks for a genome M such that $c(P_1, M) + c(P_2, M) = n + c(P_1, P_2)$ (by Theorem 1) and $c(P_3, M)$ is maximized. We define the same collection of variables as before under the same constraints (3)–(6). Since $c(P_1, M) + c(P_2, M) = n + c(P_1, P_2)$, we have the following additional constraint:

$$\sum_{k=1}^{n} (\tilde{p}_k^1 + \tilde{p}_k^2) = n + c(P_1, P_2). \tag{9}$$

Theorem 6. *Let P_1, P_2, P_3 be ordinary genomes composed of the same n genes, and $q = 3$. The linear constraints (3), (4), (5), (6), (9) together with the objective function*

$$maximize \ \sum_{k=1}^{n} \tilde{p}_k^3, \tag{10}$$

define a genome M that is a solution to the IGMP for genomes P_1, P_2, P_3.

Proof. The proof proceeds as the proof of Theorem 4 with the following difference. The constraint (9) now guarantees that M is an intermediate genome between P_1 and P_2. If the objective function in (10) is maximized, then $\sum_{k=1}^{n} \tilde{p}_k^3 = c(M, P_3)$ and the corresponding genome M is a solution to the IGMP for the genomes P_1, P_2, P_3. □

The ILP formulation for the IGMP in Theorem 6 consists of $O(n^2)$ variables and $O(n^2)$ constraints.

We remark that from the known structure of intermediate genomes [4] it further follows that a variable $r_{\{i,j\}}$ may be nonzero only if v_i, v_j belong to the same P_1P_2-cycle. In other words, we have the following additional constraints:

$$r_{\{i,j\}} = 0 \text{ if } v_i, v_j \text{ belong to different } P_1P_2\text{-cycles}, \tag{11}$$

which allows to reduce the actual number of variables in the ILP formulation given in Theorem 6.

Aliquoting Distance Problem. We consider the ADP for a given ordinary genome R on n genes and an m-duplicated genome A on the same genes present in m copies each. By the formula (2), a solution to the ADP corresponds to a maximal AR-cycle decomposition of $\hat{\mathfrak{G}}(A, mR)$. Indeed, by Theorem 2, such AR-cycle decomposition corresponds to the breakpoint graph $\mathfrak{G}(A, X)$ with the same number cycles for some genome $X \in D_m(R)$ and some labeling of gene copies in A and X. Hence, X represents a solution to the ADP. Below we show how to find a maximal AR-cycle decomposition of $\hat{\mathfrak{G}}(A, mR)$ with ILP.

Let $\bar{V} = \{\bar{v}_1, \bar{v}_2, \ldots, \bar{v}_{2mn}\}$ be the set of vertices (indexed arbitrarily) of $\mathfrak{G}(A)$. For any $X \in D_m(R)$ and any labeling of gene copies in A and X, let $\bar{G} = \mathfrak{G}(A, X)$, which we view as a graph on the same vertex set $\bar{V}$. For a vertex u in $\hat{\mathfrak{G}}(A, R)$, denote by $\text{equiv}(u) \subset [2mn]$ the set of indices of the vertices in $\bar{V}$ that are copies of u. Then we represent the genome X as the collection of binary variables $x_{\{i,j\}}$ that indicate whether an unordered pair of vertices $\{\bar{v}_i, \bar{v}_j\}$ forms an X-edge in $\bar{G}$. Since $X \in D_m(R)$, each R-edge $\{u, w\}$ in $\hat{\mathfrak{G}}(A, R)$ corresponds to an X-matching in $\bar{G}$ between the vertices with indices equiv(u) and the vertices with indices equiv(w). Hence, we have the following constraints:

$$\text{for each } R\text{-edge } \{u, v\} \text{ in } \hat{\mathfrak{G}}(A, R):$$

$$x_{\{i,j\}} \in \{0,1\}, \quad \forall i \in \text{equiv}(u), \; j \in \text{equiv}(v),$$

$$\sum_{j \in \text{equiv}(v)} x_{\{i,j\}} = 1, \quad \forall i \in \text{equiv}(u), \tag{12}$$

$$\sum_{i \in \text{equiv}(u)} x_{\{i,j\}} = 1, \quad \forall j \in \text{equiv}(v).$$

Now, let us express $c(A, X)$ via integer variables and linear constraints, similarly how the number of cycles was expressed in the ILP for the GMP. Let $e_1, e_2, \ldots, e_{mn}$ be the A-edges (indexed arbitrarily) in $\bar{G}$. For $j \in [2nm]$, let inc$(j) \in [mn]$ be the index of the A-edge incident to the vertex $\bar{v}_j$ in $\bar{G}$. For each $i \in [mn]$, we define an integer variable a_i equal the smallest index of an A-edge from the AX-cycle in $\bar{G}$ containing e_i. It follows that

$$a_i \leq i, \quad \forall i \in [mn]. \tag{13}$$

For all $s, k \in [mn]$, e_s and e_k belong to the same AX-cycle in $\bar{G}$ if and only if $a_s = a_k$. Therefore, we have

$$|a_{\text{inc}(i)} - a_{\text{inc}(j)}| \leq 2mn \left(1 - x_{\{i,j\}}\right), \quad \forall \{i,j\} \subset [2nm]. \tag{14}$$

Let $\tilde{a}_i$ be an indicator that $a_i = i$ (i.e., $\tilde{a}_i$ indicates whether e_i has the smallest index in its AX-cycle in $\bar{X}$). These variables satisfy the following constraints:

$$i \, \tilde{a}_i \leq a_i, \quad \tilde{a}_i \in \{0,1\}, \forall i \in [mn] : \tag{15}$$

Since exactly one edge has the smallest index in each AX-cycle in $\bar{G}$, it follows that $c(A, X) = \sum_{i=1}^{mn} \tilde{a}_i$.

Theorem 7. *Let A be an m-duplicated genome and R be an ordinary genome on the same n genes. Solution to the ILP with the objective function*

$$\text{maximize } \sum_{i=1}^{mn} \tilde{a}_i, \tag{16}$$

under the constraints (12), (13), (14), (15) defines a maximal AR-cycle decomposition of $\hat{\mathfrak{G}}(A, mR)$, and thus $d_{\text{DCJ}}^m(A, R) = m|R| - \max \sum_{i=1}^{mn} \tilde{a}_i$.

Proof. The constraints (12) guarantee that the collection of binary variables $x_{\{i,j\}}$ defines the breakpoint graph $\bar{G} = \mathfrak{G}(A, X)$ for some genome $X \in D_m(R)$ and some labeling of gene copies in A and X.

Similarly to the proof of Theorem 4, from the constraints (13), (14), (15), it follows that $\sum_{i=1}^{mn} \tilde{a}_i \leq c(A, X)$, where the equality is attained if and only if for all $i \in [mn]$, $\tilde{a}_i$ equals the indicator that e_i has the smallest index in its AX-cycle in $\bar{G}$. Thus, if the objective function in (16) is maximized, then $\sum_{i=1}^{mn} \tilde{a}_i = c(A, X)$, and furthermore $\mathfrak{G}(A, X)$ has the largest possible number of cycles, i.e., $c(A, X) = c_{\max}(A, R)$, the corresponding AR-cycle decomposition C is maximal, implying that

$$d_{DCJ}^m(A, R) = m|R| - c_{\max}(A, R) = m|R| - \sum_{i=1}^{mn} \tilde{a}_i.$$

$\square$

There are $O(m^2 n)$ variables and $O(m^2 n)$ constraints in the ADP ILP formulation given by Theorem 7. In the case of the double distance ($m = 2$), the size of this ILP formulation is linear in the number of genes.

Genome Aliquoting Problem. We consider the GAP for a given m-duplicated genome A on n unique genes present in m copies each. Since for $m = 2$, there exists a polynomial-time solution [1, 25, 36], we focus on the case $m \geq 3$. Recall that the GAP asks to find an ordinary genome R that maximizes $c_{\max}(A, R)$. We start to construct an ILP formulation for the GAP similarly to those for the GMP and ADP.

Let $V = \{v_1, \ldots, v_{2n}\}$ be a set of vertices (indexed arbitrarily) in $\hat{\mathfrak{G}}(A, R)$. We represent a genome R as the collection of binary variables $r_{\{i,j\}}$ ($i, j \in [2n]$) that indicate whether $\{v_i, v_j\}$ is an R-edge in $\hat{\mathfrak{G}}(A, R)$. Since R-edges form a matching on V, the constraints (3) hold.

Let $\bar{V} = \{\bar{v}_1, \bar{v}_2, \ldots, \bar{v}_{2mn}\}$ be the set of vertices (indexed arbitrarily) of $\mathfrak{G}(A)$. For any $X \in D_m(R)$ and any labeling of gene copies in A and X, let $\bar{G} = \mathfrak{G}(A, X)$, which we view as a graph on the same vertex set $\bar{V}$. For $i \in [2n]$, let equiv$(i) \subset [2mn]$ be the set of indices of the vertices from $\bar{V}$ that are glued into $v_i \in V$. Again, we represent the genome X as the collection of binary variables $x_{\{i,j\}}$ that indicate whether $\{\bar{v}_i, \bar{v}_j\}$ forms an X-edge in $\bar{G}$. Similarly to (12), we have

$$\forall i, j \in [2n]:$$
$$x_{\{k,t\}} \in \{0, 1\}, \quad \forall k \in \text{equiv}(i), t \in \text{equiv}(j),$$
$$\sum_{t \in \text{equiv}(j)} x_{\{k,t\}} = r_{\{i,j\}}, \quad \forall k \in \text{equiv}(i), \tag{17}$$
$$\sum_{k \in \text{equiv}(i)} x_{\{k,t\}} = r_{\{i,j\}}, \quad \forall t \in \text{equiv}(j).$$

We define the same collections of integer variables a_i and indicator variables $\tilde{a}_i$ as in the ILP for the ADP, under the same constraints (13) − (15), implying that $c(A, X) = \sum_{i=1}^{mn} \tilde{a}_i$.

Theorem 8. *Let A be an m-duplicated genome. The linear constraints (3), (13), (14), (15), (17) together with the objective function (16) define an ordinary genome R that is a solution to the GAP for the genome A.*[5]

Proof. The constraints (3) guarantee that the collection of binary variables $r_{\{i,j\}}$ defines a matching on the vertex set V. So, the constraints (3) define some ordinary genome R and the graph $\hat{\mathfrak{G}}(A, R)$. The constraints (17) guarantee that the collection of binary variables $x_{\{i,j\}}$ defines the breakpoint graph $\bar{G} = \mathfrak{G}(A, X)$ for some genome $X \in D_m(R)$ and some labeling of gene copies in A and X. Similarly to the proof of Theorem 7, if the objective function in (16) is maximized, it gives the maximum value of $c(A, X)$ equal $c_{\max}(A, R)$, and furthermore it maximized over all choices of R. It follows that the corresponding genome R is a solution to the GAP for the given genome A. □

There are $O(m^2 n^2)$ variables and $O(m^2 n^2)$ constraints in the GAP ILP formulation given by Theorem 8.

Guided Genome Aliquoting Problem. Let us consider the GGAP for a given ordinary genome B on n genes and m-duplicated genome A on the same genes present in m copies each. By (1) and (2), for any ordinary genome R we have

$$d_{\mathrm{DCJ}}^m(A, R) + d_{\mathrm{DCJ}}(R, B) = (m+1)n - c_{\max}(A, R) - c(R, B).$$

Hence, the GGAP asks for an ordinary genome R that maximizes $c_{\max}(A, R) + c(R, B)$. While $c_{\max}(A, R)$ equals the size of the maximal cycle decomposition of $\hat{\mathfrak{G}}(A, mR)$ and $c(R, B)$ equals the number of cycles in $\mathfrak{G}(R, B)$, we find it convenient to view both $\hat{\mathfrak{G}}(A, R)$ and $\hat{\mathfrak{G}}(R, B) = \mathfrak{G}(R, B)$ are subgraphs of the contracted breakpoint graph $\hat{\mathfrak{G}}(A, R, B)$.

Similarly to the GAP ILP formulation, for an ordinary genome R, a genome $X \in D_m(R)$ and a labeling of gene copies of X and A, we define binary variables $r_{\{i,j\}}$ ($i, j \in [2n]$), binary variables $x_{\{k,l\}}$ ($k, l \in [2mn]$), integer variables a_s and indicator variables $\tilde{a}_s$ ($s \in [mn]$) that satisfy the constraints (3), (13)–(15), and (17). Then $c(A, X) = \sum_{i=1}^{mn} \tilde{a}_i$.

Now, we express $c(R, B)$ as in the GMP ILP formulation for $m = 1$ and $P_1 = B$. Namely, we define the same collections of integer variables p_k^1 and indicator variables $\tilde{p}_k^1$ under the constraints (4)–(6). Then $c(R, B) = \sum_{k=1}^n \tilde{p}_k^1$.

Theorem 9. *Let A be an m-duplicated genome and B be an ordinary genome, $q = 1$, $P_1 = B$. The linear constraints (3)–(6), (13)–(15), (17) together with the objective function*

$$maximize \quad \sum_{i=1}^{mn} \tilde{a}_i + \sum_{i=1}^{n} \tilde{p}_i^1, \tag{18}$$

define an ordinary genome R that is a solution to the GGAP for the genomes A and B.[6]

---

[5] In fact, they also define a genome $X \in D_m(R)$ and a labeling of gene copies of A and X such that $c(A, X)$ is maximized.

[6] In fact, beside R they also define a genome $X \in D_m(R)$ and a labeling of gene copies of A and X such that $c(A, X) + c(R, B)$ is maximized.

Proof. As in the proof of the Theorem 4, the constraints (3) guarantee that the collection of binary variables $r_{\{i,j\}}$ defines some ordinary genome R and the graph $\hat{\mathfrak{G}}(A, R, B)$.

By the proof of the Theorem 4, the constraints (4), (5), (6) imply that $\sum_{i=1}^{n} \tilde{p}_i^1 \leq c(R, B)$, where equality is attained if and only if each $\tilde{p}_i^1$ equals the indicator that $p_k^1 = k$.

By the proof of the Theorem 8, the constraints (13), (14), (15), (17) imply that $\sum_{i=1}^{mn} \tilde{a}_i \leq c(A, mR)$, where equality is attained if and only if each $\tilde{a}_i$ equals the indicator that $a_k = k$.

Thus, if the objective function in (18) is maximized, then

$$\sum_{i=1}^{mn} \tilde{a}_i + \sum_{i=1}^{n} \tilde{p}_i^1 = c(A, X) + c(R, B) = c_{\max}(A, R) + c(R, B)$$

is maximized (over all choices of R), implying that the corresponding genome R is a solution to the GGAP for the genomes A and B. □

There are $O(m^2 n^2)$ variables and $O(m^2 n^2)$ constraints in the GGAP ILP formulation given by Theorem 9. For the GGHP (i.e., when $m = 2$), the size of this ILP formulation is quadratic in the number of genes.

The ILP formulation in Theorem 9 can further be adjusted for solving the CGGAP. Namely, in this case we have the following additional constraints:

$$r_{\{i,j\}} = 0 \text{ if } \{v_i, v_j\} \text{ is not a } A\text{-edge or } B\text{-edge.} \tag{19}$$

Corollary 10. *Let A be an m-duplicated genome and B be an ordinary genome, $q = 1$, $P_1 = B$. The linear constraints (3) – (6), (13) – (15), (17), (19) together with the objective function (18) define an ordinary genome R that is a solution to the CGGAP for the genomes A and B.*

Thus, we obtain a linear-size ILP formulation for the CGGAP with $O(mn)$ variables and $O(mn)$ constraints.

We remark that the ILP formulations in Theorem 9 and Corollary 10 can be easily extended to multiple ordinary genomes $B_1, B_2, \ldots, B_t$, which can further improve biological relevance of the resulting solution.

Restricted Guided Genome Halving Problem. We can easily adjust the GGAP ILP formulation for the RGGHP. For a given ordinary genome B on n genes and 2-duplicated genome A on the same genes present in two copies each, the RGGHP asks for an ordinary genome R that is a solution to the GHP for the genome A and minimizes $d_{\mathrm{DCJ}}(B, R)$. By Theorem 3, for such R, we have $c_{\max}(A, R) = n + c_{\mathrm{even}}(A)$, and $c(R, B)$ is maximized.

We let $m = 2$ and define the same collection of variables and constraints as in the GGAP ILP formulation given by Theorem 9. Since $c_{\max}(A, R) = n + c_{\mathrm{even}}(A)$, we have an additional constraint:

$$\sum_{i=1}^{mn} \tilde{a}_i = n + c_{\mathrm{even}}(A). \tag{20}$$

Theorem 11. *Let A be a 2-duplicated genome and B be an ordinary genome on the same n genes. Let $m = 2$, $q = 1$, and $P_1 = B$. Solution to the ILP with the objective function*

$$maximize: \quad \sum_{i=1}^{n} \tilde{p}_i^1 \qquad (21)$$

under the constraints (3)–(6), (13)–(15), (17), and (20) defines an ordinary genome R that is a solution to the RGGHP for the genomes A and B.

Proof. The proof proceeds as the proof of Theorem 9 with only one difference. Namely, the constraint (20) now guarantees that R is a solution to the GHP for the genome A. If the objective function in (21) is maximized, then $\sum_{i=1}^{n} \tilde{p}_i^1 = c(R, B)$, and the corresponding genome R is a solution to the RGGHP for the genomes A and B. □

The RGGHP ILP formulation given in Theorem 11 consists of $O(n^2)$ variables and $O(n^2)$ constraints.

We remark that from the known structure of the GHP solution space [4] it further follows that

$$r_{\{i,j\}} = 0 \text{ if one of } v_i, v_j \text{ belongs to an even } A\text{-cycle,}$$
$$\text{while the other vertex belong to a different } A\text{-cycle,} \qquad (22)$$

which allows to reduce the actual number of variables in the ILP formulation given in Theorem 11.

4 Experimental Results

Median Problems

We evaluate solutions to our ILP formulations for the median problems in comparison to the exact algorithm ASMedian [37] on simulated triples of ordinary unichromosomal genomes P_1, P_2, P_3. Namely, we first pick a fixed unichromosomal genome with $n = 25$ distinct "genes" as a simulated median genome, and then independently transform it into ordinary genomes P_1, P_2, P_3 with $E \in \{5, 10, 15, 20, 25\}$ random reversals. For each value of E we independently create 10 simulated datasets. The genomes P_1, P_2, P_3 are provided as an input to the different methods, and we evaluate the accuracy of the reconstructed median genome M based on the total DCJ distance (*median score*) from M to P_1, P_2, P_3, and on the set of gene adjacencies shared between M and the simulated median genome. We consider the following ILP instances for the genomes P_1, P_2, P_3:

GMP-ILP: an ILP instance for the GMP constructed as in Theorem 4 for $q = 3$;
IGMP-ILP: an ILP instance for the IGMP constructed as in Theorem 6;
CGMP-ILP: an ILP instance for the CGMP constructed as in Theorem 5 for $q = 3$.

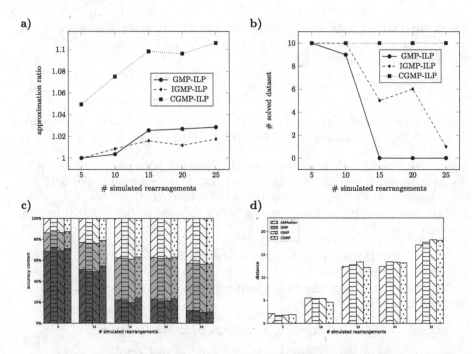

Fig. 4. Performance of GUROBI on simulated **GMP-ILP**, **IGMP-ILP**, and **CGMP-ILP** instances, with the time limit of 2 hours each. (a) Approximation ratio of the median score averaged over 10 simulated datasets. (b) The number of datasets that have been solved exactly within the time limit. (c) Proportions of true positives (dark gray bars), false negatives (gray bars), and false positives (white bars) among the reconstructed gene adjacencies in a median genome, averaged over 10 simulated datasets. (d) The average DCJ distance between the corresponding reconstructed and simulated median genome, averaged over 10 simulated datasets. The four bars correspond to ASMedian (first bar), **GMP-ILP** (second bar), **IGMP-ILP** (third bar), and **CGMP-ILP** (last bar).

We solve these ILP instances with the GUROBI solver [17]. Since in some cases obtaining an exact solution may take long time, we set a time limit of 2 h for each experiment. If the ILP solver does not terminate within this time limit, it returns the best approximate solution found so far. All computations were performed on a 4-core Intel Xeon CPU 3.70 GHz.

Figure 4a reports the median score approximation ratio in different ILP instances (with respect to the median score computed by ASMedian [37]). Figure 4b reports the number of datasets that have been solved exactly within the time limit. We observe that the approximation ratio in the **GMP-ILP** greater than 1 indicates that GUROBI was not able to find a median genome within the given time limit, which is supported by Fig. 4b. It is somewhat surprising that the approximation ratio in the **IGMP-ILP** is smaller than that in the **GMP-ILP**, however this is explained by the smaller size of the **IGMP-ILP** (thanks to the constraints (11)) and thus better performance of GUROBI within the given time limit. The **CGMP-ILP** has the worst approximation ratio, while this ILP formulation has a linear number of variables and each its instance is solved by GUROBI exactly (as shown in Fig. 4b).

Figure 4c reports the average accuracy measured as proportions of correct (true positives), missing (false negatives), and incorrect (false positives) gene adjacencies present in a solution as compared to the known simulated median genome. Figure 4d further reports the average DCJ distance (computed with the formula (1)) between the solution and the simulated median genome. Both graphs in Fig. 4c,d show that the solutions obtained by different methods (under the time limit) have similar accuracy.

The evaluated accuracy of the median genomes of GMP and IGMP (Fig. 4c) challenges the earlier empirical observation [12] that the IGMP leads to more biologically adequate median genomes than the general GMP. However, the IGMP can be viewed as a good approximation of the GMP, which maybe easier to solve due to simpler structure of the solution space [4].

The performance of the **CGMP-ILP** is somewhat interesting. While it shows a largest approximation ratio for the median score, its reconstructed median genome is often better than other median genomes in terms closeness to the simulated median genome. Since the size of the **CGMP-ILP** is linear, it can be applied for larger genomes. Thus, the **CGMP-ILP** represents a novel practical method for reconstruction of median genomes that combines information about gene adjacencies and minimizes the total distance to the input genomes (i.e., combines homology and rearrangement based methods).

Double Distance

We compare our ILP formulation for the double distance to the greedy heuristic (denoted as **GREEDY**) proposed in [28] on a simulated ordinary genome R and a 2-duplicated genome A on the same gene content. We evaluate the accuracy of two methods based on the computed double distance between A and R.

Namely, we fix an ordinary unichromosomal genome R with $n = 2000$ "genes", and apply a WGD on R to obtain a perfect 2-duplicated unichromosomal genome $2R$. Then genome $2R$ becomes a subject to a series of $E \in \{500, 1000, 1500, 2000, 2500, 3000\}$ random reversals resulting in a 2-duplicated genome A. For each value of E we independently create 10 simulated datasets.

We create the ILP instance **DD-ILP**, constructed as in Theorem 7. We solve **DD-ILP** with GUROBI under the time limit of 4 h for experiment.

Figure 5a reports inferred the double distance by **DD-ILP** and **GREEDY**. Figure 5b reports the number of datasets for which **DD-ILP** have been solved within the time limit. We remark that the accuracy of **GREEDY** was previously evaluated on simulated genomes with just 25 distinct genes, and it was shown to produce almost exact results when the breakpoint reuse (i.e., twice the number of rearrangements divided by the genome size) is below 1 [28]. Since **DD-ILP** computed the double distance exactly in the experiments with $E \in \{500, 1000, 1500\}$

a)   b)

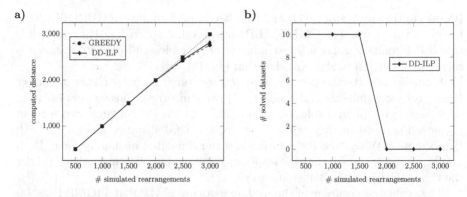

Fig. 5. (a) The double distances computed by **DD-ILP** (black lines) and **GREEDY** (dashed line) averaged over 10 datasets. (b) The number of datasets for **DD-ILP** that have been solved exactly within the time limit.

(the breakpoint reuse up to 0.75), it shows that **GREEDY** works well even for the bigger genomes with bounded breakpoint reuse. We further remark that for datasets with $E \in \{2000, 2500, 3000\}$, **GREEDY** outperforms **DD-ILP** since GUROBI was not able to find an exact solution within the given time limit.

Guided Halving Problems

To the best of knowledge there is no exact or heuristic algorithms for the GGHP available online. We therefore compare solutions produced to our ILP formulations for guided halving problems between themselves. Namely, we fix a simulated unichromosomal genome B with $n = 25$ distinct "genes" and transform it into an ordinary genome R with $E \in \{5, 10, 25, 20, 25\}$ random reversals. Then we apply a WGD for ordinary genome R to obtain a perfect unichromosomal 2-duplicated genome $2R$. Finally, we transform $2R$ into a 2-duplicated genome A with E random reversals. For each value of E we independently create 10 simulated datasets. The genomes A and B are provided as an input to the different methods, and we evaluate their accuracy based on the computed ordinary genome R and the total DCJ distance (*GGH-score*) from R to A and B. Namely, we consider the following ILP instances for the genomes A and B:

GGHP-ILP: an ILP instance for the GGHP constructed as in Theorem 9 for $m = 2$;

RGGHP-ILP: an ILP instance for the RGGHP constructed as in Theorem 11;

CGGHP-ILP: an ILP instance for the CGGHP constructed as in Corollary 10 for $m = 2$.

We solve these ILP instances with GUROBI with the time limit of 2 h.

Figure 6a reports the GGH-score for different ILP instances. Figure 6b reports the number of datasets that have been solved exactly within the time limit. We

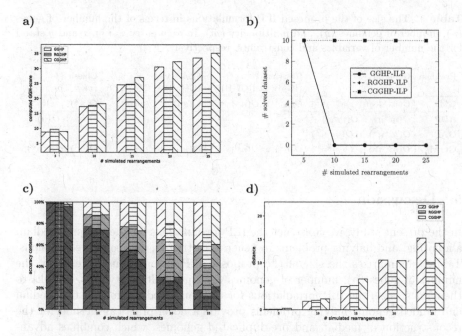

Fig. 6. (a) The GGH-score by **GGHP-ILP** (left bar), **RGGHP-ILP** (middle bar), and **CGGHP-ILP** (right bar) computed within the 2-hour time limit, averaged over 10 datasets, (b) The number of datasets that have been solved exactly within the time limit. (c) Normalized true positives (dark gray bars), false negatives (gray bars), and false positives (white bars) among the reconstructed gene adjacencies in an ordinary genome, averaged over 10 simulated datasets. (d) The average DCJ distance between the corresponding reconstructed and simulated ordinary genome, averaged over 10 simulated datasets with parameters.

observe that the **CGGHP-ILP** shows the worst GGH-score among the ILP instances, while this ILP formulation has a linear number of variables and each its instance is solved by GUROBI exactly (as shown in Fig. 6b). It is somewhat surprising that the GGH-score of the **RGGHP-ILP** is smaller than that of **GGHP-ILP** on the datasets with $E \in 20, 25$. However, similarly to the case of **GMP-ILP** and **IGMP-ILP**, this is explained by the smaller size of the **RGGHP-ILP** (thanks to the constraints (22)) and thus better performance of GUROBI within the given time limit.

Figure 6c reports the average accuracy of the reconstructed pre-duplicated genome R' measured in terms of correct (true positives), missing (false negatives), and incorrect (false positives) gene adjacencies present in R' as compared to the known simulated pre-duplicated genome R. Figure 6d further reports the average DCJ distance (computed with the formula (1)) between R' and R. Both graphs show that the solutions obtained by different methods have similar accuracy. Hence, **CGGHP-ILP** can be used as practical approximation method for solving the GGHP for large genomes.

Table 1. The size of the proposed ILP formulations in terms of the number of genes (n), number of genomes (q), and multiplicity (m). In each entry $x : y$, x and y stand for the number of variables and constraints, respectively.

Problem	General	Restricted (IGMP & RGGHP)	Conserved (CGMP & CGGHP)	Classic ($m = 2$ or $q = 3$)
GMP	$O(n^2 + qn)$: $O(qn^2)$	$O(n^2)$: $O(n^2)$	$O(qn)$: $O(qn)$	$O(n^2)$: $O(n^2)$
ADP	$O(m^2 n)$: $O(m^2 n)$	–	–	$O(n)$: $O(n)$
GAP	$O(m^2 n^2)$: $O(m^2 n^2)$	–	–	–
GGAP	$O(m^2 n^2)$: $O(m^2 n^2)$	$O(n^2)$: $O(n^2)$	$O(mn)$: $O(mn)$	$O(n^2)$: $O(n^2)$

5 Discussion

In the present study, we construct the ILP formulations for the genome median, aliquoting, and halving problems in general, restricted, and conserved versions. Table 1 summarizes the size of the proposed ILP formulations in terms of the number of genes (n), number of genomes (q), and multiplicity (m). Thanks to their linear size, the ILP formulations for the conserved versions of the median and guided genome halving problems provide a novel practical method for the reconstruction of median and pre-duplicated genomes, which combines advantages of homology and rearrangement based methods.

We provide extensive comparison between ILP solutions to the conserved, restricted, and general versions of different problems and other existing algorithms. In particular, our results challenge the earlier empirical observation [12] that the IGMP leads to more biological adequate median genomes than the GMP. We also show that the greedy heuristic for the double distance [28] works well for even bigger genomes than previously tested.

In future work, we plan to extend our ILP formulations to support genomes with unequal gene content, genomes with linear chromosomes, and median genomes with duplicated genes.

Acknowledgements. The work of PA and MAA is supported by the National Science Foundation under the grant No. IIS-1462107. The work of NA and YR is partially supported by the National Science Foundation under the grant No. DMS-1406984.

References

1. Alekseyev, M.A., Pevzner, P.A.: Colored de Bruijn graphs and the genome halving problem. IEEE/ACM Trans. Comput. Biol. Bioinf. (TCBB) **4**(1), 98–107 (2007)
2. Alekseyev, M.A., Pevzner, P.A.: Whole genome duplications, multi-break rearrangements, and genome halving problem. In: Proceedings of the Eighteenth Annual ACM-SIAM Symposium on Discrete Algorithms (SODA 2007), pp. 665–679. Society for Industrial and Applied Mathematics, Philadelphia, PA, USA (2007)
3. Alekseyev, M.A., Pevzner, P.A.: Multi-break rearrangements and chromosomal evolution. Theoret. Comput. Sci. **395**(2), 193–202 (2008)

4. Alexeev, N., Avdeyev, P., Alekseyev, M.A.: Comparative genomics meets topology: a novel view on genome median and halving problems. BMC Bioinf. **17**(14), 418 (2016)
5. Avdeyev, P., Jiang, S., Aganezov, S., Hu, F., Alekseyev, M.A.: Reconstruction of ancestral genomes in presence of gene gain and loss. J. Comput. Biol. **23**(3), 150–164 (2016)
6. Bergeron, A., Mixtacki, J., Stoye, J.: A unifying view of genome rearrangements. In: Bücher, P., Moret, B.M.E. (eds.) WABI 2006. LNCS, vol. 4175, pp. 163–173. Springer, Heidelberg (2006). doi:10.1007/11851561_16
7. Caprara, A.: The reversal median problem. INFORMS J. Comput. **15**(1), 93–113 (2003)
8. Caprara, A., Lancia, G., Ng, S.K.: Fast practical solution of sorting by reversals. In: Proceedings of the Eleventh Annual ACM-SIAM Symposium on Discrete Algorithms (SODA 2000), pp. 12–21. Society for Industrial and Applied Mathematics (2000)
9. Dehal, P., Boore, J.L.: Two rounds of whole genome duplication in the ancestral vertebrate. PLoS Biol. **3**(10), e314 (2005)
10. Dias, Z., de Souza, C.C.: Polynomial-sized ILP models for rearrangement distance problems. In: Brazilian Symposium On Bioinformatics, p. 74 (2007)
11. El-Mabrouk, N., Sankoff, D.: The reconstruction of doubled genomes. SIAM J. Comput. **32**(3), 754–792 (2003)
12. Feijão, P.: Reconstruction of ancestral gene orders using intermediate genomes. BMC Bioinf. **16**(Suppl 14), S3 (2015)
13. Feijão, P., Araujo, E.: Fast ancestral gene order reconstruction of genomes with unequal gene content. BMC Bioinf. **17**(14), 413 (2016)
14. Gagnon, Y., Savard, O.T., Bertrand, D., El-Mabrouk, N.: Advances on genome duplication distances. In: Tannier, E. (ed.) RECOMB-CG 2010. LNCS, vol. 6398, pp. 25–38. Springer, Heidelberg (2010). doi:10.1007/978-3-642-16181-0_3
15. Gao, N., Yang, N., Tang, J.: Ancestral genome inference using a genetic algorithm approach. PLoS ONE **8**(5), 1–6 (2013)
16. Gavranović, H., Tannier, E.: Guided genome halving: provably optimal solutions provide good insights into the preduplication ancestral genome of saccharomyces cerevisiae. Pac. Symp. Biocomput. **15**, 21–30 (2010)
17. Gurobi Optimization Inc: Gurobi optimizer reference manual (2016). http://www.gurobi.com
18. Guyot, R., Keller, B.: Ancestral genome duplication in rice. Genome **47**(3), 610–614 (2004)
19. Haghighi, M., Sankoff, D.: Medians seek the corners, and other conjectures. BMC Bioinform. **13**(19), 1 (2012)
20. Hartmann, T., Wieseke, N., Sharan, R., Middendorf, M., Bernt, M.: Genome Rearrangement with ILP. IEEE/ACM Trans. Comput. Biol. Bioinform. (2017, in press). doi:10.1109/TCBB.2017.2708121
21. Kellis, M., Birren, B.W., Lander, E.S.: Proof and evolutionary analysis of ancient genome duplication in the yeast saccharomyces cerevisiae. Nature **428**(6983), 617–624 (2004)
22. Lancia, A.C.G., Ng, S.K.: A column-generation based branch-and-bound algorithm for sorting by reversals. Math. Support Mol. Biol. **47**, 213 (1999)
23. Lancia, G., Rinaldi, F., Serafini, P.: A unified integer programming model for genome rearrangement problems. In: Ortuño, F., Rojas, I. (eds.) IWB-BIO 2015. LNCS, vol. 9043, pp. 491–502. Springer, Cham (2015). doi:10.1007/978-3-319-16483-0_48

24. Laohakiat, S., Lursinsap, C., Suksawatchon, J.: Duplicated genes reversal distance under gene deletion constraint by integer programming. In: 2008 2nd International Conference on Bioinformatics and Biomedical Engineering, pp. 527–530, May 2008
25. Mixtacki, J.: Genome halving under DCJ revisited. In: Hu, X., Wang, J. (eds.) COCOON 2008. LNCS, vol. 5092, pp. 276–286. Springer, Heidelberg (2008). doi:10. 1007/978-3-540-69733-6_28
26. Postlethwait, J.H., Yan, Y.L., Gates, M.A., Horne, S., Amores, A., Brownlie, A., Donovan, A., Egan, E.S., Force, A., Gong, Z., et al.: Vertebrate genome evolution and the zebrafish gene map. Nat. Genet. **18**(4), 345–349 (1998)
27. Rajan, V., Xu, A.W., Lin, Y., Swenson, K.M., Moret, B.M.: Heuristics for the inversion median problem. BMC Bioinf. **11**(1), S30 (2010)
28. Savard, O.T., Gagnon, Y., Bertrand, D., El-Mabrouk, N.: Genome halving and double distance with losses. J. Comput. Biol. **18**(9), 1185–1199 (2011)
29. Shao, M., Lin, Y., Moret, B.M.: An exact algorithm to compute the double-cut-and-join distance for genomes with duplicate genes. J. Comput. Biol. **22**(5), 425–435 (2015)
30. Shao, M., Moret, B.M.: Comparing genomes with rearrangements and segmental duplications. Bioinformatics **31**(12), i329 (2015)
31. Suksawatchon, J., Lursinsap, C., Boden, M.: Computing the reversal distance between genomes in the presence of multi-gene families via binary integer programming. J. Bioinf. Comput. Biol. **05**(01), 117–133 (2007)
32. Swenson, K.M., Moret, B.M.: Inversion-based genomic signatures. BMC Bioinf. **10**(1), 1 (2009)
33. Tannier, E., Zheng, C., Sankoff, D.: Multichromosomal median and halving problems under different genomic distances. BMC Bioinf. **10**(1), 1 (2009)
34. The OEIS Foundation: The On-Line Encyclopedia of Integer Sequences. Published electronically at http://oeis.org (2017)
35. Warren, R., Sankoff, D.: Genome aliquoting with double cut and join. BMC Bioinf. **10**(1), S2 (2009)
36. Warren, R., Sankoff, D.: Genome halving with double cut and join. J. Bioinf. Comput. Biol. **7**(02), 357–371 (2009)
37. Xu, A.W.: A fast and exact algorithm for the median of three problem: A graph decomposition approach. J. Comput. Biol. **16**(10), 1369–1381 (2009)
38. Yancopoulos, S., Attie, O., Friedberg, R.: Efficient sorting of genomic permutations by translocation, inversion and block interchange. Bioinformatics **21**(16), 3340–3346 (2005)
39. Zhang, M., Arndt, W., Tang, J.: An exact solver for the DCJ median problem. In: Pacific Symposium on Biocomputing. Pacific Symposium on Biocomputing, p. 138. NIH Public Access (2009)
40. Zheng, C., Zhu, Q., Adam, Z., Sankoff, D.: Guided genome halving: hardness, heuristics and the history of the hemiascomycetes. Bioinformatics **24**(13), i96 (2008)
41. Zheng, C., Zhu, Q., Sankoff, D.: Genome halving with an outgroup. Evol. Bioinf. **2**, 295–302 (2006)

Orientation of Ordered Scaffolds

Sergey Aganezov[1,2(✉)] and Max A. Alekseyev[3]

[1] Princeton University, Princeton, NJ, USA
aganezov@cs.princeton.edu
[2] ITMO University, St. Petersburg, Russia
[3] The George Washington University, Washington, DC, USA

Abstract. Despite the recent progress in genome sequencing and assembly, many of the currently available assembled genomes come in a draft form. Such draft genomes consist of a large number of genomic fragments (*scaffolds*), whose order and/or orientation (i.e., strand) in the genome are unknown. There exist various scaffold assembly methods, which attempt to determine the order and orientation of scaffolds along the genome chromosomes. Some of these methods (e.g., based on FISH physical mapping, chromatin conformation capture, etc.) can infer the order of scaffolds, but not necessarily their orientation. This leads to a special case of the *scaffold orientation problem* (i.e., deducing the orientation of each scaffold) with a known order of the scaffolds.

We address the problem of orientation of ordered scaffolds as an optimization problem based on given weighted orientations of scaffolds and their pairs (e.g., coming from pair-end sequencing reads, long reads, or homologous relations). We formalize this problem within the earlier introduced framework for comparative analysis and merging of scaffold assemblies (CAMSA). We prove that this problem is NP-hard, and further present a polynomial-time algorithm for solving its special case, where orientation of each scaffold is imposed relatively to at most two other scaffolds. This lays the foundation for a follow-up FPT algorithm for the general case. The proposed algorithms are implemented in the CAMSA software version 2.

Keywords: Genome assembly · Genome scaffolding · Scaffold orientation · Computational complexity · Algorithms

1 Introduction

While genome sequencing technologies are constantly evolving, they are still unable to read at once complete genomic sequences from organisms of interest. Instead, they produce a large number of rather short genomic fragments, called *reads*, originating from unknown locations and strands of the genome. The problem then becomes to assemble the reads into the complete genome. Existing genome assemblers usually assemble reads based on their overlap patterns and produce longer genomic fragments, called *contigs*, which are typically interweaved with highly polymorphic and/or repetitive regions in the genome.

© Springer International Publishing AG 2017
J. Meidanis and L. Nakhleh (Eds.): RECOMB CG 2017, LNBI 10562, pp. 179–196, 2017.
DOI: 10.1007/978-3-319-67979-2_10

Contigs are further assembled into *scaffolds*, i.e., sequences of contigs interspaced with gaps.[1] Assembling scaffolds into larger scaffolds (ideally representing complete chromosomes) is called the *scaffold assembly problem*.

The scaffold assembly problem is known to be NP-hard [13, 16, 22, 28, 34], but there still exists a number of methods that use heuristic and/or exact algorithmic approaches to address it. The *scaffold assembly problem* consists of two subproblems:

1. determine the order of scaffolds (*scaffold order problem*); and
2. determine the orientation (i.e., strand of origin) of scaffolds (*scaffold orientation problem*).

Some methods attempt to solve these subproblems jointly by using various types of additional data including jumping libraries [10, 14, 19, 20, 24, 26, 31], long error-prone reads [5, 6, 11, 25, 33], homology relationships between genomes [1, 3, 4, 23], etc. Other methods (typically based on wet-lab experiments [12, 21, 27, 29, 30, 32]) can often reliably reconstruct the order of scaffolds, but may fail to impose their orientation.

The scaffold orientation problem is also known to be NP-hard [9, 22]. Since the scaffold order problem can often be reliably solved with wet-lab based methods, this inspires us to consider the special case of the scaffold orientation problem with the given order of scaffolds, which we refer to as the *orientation of ordered scaffolds* (OOS) problem. We formulate the OOS as an optimization problem based on given weighted orientations of scaffolds and their pairs (e.g., coming from pair-end sequencing reads, long reads, or homologous relations) and further address it within the previously introduced CAMSA framework [2] for comparative analysis and merging of scaffold assemblies. We prove that the OOS is NP-hard both in the case of linear genomes and in the case of circular genomes. We also present a polynomial-time algorithm for solving the special case of the OOS, where the orientation of each scaffold is imposed relatively to at most two other scaffolds.

2   Background

We start with a brief description of the CAMSA framework, which provides a unifying way to represent scaffold assemblies obtained by different methods.

2.1   CAMSA Framework

We represent an *assembly* of scaffolds from a set $\mathbb{S} = \{s_i\}_{i=1}^n$ as a set of *assembly points*. Each assembly point is formed by an adjacency between two scaffolds from $\mathbb{S}$. Namely, an assembly point $p = (s_i, s_j)$ tells that the scaffolds s_i and s_j are adjacent in the assembly. Additionally, we may know the orientation of either or both of the scaffolds and thus distinguish between three types of assembly points:

---

[1] We remark that contigs can be viewed as a special type of scaffolds with no gaps.

1. p is *oriented* if the orientation of both scaffolds s_i and s_j is known;
2. p is *semi-oriented* if the orientation of only one scaffold among s_i and s_j is known;
3. p is *unoriented* if the orientation of neither of s_i and s_j is known.

We denote the known orientation of scaffolds in assembly points by overhead arrows, where the right arrow corresponds to the original genomic sequence representing a scaffold, while the left arrow corresponds to the reverse complement of this sequence. For example, $(\overrightarrow{s_i}, \overleftarrow{s_j})$, $(\overrightarrow{s_i}, s_j)$, and (s_i, s_j) are oriented, semi-oriented, and unoriented assembly points, respectively. We remark that assembly points $(\overrightarrow{s_i}, \overrightarrow{s_j})$ and $(\overleftarrow{s_j}, \overleftarrow{s_i})$ represent the same adjacency between oriented scaffolds; to make this representation unique we will require that in all assembly points (s_i, s_j) we have $i < j$. Another way to represent the orientation of the scaffolds in an assembly point is by using superscripts h and t denoting the head and tail extremities of the scaffold's genomic sequence, e.g., $(\overrightarrow{s_i}, \overrightarrow{s_j})$ can also be written as (s_i^h, s_j^t).

We will need an auxiliary function $\mathrm{sn}(p, i)$ defined on an assembly point p and an index $i \in \{1, 2\}$ that returns the scaffold corresponding to the component i of p (e.g., $\mathrm{sn}((\overrightarrow{s_i}, \overrightarrow{s_j}), 2) = s_j$).

We define a *realization* of an assembly point p as any oriented assembly point that can be obtained from p by orienting the unoriented scaffolds. We denote the set of realizations of p by $\mathrm{R}(p)$. If p is oriented, than it has a single realization equal p itself (i.e., $\mathrm{R}(p) = \{p\}$); if p is semi-oriented, then it has two realizations (i.e., $|\mathrm{R}(p)| = 2$); and if p is unoriented, then it has four realizations (i.e., $|\mathrm{R}(p)| = 4$). For example,

$$\mathrm{R}((s_i, s_j)) = \{(\overrightarrow{s_i}, \overrightarrow{s_j}), (\overrightarrow{s_i}, \overleftarrow{s_j}), (\overleftarrow{s_i}, \overrightarrow{s_j}), (\overleftarrow{s_i}, \overleftarrow{s_j})\}. \tag{1}$$

An assembly point p is called a *refinement* of an assembly point q if $\mathrm{R}(p) \subset \mathrm{R}(q)$. From now on, we assume that no assembly point in a given assembly is a refinement of another assembly point (otherwise we simply discard the latter assembly point as less informative). We refer to an assembly with no assembly point refinements as a *proper assembly*.

For a given assembly $\mathbb{A}$ we will use subscripts $u/s/o$ to denote the sets of unoriented/semi-oriented/oriented assembly points in $\mathbb{A}$ (e.g., $\mathbb{A}_u \subset \mathbb{A}$ is the set of all unoriented assembly points from $\mathbb{A}$). We also denote by $\mathbb{S}(\mathbb{A})$ the set of scaffolds appearing in the assembly points from $\mathbb{A}$.

We call two assembly points *overlapping* if they involve the same scaffold, and further call them *conflicting* if they involve the same extremity of this scaffold. We generalize this notion for semi-oriented and unoriented assembly points: two assembly points p and q are *conflicting* if all pairs of their realizations $\{p', q'\} \in \mathrm{R}(p) \times \mathrm{R}(p)$ are conflicting. If some, but not all, pairs of the realizations are conflicting, p and q are called *semi-conflicting*. Otherwise, p and q are called *non-conflicting*.

We extend the notion of non-/semi- conflictness to entire assemblies as follows. A scaffold assembly $\mathbb{A}$ is *non-conflicting* if all pairs of assembly points in it are non-conflicting, and $\mathbb{A}$ is *semi-conflicting* if all pairs of assembly points are non-conflicting or semi-conflicting with at least one pair being semi-conflicting.

2.2   Assembly Realizations

For an assembly $\mathbb{A} = \{p_i\}_{i=1}^k$, an assembly $\mathbb{A}' = \{q_i\}_{i=1}^k$ is called a *realization*[2] of $\mathbb{A}$ if there exists a permutation π of order k such that $q_{\pi_i} \in R(p_i)$ for all $i = 1, 2, \ldots, k$. We denote by $R(\mathbb{A})$ the set of realizations of assembly $\mathbb{A}$, and by $NR(\mathbb{A})$ the set of non-conflicting realizations among them.

We define the *scaffold assembly graph* $SAG(\mathbb{A})$ on the set of vertices $\{s^h, s^t : s \in \mathbb{S}(\mathbb{A})\}$ and edges of two types: directed edges (s^t, s^h) that encode scaffolds from $\mathbb{S}(\mathbb{A})$, and undirected edges that encode all possible realizations of all assembly points in $\mathbb{A}$ (Fig. 1a). We further define the *order (multi)graph* $OG(\mathbb{A})$ formed by the set of vertices $\mathbb{S}(\mathbb{A})$ and the set of undirected edges $\{\{sn(p, 1), sn(p, 2)\} : p \in \mathbb{A}\}$ (Fig. 1b). The order graph can also be obtained from $SAG(\mathbb{A})$ by first contracting the directed edges, and then by substituting all edges that encode realizations of the same assembly point with a single edge (Fig. 1b).

Lemma 1. *For a non-conflicting realization $\mathbb{A}'$ of an assembly $\mathbb{A}$, $OG(\mathbb{A}')$ is non-branching, i.e., $\deg(v) \leq 2$ for all vertices v of $OG(\mathbb{A}')$.*[3]

Proof. Each vertex v in $OG(\mathbb{A}')$ represents a scaffold, which has two extremities and thus can participate in at most two non-conflicting assembly points in $\mathbb{A}'$. Hence, $\deg(v) \leq 2$. □

We notice that any non-conflicting realization $\mathbb{A}'$ of an assembly $\mathbb{A}$ provides orientation for all scaffolds involved in each connected component of $SAG(\mathbb{A}')$ (as well as of $OG(\mathbb{A}')$) relatively to each other.

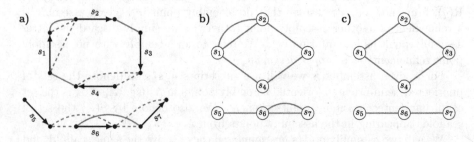

Fig. 1. For an assembly $A = \{(s_1, \overrightarrow{s_2}), (\overrightarrow{s_1}, \overrightarrow{s_2}), (\overrightarrow{s_2}, \overrightarrow{s_3}), (\overrightarrow{s_3}, s_4), (\overleftarrow{s_1}, \overleftarrow{s_4}), (\overleftarrow{s_5}, s_6),$ $(\overleftarrow{s_6}, \overrightarrow{s_7}), (\overrightarrow{s_6}, s_7)\}$, (a) the scaffold assembly graph $SAG(A)$, where semi-oriented assembly points, oriented assembly points, and scaffolds are represented by dashed red edges, solid red edges, and directed black edges, respectively. (b) The order graph $OG(A)$. (c) The contracted order graph $COG(A)$. (Color figure online)

---

[2] It can be easily seen that a realization of $\mathbb{A}$ may exist only if $\mathbb{A}$ is proper.

[3] $\deg(v)$ denotes the degree of a vertex v, i.e., the number of edges (counted with multiplicity) incident to v.

Theorem 2. *An assembly* $\mathbb{A}$ *has at least one non-conflicting realization (i.e., $|\mathrm{NR}(\mathbb{A})| \geq 1$) if and only if $\mathbb{A}$ is non-conflicting or semi-conflicting and $\mathrm{OG}(\mathbb{A})$ is non-branching.*

Proof. Suppose that $|\mathrm{NR}(\mathbb{A})| \geq 1$ and pick any $\mathbb{A}' \in \mathrm{NR}(\mathbb{A})$. Then for every pair of assembly points $p, q \in \mathbb{A}$, their realizations in $\mathbb{A}'$ are non-conflicting, implying that p and q are either non-conflicting or semi-conflicting. Hence, $\mathbb{A}$ is non-conflicting or semi-conflicting. Since $\mathrm{OG}(A) = \mathrm{OG}(A')$ and A' is non-conflicting, Lemma 1 implies that $\mathrm{OG}(A)$ is non-branching.

Vice versa, suppose that $\mathbb{A}$ is non-conflicting or semi-conflicting and $\mathrm{OG}(\mathbb{A})$ is non-branching. To prove that $|\mathrm{NR}(\mathbb{A})| \geq 1$, we will orient unoriented scaffolds in all assembly points in $\mathbb{A}$ without creating conflicts. Every scaffold s corresponds to a vertex v in $\mathrm{OG}(\mathbb{A})$ of degree at most 2. If $\deg(v) = 1$, then s participates in one assembly point p, and s is either already oriented in p or we pick an arbitrary orientation for it. If $\deg(v) = 2$, then s participates in two overlapping assembly points p and q. If s is not oriented in either of p, q, we pick an arbitrary orientation for it consistently across p and q (i.e., keeping them non-conflicting). If s is oriented in exactly one assembly point, we orient the unoriented instance of s consistently with its orientation in the other assembly point. Since conflicts may appear only between assembly points that share a vertex in $\mathrm{OG}(\mathbb{A})$, the constructed orientations produce no new conflicts. On other hand, the scaffolds that are already oriented in $\mathbb{A}$ impose no conflicts since $\mathbb{A}$ is non-conflicting or semi-conflicting. Hence, the resulting oriented assembly points form a non-conflicting assembly from $\mathrm{NR}(\mathbb{A})$, i.e., $|\mathrm{NR}(\mathbb{A})| \geq 1$. ☐

We remark that if $\mathrm{OG}(\mathbb{A})$ is branching, the assembly $\mathbb{A}$ may be semi-conflicting but have $|\mathrm{NR}(\mathbb{A})| = 0$. An example is given by $\mathbb{A} = \{(s_1, s_{i+1})\}_{i=1}^k$ with $k > 2$, which contains no conflicting assembly points (in fact, all assembly points in $\mathbb{A}$ are semi-conflicting), but $|\mathrm{NR}(\mathbb{A})| = 0$.

For an assembly $\mathbb{A}$ with $|\mathrm{NR}(\mathbb{A})| \geq 1$, the orientation of some scaffolds from $\mathbb{S}(\mathbb{A})$ does not depend on the choice of a realization from $\mathrm{NR}(\mathbb{A})$ (we denote the set of such scaffolds by $\mathbb{S}_o(\mathbb{A})$), while the orientation of other scaffolds within some assembly points varies across realizations from $\mathrm{NR}(\mathbb{A})$ (we denote the set of such scaffolds by $\mathbb{S}_u(\mathbb{A})$). Trivially, we have $\mathbb{S}_u(\mathbb{A}) \cup \mathbb{S}_o(\mathbb{A}) = \mathbb{S}(\mathbb{A})$. It can be easily seen that the set $\mathbb{S}_u(\mathbb{A})$ is formed by the scaffolds for which the orientation in the proof of Theorem 2 was chosen arbitrarily, implying the following statement.

Corollary 3. *For a given assembly $\mathbb{A}$ with $|\mathrm{NR}(\mathbb{A})| \geq 1$, we have $|\mathrm{NR}(\mathbb{A})| = 2^{|\mathbb{S}_u(\mathbb{A})|}$.*

Lemma 4. *Testing whether a given assembly $\mathbb{A}$ has a non-conflicting realization can be done in $\mathcal{O}(k \cdot \log(k))$ time, where $k = |\mathbb{S}(\mathbb{A})|$.*

Proof. To test whether $\mathbb{A}$ has a non-conflicting realization, we first create a hash table indexed by $\mathbb{S}(\mathbb{A})$ that for every scaffold $s \in \mathbb{S}(\mathbb{A})$ will contain a list of assembly points that involve s. We iterate over all assembly points $p \in \mathbb{A}$ and add p to two lists in the hash table indexed by the scaffolds participating in p.

If the length of some list becomes greater than 2, then $\mathbb{A}$ is conflicting and we stop. If we successfully complete the iterations, then every scaffold from $\mathbb{S}(\mathbb{A})$ participates in at most two assembly points in $\mathbb{A}$, and thus we made $\mathcal{O}(k)$ steps of $\mathcal{O}(\log(k))$ time each.

Next, for every scaffold whose list in the hash table has length 2, we check whether the corresponding assembly points are either non-conflicting or semi-conflicting. If not, then $\mathbb{A}$ is conflicting and we stop. If the check completes successfully, then $\mathbb{A}$ has a non-conflicting realization by Theorem 2. The check takes $\mathcal{O}(k)$ steps of $\mathcal{O}(\log(k))$ time each, and thus the total running time comes to $\mathcal{O}(k \cdot \log(k))$. $\qquad\square$

A pseudocode for the test described in the proof of Lemma 4 is given in Algorithm 2 in the Appendix.

Lemma 5. *For a given assembly* $\mathbb{A}$ *with* $|\mathrm{NR}(\mathbb{A})| \geq 1$, *the set* $\mathbb{S}_u(\mathbb{A})$ *can be computed in* $\mathcal{O}(k \cdot \log(k))$ *time, where* $k = |\mathbb{S}(\mathbb{A})|$.

Proof. We will construct the set $S = \mathbb{S}_u(\mathbb{A})$ iteratively. Initially we let $S = \emptyset$. Following the algorithm described in the proof for Lemma 4, we construct a hash table that for every scaffold $s \in \mathbb{S}(\mathbb{A})$ contains a list of assembly points that involve s (which takes $\mathcal{O}(k \cdot \log(k))$ time). Then for every $s \in \mathbb{S}(\mathbb{A})$, we check if either of the corresponding assembly points provides an orientation for s; if not, we add s to S. This check for each scaffolds takes $\mathcal{O}(1)$ time, bringing the total running time to $\mathcal{O}(k \cdot \log(k))$. $\qquad\square$

A pseudocode for the computation of $\mathbb{S}_u(\mathbb{A})$ described in the proof of Lemma 5 is given in Algorithm 3 in the Appendix.

3   Orientation of Ordered Scaffolds

For a non-conflicting assembly $\mathbb{A}$ composed only of oriented assembly points, an assembly point p on scaffolds $s_i, s_j \in \mathbb{S}(\mathbb{A})$ has a *consistent orientation* with $\mathbb{A}$ if for some $p' \in \mathrm{R}(p)$ there exists a path connecting edges s_i and s_j in $\mathsf{SAG}(\mathbb{A})$ such that direction of edges s_i and s_j at the path ends is consistent with p' (e.g., in Fig. 1a, the assembly point $(\overrightarrow{s_1}, \overrightarrow{s_3})$ has a consistent orientation with the assembly A).

We formulate the orientation of ordered scaffolds problem as follows.

Problem 1 (Orientation of Ordered Scaffolds, OOS). Let $\mathbb{A}$ be an assembly and $\mathbb{O}$ be a set[4] of assembly points such that $|\mathrm{NR}(\mathbb{A})| \geq 1$ and $\mathbb{S}(\mathbb{O}) \subset \mathbb{S}(\mathbb{A})$. Find a non-conflicting realization $\mathbb{A}' \in \mathrm{NR}(\mathbb{A})$ that maximizes the number (total weight) of assembly points from $\mathbb{O}$ having consistent orientations with $\mathbb{A}'$.

---

[4] More generally, $\mathbb{O}$ may be a multiset whose elements have real positive multiplicities (weights).

From the biological perspective, the OOS can be viewed as a formalization of the case where (sub)orders of scaffolds have been determined (which defines $\mathbb{A}$), while there exists some information (possibly coming from different sources and conflicting) about their relative orientation (which defines $\mathbb{O}$). The OOS asks to orient unoriented scaffolds in the given scaffold orders in a way that is most consistent with the given orientation information.

We also remark that the OOS can be viewed as a fine-grained variant of the scaffold orientation problem studied in [9]. In our terminology, the latter problem concerns an artificial circular genome $\mathbb{A}$ formed by the given scaffolds in an arbitrary order (so that there is a path connecting any scaffold or its reverse complement to any other scaffold in $OG(\mathbb{A})$), and $\mathbb{O}$ formed by unordered pairs of scaffolds supplemented with the binary information on whether each such pair come from the same or different strands of the genome. In contrast, in the OOS, the assembly $\mathbb{A}$ is given and $OG(\mathbb{A})$ does not have to be connected or non-branching, while $\mathbb{O}$ may provide a pair of scaffolds with up to four options (as in (1)) about their relative orientation.

3.1   NP-hardness of the OOS

We consider two important partial cases of the OOS, where the assembly $\mathbb{A}$ represents a linear or circular genome up to unknown orientations of the scaffolds. In these cases, the graph $OG(\mathbb{A})$ forms a collection of paths or cycles, respectively. Below we prove that the OOS in both these cases is NP-hard.

Theorem 6. *The OOS for linear genomes is* NP-*hard.*

Proof. We will construct a polynomial-time reduction from the MAX 2-DNF problem, which is known to be NP-hard [7,15]. Given an instance I of MAX 2-DNF consisting of conjunctions $C = \{c_i\}_{i=1}^k$ on variables $X = \{x_i\}_{i=1}^n$, we define an assembly

$$\mathbb{A} = \{(0, x_1)\} \cup \{(x_i, x_{i+1}) \; : \; i = 1, 2, \ldots, n - 1\}.$$

We then construct a set of assembly points $\mathbb{O}$ from the clauses in C as follows. For each clause $c \in C$ with two variables x_i and x_j $(i < j)$, we add an oriented assembly point on scaffolds x_i, x_j to $\mathbb{O}$ with the orientation depending on the negation of these variables in c (i.e., a clause $x_i \wedge \overline{x_j}$ is translated into an assembly point $(\overrightarrow{x_i}, \overleftarrow{x_j})$). For each clause from C with a single variable x, we add an assembly point $(\overrightarrow{0}, \overrightarrow{x})$ or $(\overrightarrow{0}, \overleftarrow{x})$ depending whether x is negated in the clause.

It is easy to see that the constructed assembly $\mathbb{A}$ is semi-conflicting and $OG(\mathbb{A})$ is a path, and thus by Theorem 2 $\mathbb{A}$ has a non-conflicting realization. Hence, $\mathbb{A}$ and $\mathbb{O}$ form an instance of the OOS for linear genomes. A solution $\mathbb{A}'$ to this OOS provides an orientation for each $x \in \mathbb{S}$ that maximizes the number of assembly points from $\mathbb{O}$ having consistent orientations with A'. A solution to I is obtained from A' as the assignment of 0 or 1 to each variable x depending on whether the orientation of scaffold x in A' is forward or reverse. Indeed, since each assembly point in $\mathbb{O}$ having consistent orientation with $\mathbb{A}'$ corresponds to a truthful clause in I, the number of such clauses is maximized.

It is easy to see that the OOS instance and the solution to I can be computed in polynomial time, thus we constructed a polynomial-time reduction from the MAX 2-DNF to the OOS for linear genomes. $\qquad\square$

Theorem 7. *The OOS for circular genomes is* NP-*hard.*

Proof. We construct a polynomial-time reduction from the MAX-CUT problem, which is known to be NP-hard [17,18]. An instance I of MAX-CUT for a given a graph (V, E) asks to partition the set of vertices $V = \{v_i\}_{i=1}^n$ into two disjoint subsets V_1 and V_2 such that the number of edges $\{u, v\} \in E$ with $u \in V_1$ and $v \in V_2$ is maximized. For a given instance I of MAX-CUT problem, we define the assembly

$$\mathbb{A} = \{(v_i, v_{i+1}) \ : \ i = 1, 2, \ldots, n-1\} \cup \{(v_1, v_n)\}$$

and the set of assembly points

$$\mathbb{O} = \{(\overrightarrow{v_i}, \overleftarrow{v_j}) \ : \ \{v_i, v_j\} \in E\}.$$

It is easy to see that $\mathbb{A}$ has a non-conflicting realization and $\mathrm{OG}(\mathbb{A})$ is a cycle, i.e., $\mathbb{A}$ and $\mathbb{O}$ form an instance of the OOS for circular genomes. A solution $\mathbb{A}'$ to this OOS instance provides orientations for all elements $\mathbb{S}(\mathbb{A}) = V$ that maximizes the number of assembly points from $\mathbb{O}$ having consistent orientations with $\mathbb{A}'$. A solution to I is obtained as the partition of V into two disjoint subsets, depending on the orientation of scaffolds in $\mathbb{A}'$ (forward vs reverse). Indeed, since each assembly point in $\mathbb{O}$ having a consistent orientation with $\mathbb{A}'$ corresponds to an edge from E whose endpoints belong to distinct subsets in the partition, the number of such edges is maximized.

It is easy to see that the OOS instance and the solution to I can be computed in polynomial time, thus we constructed a polynomial-time reduction from the MAX-CUT to the OOS for circular genomes. $\qquad\square$

As a trivial consequence of Theorems 6 and 7, we obtain that the general OOS problem is NP-hard.

Corollary 8. *The OOS is* NP-*hard.*

3.2 Properties of the OOS

In this subsection, we formulate and prove some important properties of the OOS. We start with the following lemma that trivially follows from the definition of consistent orientation.

Lemma 9. *Let $\mathbb{A}$ be an assembly. An assembly point on scaffolds $s_i, s_j \in \mathbb{S}(\mathbb{A})$ may have a consistent orientation with $\mathbb{A}$ only if both s_i and s_j belong to the same connected component in* $\mathrm{OG}(\mathbb{A})$.

Theorem 10. *Let* $(\mathbb{A}, \mathbb{O})$ *be an OOS instance, and* $\mathbb{A} = \mathbb{A}_1 \cup \cdots \cup \mathbb{A}_k$ *be the partition such that* $\mathsf{OG}(\mathbb{A}_1), \ldots, \mathsf{OG}(\mathbb{A}_k)$ *represent the connected components of* $\mathsf{OG}(\mathbb{A})$. *For each* $i = 1, 2, \ldots, k$, *define* $\mathbb{O}_i = \{p \in \mathbb{O} : \mathrm{sn}(p, 1), \mathrm{sn}(p, 2) \in \mathbb{S}(\mathbb{A}_i)\}$ *and let* $\mathbb{A}'_i$ *be a solution to the OOS instance* $(\mathbb{A}_i, \mathbb{O}_i)$. *Then* $\mathbb{A}'_1 \cup \cdots \cup \mathbb{A}'_k$ *is a solution to the OOS instance* $(\mathbb{A}, \mathbb{O})$.

Proof. Lemma 9 implies that we can discard from $\mathbb{O}$ all assembly points that are formed by scaffolds from different connected components in $\mathsf{OG}(\mathbb{A})$. Hence, we may assume that $\mathbb{O} = \mathbb{O}_1 \cup \cdots \cup \mathbb{O}_k\}$.

Lemma 9 further implies that an assembly point from $\mathbb{O}_i$ may have a consistent orientation with $\mathbb{A}_j$ only if $i = j$. Therefore, any solution to the OOS instance $(\mathbb{A}, \mathbb{O})$ is formed by the union of solutions to the OOS instances $(\mathbb{A}_i, \mathbb{O}_i)$. □

Theorem 10 allows us to focus on instances of the OOS, where $\mathsf{OG}(\mathbb{A})$ is connected and thus forms a path or a cycle (by Theorem 2).

The following lemma is almost trivial.

Lemma 11. *Let* $\mathbb{A}$ *be an assembly, and* s_i, s_j *be scaffolds from the same connected component* C *in* $\mathsf{SAG}(\mathbb{A})$. *Then an unoriented assembly point* (s_i, s_j) *has a consistent orientation with* $\mathbb{A}$. *Furthermore, if* C *is a cycle, then any semi-oriented assembly point on* s_i, s_j *has a consistent orientation with* $\mathbb{A}$.

By Lemma 11, we can assume that $\mathbb{O}$ does not contain any unoriented assembly points (i.e., $\mathbb{O} = \mathbb{O}_o \cup \mathbb{O}_s$). Furthermore, if $\mathsf{OG}(\mathbb{A})$ is a cycle, we can assume that $\mathbb{O} = \mathbb{O}_o$ (i.e., $\mathbb{O}$ consists of oriented assembly points only).

Below we show that an OOS instance can also be solved independently for each connected component of $\mathsf{OG}(\mathbb{O}_o)$. We consider two cases depending on whether $\mathsf{OG}(\mathbb{A})$ forms a path or a cycle.

Case 1: $\mathsf{OG}(\mathbb{A})$ *is a path.* Let $(\mathbb{A}, \mathbb{O})$ be an OOS instance such that $\mathsf{OG}(\mathbb{A}) = (s_1, s_2, \ldots, s_n)$ is a path and $\mathbb{O} = \mathbb{O}_o \cup \mathbb{O}_s$. First we notice that since $|\mathrm{NR}(\mathbb{A})| \geq 1$, a solution $\mathbb{A}'$ to this OOS can be viewed as a sequence of the same scaffolds $(s_1, s_2, \ldots, s_n)$ where every element is oriented.

Let $C_1, C_2, \ldots, C_k$ be the connected components of $\mathsf{OG}(\mathbb{O}_o)$ and, for $i = 1, 2, \ldots, k$, $(s_{j_{i,1}}, \ldots, s_{j_{i,m_i}})$ be a sequence of vertices of C_i such that $j_{i,1} < j_{i,2} < \cdots < j_{i,m_i}$.

We define an assembly $\mathbb{A}_i$ $(i = 1, 2, \ldots, k)$ such that $\mathsf{OG}(\mathbb{A}_i)$ is the path $(x_i, s_{j_{i,1}}, \ldots, s_{j_{i,m_i}}, y_i)$, where x_i and y_i are artificial vertices, and the assembly points in $\mathbb{A}_i$ (corresponding to the edges in $\mathsf{OG}(\mathbb{A}_i)$) are oriented or semi-oriented. Namely, the edges $\{x_i, s_{j_{i,1}}\}$ and $\{s_{j_{i,m_i}}, y_i\}$ correspond to semi-oriented assembly points $(\overrightarrow{x_i}, s_{j_{i,1}})$ and $(s_{j_{i,m_i}}, \overrightarrow{y_i})$, respectively. The orientation of the scaffolds in the assembly point corresponding to the edge $\{s_{j_{i,l}}, s_{j_{i,l+1}}\}$ is inherited from the assembly points in $\mathbb{A}$ formed by $(s_{j_{i,l}}, s_{j_{i,l}+1})$ and $(s_{j_{i,l+1}-1}, s_{j_{i,l+1}})$, respectively.

We further define $\mathbb{O}_i$ $(i = 1, 2, \ldots, k)$ formed by the oriented assembly points from C_i and the following oriented assembly points. For each semi-oriented assembly point $p \in \mathbb{O}$ formed by scaffolds s_m and s_l $(m < l)$, $\mathbb{O}_i$ contains

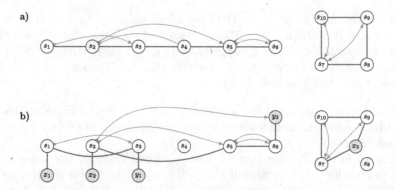

Fig. 2. Decomposition of an OOS problem instance $(\mathbb{A}, \mathbb{O})$ based on the connected components of $OG(\mathbb{O}_o)$. (a) The superposition of $OG(\mathbb{A})$ (red edges) and $OG(\mathbb{O})$ (green edges), where arrows (if present) at the ends of green edges encode the orientation of the scaffolds in the corresponding assembly points. (b) The superposition of five graphs $OG(\mathbb{A}_i)$ (red edges) and three graphs $OG(\mathbb{O}_j)$ (green edges) constructed based on the connected components of $OG(\mathbb{O}_o)$. Unless $OG(\mathbb{A}_i)$ is formed by an isolated vertex, it contains artificial vertices x_i and y_i, which coincide if $OG(\mathbb{A}_i)$ is a cycle. (Color figure online)

(i) an oriented point p' formed by s_m and $\overrightarrow{y_i}$ to $\mathbb{O}_i$ whenever s_m is oriented in p and belongs to C_i (and its orientation in p' is inherited from p); and (ii) an oriented point p'' formed by $\overrightarrow{x_i}$ and s_l whenever s_l is oriented in p and belongs to C_i (and its orientation in p'' is inherited from p) (Fig. 2).

Theorem 12. *Let $(\mathbb{A}, \mathbb{O})$ be an OOS instance, and $\mathbb{A}_i$ and $\mathbb{O}_i$ ($i = 1, 2, \ldots, k$) be defined as above. For each $i = 1, 2, \ldots, k$, let $\mathbb{A}'_i$ be a solution to the OOS instance $(\mathbb{A}_i, \mathbb{O}_i)$. Then a solution $\mathbb{A}'$ to the OOS instance $(\mathbb{A}, \mathbb{O})$ can be constructed as follows. For a scaffold $s \in \mathbb{S}(\mathbb{A})$ present in some $\mathbb{A}'_i$, $\mathbb{A}'$ inherits the orientation of s from $\mathbb{A}'_i$. For a scaffold $s \in \mathbb{S}(\mathbb{A})$ not present in any $\mathbb{A}'_i$, $\mathbb{A}'$ either inherits the orientation of s from $\mathbb{A}$ (if s is oriented in any assembly point), or orients s arbitrarily.*

Proof. The graph $SAG(\mathbb{A}')$ can be viewed as an ordered sequence of directed scaffold edges (interweaved with undirected edges encoding assembly points). Then each $SAG(\mathbb{A}'_i)$, with the exception of scaffold edges x_i and y_i, corresponds to a subsequence of this sequence.

Each oriented assembly point $p \in \mathbb{O}$ is formed by scaffolds u, v from C_i for some $i \in \{1, \ldots, k\}$. Then $p \in \mathbb{O} \cap \mathbb{O}_i$ and there exist a unique path in $SAG(\mathbb{A}'_i)$ and a unique path in $SAG(\mathbb{A}')$ having the same directed edges u, v at the ends. Hence, if p has a consistent orientation with one of assemblies $\mathbb{A}'$ or $\mathbb{A}'_i$, then it has a consistent orientation with the other.

Each semi-oriented assembly point $p \in \mathbb{O}$ formed by scaffolds u, v corresponds to an oriented assembly point $q \in \mathbb{O}_i$ (for some i) formed by u and y_i (in which case $u \in C_i$ and u is oriented in p), or by x_i and v (in which case $v \in C_i$ and

v is oriented in p). Without loss of generality, we assume the former case. Then there exists a unique path Q in $\mathsf{SAG}(\mathbb{A}'_i)$ connecting directed edges u and y_i, and there exists a unique path P in $\mathsf{SAG}(\mathbb{A}')$ connecting directed edges u and v, where the orientation of u is the same in the two paths. By construction, the orientation of y_i in q matches that in Q. Hence, q has a consistent orientation with $\mathbb{A}'_i$ if and only if the orientation of u in q matches that in Q, which happens if and only if the orientation of u in p matches its orientation in P, i.e., p has a consistent orientation with $\mathbb{A}'$.

We proved that the number of assembly points from $\mathbb{O}$ having consistent orientation with $\mathbb{A}'$ equals the total number of assembly points from $\mathbb{O}_i$ having consistent orientation with $\mathbb{A}'_i$ for all $i = 1, 2, \ldots, k$. It remains to notice that this number is maximum possible, i.e., $\mathbb{A}'$ is indeed a solution to the OOS instance $(\mathbb{A}, \mathbb{O})$ (if it is not, then the sets $\mathbb{A}_i$ constructed from $\mathbb{A}$ being an actual solution to the OOS will give a better solution to at least one of the subproblems).   □

Case 2: $\mathsf{OG}(\mathbb{A})$ *is a cycle.* In this case, we can construct subproblems based on the connected components of $\mathsf{OG}(\mathbb{O})$ similarly to Case 1, with the following differences. First, we assume that $\mathbb{O} = \mathbb{O}_o$ (discarding all unoriented and oriented assembly points from $\mathbb{O}$). Second, we assume that $x_i = y_i$ and thus $\mathsf{OG}(\mathbb{A}_i)$ forms a cycle. Theorem 12 still holds in this case.

Articulation Vertices in $\mathsf{OG}(\mathbb{A})$. While Theorem 12 allows us to divide the OOS problems into subproblems based on the connected components of $\mathsf{OG}(\mathbb{O}_o)$, we show below that similar division is possible when $\mathsf{OG}(\mathbb{O}_o)$ is connected but contains an articulation vertex.[5]

Any articulation vertex v in $\mathsf{OG}(\mathbb{O}_o)$ defines a partition of $\mathbb{S}(\mathbb{O}_o)$ into subsets:

$$\mathbb{S}(\mathbb{O}_o) = \{v\} \cup V_1 \cup V_2 \cup \cdots \cup V_k, \tag{2}$$

where $k > 1$ and the V_i represent the vertex sets of the connected components resulted from removal of v from $\mathsf{OG}(\mathbb{O}_o)$.

The following theorem shows that an articulation vertex in $\mathsf{OG}(\mathbb{O})$ enables application of Theorem 12.

Theorem 13. *Let* $(\mathbb{A}, \mathbb{O})$ *be an instance of the OOS problem such that* $\mathsf{OG}(\mathbb{A})$ *and* $\mathsf{OG}(\mathbb{O}_o)$ *are connected. Let* v *be an articulation vertex in* $\mathsf{OG}(\mathbb{O}_o)$*, defining a partition* (2).

If $v \notin \mathbb{S}_o(\mathbb{A})$*, we introduce copies* $v_1, \ldots, v_k$ *of* v*, and construct* $\hat{\mathbb{A}}$ *from* $\mathbb{A}$ *by replacing a path* (u, v, w) *in* $\mathsf{OG}(\mathbb{A})$ *with a path* $(u, v_1, v_2, \ldots, v_k, w)$ *where all* v_i *inherit the orientation from* v*. Then we construct* $\hat{\mathbb{O}}$ *from* $\mathbb{O}$ *by replacing in each assembly point* p *formed by* v *and* $u \in V_i$ *(for some* $i \in \{1, 2, \ldots, k\}$*) with an assembly point formed by* v_i *and* u *(keeping their orientations intact). Then a solution to the OOS instance* $(\mathbb{A}, \mathbb{O})$ *can be obtained from a solution to the OOS instance* $(\hat{\mathbb{A}}, \hat{\mathbb{O}})$ *by replacing every scaffold* v_i *with* v.

---

[5] We remind that a vertex is *articulation* if its removal from the graph increases the number of connected components.

If $v \notin \mathbb{S}_o(\mathbb{A})$, we iteratively fix the two possible orientations of v, and proceed with construction and solution of the OOS instance $(\hat{\mathbb{A}}, \hat{\mathbb{O}})$ as above. Then a solution to the OOS instance $(\mathbb{A}, \mathbb{O})$ can be obtained from a better solution among the two.

Proof. Let $\hat{\mathbb{A}}'$ be a solution to the OOS instance $(\hat{\mathbb{A}}, \hat{\mathbb{O}})$, and $\mathbb{A}'$ be obtained from $\hat{\mathbb{A}}'$ by replacing every v_i with v. We remark that $\mathbb{O}$ can be obtained from $\hat{\mathbb{O}}$ by similar replacement.

This establishes an one-to-one correspondence between the assembly points in $\hat{\mathbb{A}}'$ and $\mathbb{A}'$, as well as between the assembly points in $\hat{\mathbb{O}}'$ and $\mathbb{O}'$. It remains to show that consistent orientations are invariant under this correspondence.

We remark that $\mathsf{SAG}(\mathbb{A}')$ can be obtained from $\mathsf{SAG}(\hat{\mathbb{A}}')$ by replacing a sequence of edges $(r_1, v_1, r_2, v_2, \ldots, r_k, v_k, r_{k+1})$, where r_i are assembly edges, with a sequence of edges (r_1, v, r_2). Therefore, if there exists a path in one graph proving existence of consistent orientation for some assembly point, then there exists a corresponding path in the other graph (having the same orientations of the end edges). □

3.3   Non-branching Orientation of Ordered Scaffolds

At the latest stages of genome assembly, the constructed scaffolds are usually of significant length. If (sub)orders for these scaffolds are known, it is rather rare to have orientation-imposing information that would involve non-neighboring scaffolds. Or, more generally, it is rather rare to have orientation imposing information for one scaffold with respect to more than two other scaffolds. This inspires us to consider a special case of the OOS problem.

For a set of assembly points $\mathbb{A}$, we define the *contracted order graph* $\mathsf{COG}(\mathbb{A})$ obtained from $\mathsf{OG}(\mathbb{A})$ by replacing all multi-edges edges with single edges (Fig. 1c). We now consider a special type of the OOS problem:

Problem 2 (Non-branching Orientation of Ordered Scaffolds, NOOS). Given an OOS instance $(\mathbb{A}, \mathbb{O})$ such that the graph $\mathsf{COG}(\mathbb{O}_o)$ is non-branching. Find $\mathbb{A}' \in \mathsf{NR}(\mathbb{A})$ that maximizes the number of assembly points from $\mathbb{O}$ having consistent orientations with $\mathbb{A}'$.

Theorem 14. *The NOOS is in* P.

Proof. By Theorems 10 and 12, we can assume that both $\mathsf{OG}(\mathbb{A})$ and $\mathsf{OG}(\mathbb{O}_o)$ are connected. Since $\mathsf{COG}(\mathbb{O}_o)$ is non-branching, and we consider two cases depending on whether it is a path or a cycle.

If $\mathsf{COG}(\mathbb{O}_o)$ is a path, then every vertex in it is an articulation vertex in both $\mathsf{COG}(\mathbb{O}_o)$ and $\mathsf{OG}(\mathbb{O}_o)$. Our algorithm will process this path in a divide and conquer manner. Namely, for a path of length greater than 2, we pick a vertex v closest to the path middle and proceed as in Theorem 13. A path of length at most 2 can be solved in $\mathcal{O}(|\mathbb{O}|)$ time by brute-forcing all possible orientations of the scaffolds in the path and counting how many assembly points in $\mathbb{O}$ get consistent orientations.

The number of operations for the preprocessing stage of the algorithm (dominated by construction of hash tables) is $\mathcal{O}\left(k \cdot \log(k)\right)$, where $k = \max\{|\mathbb{O}|, |\mathbb{S}(\mathbb{A})|\}$. The running time for recursive part of the algorithm can described as

$$T(l) = \begin{cases} 4 \cdot T\left(\frac{l}{2}\right) + \mathcal{O}\left(1\right), & \text{if } l > 2; \\ \mathcal{O}\left(|\mathbb{O}|\right), & \text{if } l \leq 2. \end{cases}$$

From the Master theorem [8], we conclude that the total running time for the proposed recursive algorithm is $\mathcal{O}\left(l^2 + k \cdot \log(k)\right)$, where $l = |\mathbb{O}|$.

If $\mathsf{COG}(\mathbb{O}_o)$ is a cycle, we can reduce the corresponding NOOS instance to the case of a path as follows. First, we pick a random vertex w in $\mathsf{COG}(\mathbb{O}_o)$ and replace it with new vertices w_1 and w_2 such that the edges $\{u, w\}$, $\{w, v\}$ in $\mathsf{COG}(\mathbb{O}_o)$ are replaced with $\{u, w_1\}$, $\{w_2, v\}$. Then we solve the NOOS for the resulting path one or two times (depending on whether $w \in \mathbb{S}_o(\mathbb{A})$): once for each of possible orientations of scaffold w (inherited by w_1 and w_2), and then select the orientation for w that produces a better result. $\qquad\square$

A pseudocode for the algorithm described in the proof of Theorem 14 is given in Algorithm 1 in the Appendix.

4   Conclusion

In the present study, we posed the orientation of ordered scaffolds (OOS) problem as an optimization problem based on given weighted orientations of scaffolds and their pairs. We further addressed it within the earlier introduced CAMSA framework, taking advantage of the simple yet powerful concept of assembly points describing (semi-/un-) oriented adjacencies between scaffolds. This approach allows one to uniformly represent both orders of (oriented and/or unoriented) scaffolds and orientation-imposing data.

We proved that the OOS problem is NP-hard when the given scaffold order represents a linear or circular genome. We also described a polynomial-time algorithm for the special case of non-branching OOS (NOOS), where the orientation of each scaffold is imposed relatively to at most two other scaffolds. Our algorithm for the NOOS problem and Theorems 10, 12, and 13 further enable us to develop an FPT algorithm for the general OOS problem (to be described elsewhere).

Acknowledgements. The authors thank the anonymous reviewers for their suggestions and comments that helped to improve the exposition.

The work is supported by the National Science Foundation under the grant No. IIS-1462107. The work of SA is also partially supported by the National Science Foundation under the grant No. CCF-1053753 and by the National Institute of Health under the grant No. U24CA211000.

Appendix. Pseudocodes

In the algorithms below we do not explicitly describe the function ORCON-SCOUNT, which takes 4 arguments:

1. a subgraph c from $COG(\mathbb{O}_o)$ with 1 or 2 vertices;
2. a hash table so with scaffolds as keys and their orientations as values;

Algorithm 1. Solving the NOOS

```
 1: function SolveNOOSForCC(c, 𝕆, so, 𝔸)
 2:     score ← 0
 3:     if c has less than 3 vertices then
 4:         return ORCONSCOUNT(c, so, 𝕆, 𝔸)
 5:     end if
 6:     p₁, p₂ ← split c into two paths of equal length, at vertex s
 7:     var ← empty hash table
 8:     for or in {→, ←} do
 9:         so[s] ← or
10:         var[or] = SolveNOOSForCC(p₁, 𝕆, so)
11:         var[or] += SolveNOOSForCC(p₂, 𝕆, so)
12:         var[or] += ORCONSCOUNT(s, so, 𝕆, 𝔸)
13:     end for
14:     score ← maximum value in var
15:     or ← key corresponding to the maximum value in var
16:     so[s] ← or
17:     return score
18: end function
19:
20: function SolveNOOS(𝔸, 𝕆)
21:     so ← hash table with 𝕊(𝔸) as keys and their orientation as values
22:     cog ← COG(𝕆ₒ)
23:     var ← empty hash table
24:     for each connected component c in cog do
25:         if c is a cycle then
26:             pick a random vertex w in c
27:             remove two edges {u, w}, {w, v} from c
28:             add two edges {u, w₁}, {w₂, u} to c
29:             for or in {→, ←} do
30:                 so[w₁], so[w₂] ← or, or
31:                 var[(w, or)] ← SolveNOOSForCC(c, 𝕆, so, 𝔸)
32:             end for
33:         else
34:             s₁, s₂ ← scaffolds, corresponding extremities of c
35:             for or₁ in {→, ←} do
36:                 for or₂ in {→, ←} do
37:                     so[s₁], so[s₂] ← or₁, or₂
38:                     r ← SolveNOOSForCC(c, 𝕆, so, 𝔸)
39:                     var[(s₁, or₁)], var[(s₂, or₂)] ← r, r
40:                 end for
41:             end for
42:         end if
43:     end for
44:     return var
45: end function
```

Algorithm 2. Checking if a given assembly has a non-conflicting realization

---

1: **procedure** HASNONCONFLICTINGREALIZATION($\mathbb{A}$)
2:     $S \leftarrow \mathbb{S}(\mathbb{A})$ ▷ retrieving scaffolds that $\mathbb{A}$ is built on
3:     sort S
4:     $t \leftarrow$ hash table with S as keys and empty linked lists as values
5:     **for** p in $\mathbb{A}$ **do**
6:         append p to the list at $t[\text{sn}(p, 1)]$
7:         append p to the list at $t[\text{sn}(p, 2)]$
8:         **if** length of list at $t[\text{sn}(p, 1)] > 2$ **or** length of list at $t[\text{sn}(p, 2)] > 2$ **then**
9:             **return** False
10:         **end if**
11:     **end for**
12:     **for** s in S **do**
13:         **if** length of list at $t[s] = 2$ **and** assembly points in $t[s]$ are conflicting **then**
14:             **return** False
15:         **end if**
16:     **end for**
17:     **return** True
18: **end procedure**

Algorithm 3. Computing $\mathbb{S}_u(\mathbb{A})$

1: **procedure** UNORIENTEDSCAFFOLDS($\mathbb{A}$)
2:     $r \leftarrow$ empty set
3:     **if** not HasNonConflictingRealization($\mathbb{A}$) **then**
4:         **return** r
5:     **end if**
6:     $S \leftarrow \mathbb{S}(\mathbb{A})$
7:     sort S
8:     $t \leftarrow$ hash table with S as keys and empty linked lists as values
9:     **for** p in $\mathbb{A}$ **do**
10:         append p to the list at $t[\text{sn}(p, 1)]$
11:         append p to the list at $t[\text{sn}(p, 2)]$
12:     **end for**
13:     **for** s in S **do**
14:         $flag \leftarrow$ True
15:         **for** p in the list at $t[s]$ **do**
16:             **if** p involves s with orientation **then**
17:                 $flag \leftarrow$ False
18:             **end if**
19:         **end for**
20:         **if** $flag$ **then**
21:             add s to r
22:         **end if**
23:     **end for**
24:     **return** r
25: **end procedure**

---

3. a set of orientation imposing assembly points $\mathbb{O}_o$;
4. an assembly $\mathbb{A}$

and counts the assembly points from $\mathbb{O}$ that have consistent orientation with $\mathbb{A}$ in the case where scaffold(s) corresponding to vertices from c were to have orientation from so in $\mathbb{A}$. With simple hash-table based preprocessing of $\mathbb{A}$ and $\mathbb{O}$ (can be done in $\mathcal{O}\left(k\log(k)\right)$ time, where $k = \max\{|\mathbb{O}|, \mathbb{S}(\mathbb{A})\}$) this function runs in $\mathcal{O}\left(n\right)$ time, where n is a number of assembly points in $\mathbb{O}$ involving scaffolds that correspond to vertices in c. So, total running time for all invocations of this function will be $\mathcal{O}\left(|\mathbb{O}|\right)$.

References

1. Aganezov, S., Alekseyev, M.A.: Multi-genome scaffold co-assembly based on the analysis of gene orders and genomic repeats. In: Bourgeois, A., Skums, P., Wan, X., Zelikovsky, A. (eds.) ISBRA 2016. LNCS, vol. 9683, pp. 237–249. Springer, Cham (2016). doi:10.1007/978-3-319-38782-6_20

2. Aganezov, S.S., Alekseyev, M.A.: CAMSA: A Tool for Comparative Analysis and Merging of Scaffold Assemblies. Preprint bioRrxiv:10.1101/069153 (2016)

3. Anselmetti, Y., Berry, V., Chauve, C., Chateau, A., Tannier, E., Bérard, S.: Ancestral gene synteny reconstruction improves extant species scaffolding. BMC Genom. **16**(Suppl 10), S11 (2015)

4. Assour, L.A., Emrich, S.J.: Multi-genome synteny for assembly improvement multi-genome synteny for assembly improvement. In: Proceedings of 7th International Conference on Bioinformatics and Computational Biology, pp. 193–199 (2015)

5. Bankevich, A., Nurk, S., Antipov, D., Gurevich, A.A., Dvorkin, M., Kulikov, A.S., Lesin, V.M., Nikolenko, S.I., Pham, S., Prjibelski, A.D., Pyshkin, A.V., Sirotkin, A.V., Vyahhi, N., Tesler, G., Alekseyev, M.A., Pevzner, P.A.: SPAdes: a new genome assembly algorithm and its applications to single-cell sequencing. J. Comput. Biol. **19**(5), 455–477 (2012)

6. Bashir, A., Klammer, A.A., Robins, W.P., Chin, C.S., Webster, D., Paxinos, E., Hsu, D., Ashby, M., Wang, S., Peluso, P., Sebra, R., Sorenson, J., Bullard, J., Yen, J., Valdovino, M., Mollova, E., Luong, K., Lin, S., LaMay, B., Joshi, A., Rowe, L., Frace, M., Tarr, C.L., Turnsek, M., Davis, B.M., Kasarskis, A., Mekalanos, J.J., Waldor, M.K., Schadt, E.E.: A hybrid approach for the automated finishing of bacterial genomes. Nat. Biotech. **30**(7), 701–707 (2012)

7. Bazgan, C., Paschos, V.T.: Differential approximation for optimal satisfiability and related problems. Eur. J. Oper. Res. **147**(2), 397–404 (2003)

8. Bentley, J.L., Haken, D., Saxe, J.B.: A general method for solving divide-and-conquer recurrences. ACM SIGACT News **12**(3), 36–44 (1980)

9. Bodily, P.M., Fujimoto, M.S., Snell, Q., Ventura, D., Clement, M.J.: ScaffoldScaffolder: solving contig orientation via bidirected to directed graph reduction. Bioinformatics **32**(1), 17–24 (2015)

10. Boetzer, M., Henkel, C.V., Jansen, H.J., Butler, D., Pirovano, W.: Scaffolding pre-assembled contigs using SSPACE. Bioinformatics **27**(4), 578–579 (2011)

11. Boetzer, M., Pirovano, W.: SSPACE-LongRead: scaffolding bacterial draft genomes using long read sequence information. BMC Bioinf. **15**(1), 211 (2014)

12. Burton, J.N., Adey, A., Patwardhan, R.P., Qiu, R., Kitzman, J.O., Shendure, J.: Chromosome-scale scaffolding of de novo genome assemblies based on chromatin interactions. Nat. Biotechnol. **31**(12), 1119–1125 (2013)

13. Chen, Z.Z., Harada, Y., Guo, F., Wang, L.: Approximation algorithms for the scaffolding problem and its generalizations. Theoret. Comput. Sci. (2017). http://www.sciencedirect.com/science/article/pii/S0304397517302815

14. Dayarian, A., Michael, T.P., Sengupta, A.M.: SOPRA: scaffolding algorithm for paired reads via statistical optimization. BMC Bioinf. **11**, 345 (2010)
15. Escoffier, B., Paschos, V.T.: Differential approximation of min sat, max sat and related problems. Eur. J. Oper. Res. **181**(2), 620–633 (2007)
16. Gao, S., Nagarajan, N., Sung, W.-K.: Opera: reconstructing optimal genomic scaffolds with high-throughput paired-end sequences. In: Bafna, V., Sahinalp, S.C. (eds.) RECOMB 2011. LNCS, vol. 6577, pp. 437–451. Springer, Heidelberg (2011). doi:10.1007/978-3-642-20036-6_40
17. Garey, M.R., Johnson, D.S.: Computers and Intractability: A Guide To The Theory of Np-completeness, vol. 58. Freeman, San Francisco (1979)
18. Garey, M.R., Johnson, D.S., Stockmeyer, L.: Some simplified NP-complete graph problems. Theoret. Comput. Sci. **1**(3), 237–267 (1976)
19. Gritsenko, A.A., Nijkamp, J.F., Reinders, M.J.T., de Ridder, D.: GRASS: a generic algorithm for scaffolding next-generation sequencing assemblies. Bioinformatics **28**(11), 1429–1437 (2012)
20. Hunt, M., Newbold, C., Berriman, M., Otto, T.D.: A comprehensive evaluation of assembly scaffolding tools. Genome Biol. **15**(3), R42 (2014)
21. Jiao, W.B., Garcia Accinelli, G., Hartwig, B., Kiefer, C., Baker, D., Severing, E., Willing, E.M., Piednoel, M., Woetzel, S., Madrid-Herrero, E., Huettel, B., Hümann, U., Reinhard, R., Koch, M.A., Swan, D., Clavijo, B., Coupland, G., Schneeberger, K.: Improving and correcting the contiguity of long-read genome assemblies of three plant species using optical mapping and chromosome conformation capture data. Genome Res. **27**(5), 116 (2017)
22. Kececioglu, J.D., Myers, E.W.: Combinatorial algorithms for DNA sequence assembly. Algorithmica **13**(1–2), 7–51 (1995)
23. Kolmogorov, M., Armstrong, J., Raney, B.J., Streeter, I., Dunn, M., Yang, F., Odom, D., Flicek, P., Keane, T., Thybert, D., Paten, B., Pham, S.: Chromosome assembly of large and complex genomes using multiple references. Preprint bioRxiv:10.1101/088435 (2016)
24. Koren, S., Treangen, T.J., Pop, M.: Bambus 2: scaffolding metagenomes. Bioinformatics **27**(21), 2964–2971 (2011)
25. Lam, K.K., Labutti, K., Khalak, A., Tse, D.: FinisherSC: a repeat-aware tool for upgrading de novo assembly using long reads. Bioinformatics **31**(19), 3207–3209 (2015)
26. Luo, R., Liu, B., Xie, Y., Li, Z., Huang, W., Yuan, J., Wang, J.: SOAPdenovo2: an empirically improved memory-efficient short-read de novo assembler. Gigascience **1**(1), 18 (2012)
27. Nagarajan, N., Read, T.D., Pop, M.: Scaffolding and validation of bacterial genome assemblies using optical restriction maps. Bioinformatics **24**(10), 1229–1235 (2008)
28. Pop, M., Kosack, D.S., Salzberg, S.L.: Hierarchical scaffolding with Bambus. Genome Res. **14**(1), 149–159 (2004)
29. Putnam, N.H., O'Connell, B.L., Stites, J.C., Rice, B.J., Blanchette, M., Calef, R., Troll, C.J., Fields, A., Hartley, P.D., Sugnet, C.W., Haussler, D., Rokhsar, D.S., Green, R.E.: Chromosome-scale shotgun assembly using an in vitro method for long-range linkage. Genome Res. **26**(3), 342–350 (2016)
30. Reyes-Chin-Wo, S., Wang, Z., Yang, X., Kozik, A., Arikit, S., Song, C., Xia, L., Froenicke, L., Lavelle, D.O., Truco, M.J., Xia, R., Zhu, S., Xu, C., Xu, H., Xu, X., Cox, K., Korf, I., Meyers, B.C., Michelmore, R.W.: Genome assembly with in vitro proximity ligation data and whole-genome triplication in lettuce. Nat. Commun. **8**, Article no. 14953 (2017). https://www.nature.com/articles/ncomms14953

31. Simpson, J.T., Wong, K., Jackman, S.D., Schein, J.E., Jones, S.J., Birol, I.: ABySS: a parallel assembler for short read sequence data. Genome Res. **19**(6), 1117–1123 (2009)
32. Tang, H., Zhang, X., Miao, C., Zhang, J., Ming, R., Schnable, J.C., Schnable, P.S., Lyons, E., Lu, J.: ALLMAPS: robust scaffold ordering based on multiple maps. Genome Biol. **16**(1), 3 (2015)
33. Warren, R.L., Yang, C., Vandervalk, B.P., Behsaz, B., Lagman, A., Jones, S.J.M., Birol, I.: LINKS: scalable, alignment-free scaffolding of draft genomes with long reads. GigaScience **4**(1), 35 (2015)
34. Zimin, A.V., Smith, D.R., Sutton, G., Yorke, J.A.: Assembly reconciliation. Bioinformatics **24**(1), 42–45 (2008)

Finding Teams in Graphs and Its Application to Spatial Gene Cluster Discovery

Tizian Schulz[1,2], Jens Stoye[1], and Daniel Doerr[1(✉)]

[1] Faculty of Technology and CeBiTec, Bielefeld University, Bielefeld, Germany
{tizian.schulz,jens.stoye,daniel.doerr}@uni-bielefeld.de
[2] International Research Training Group 1906 "Computational Methods for the Analysis of the Diversity and Dynamics of Genomes", Bielefeld University, Bielefeld, Germany

Abstract. Gene clusters are sets of genes in a genome with associated functionality. Often, they exhibit close proximity to each other on the chromosome which can be beneficial for their common regulation. A popular strategy for finding gene clusters is to exploit the close proximity by identifying sets of genes that are consistently close to each other on their respective chromosomal sequences across several related species.

Yet, even more than gene proximity on linear DNA sequences, the spatial conformation of chromosomes may provide a pivotal indicator for common regulation and/or associated function of sets of genes.

We present the first gene cluster model capable of handling spatial data. Our model extends a popular computational model for gene cluster prediction, called δ-teams, from sequences to general graphs. In doing so, δ-teams are single-linkage clusters of a set of shared vertices between two or more undirected weighted graphs such that the largest link in the cluster does not exceed a given threshold δ in any input graph.

We apply our model to human and mouse data to find *spatial gene clusters*, i.e., gene sets with functional associations that exhibit close neighborhood in the spatial conformation of the chromosome across species.

Keywords: Spatial gene cluster · Gene teams · Single-linkage clustering · Graph teams · Hi-C data

1 Introduction

Distance-based clustering algorithms are paramount to approach various questions across all data-driven fields including comparative genomics. Here, we study the problem of discovering single-linkage clusters of a set of corresponding vertices (where correspondence is either provided through a bijective mapping or equivalence classes) between two or more undirected weighted graphs $G_1, \ldots, G_k$ such that the largest link in the cluster (measured in terms of the weighted shortest path) does not exceed a given threshold δ in either graph G_i, $1 \leq i \leq k$.

© Springer International Publishing AG 2017
J. Meidanis and L. Nakhleh (Eds.): RECOMB CG 2017, LNBI 10562, pp. 197–212, 2017.
DOI: 10.1007/978-3-319-67979-2_11

We call such clusters (δ-) teams, thereby adopting notation used by an extensive trail of literature that studies the equivalent problem on permutations and sequences [2,9,19,21].

A prominent use case of δ-teams in comparative genomics is the detection of *gene clusters*, which are sets of genes with associated functionality such as the encoding of different enzymes used in the same metabolic pathway. In many organisms instances exist where such genes are also locally close to each other in the genome, i.e., their positions fall within a narrow region on the same chromosome. They may even remain in close proximity over a longer evolutionary period, despite the fact that genomes regularly undergo mutations such as genome rearrangements, gene- or segmental duplications, as well as gene insertions and deletions. Such mutations may also affect the order and copy number of genes within a gene cluster. Molecular biologists argue that a conserved neighborhood is beneficial for co-regulation, as is true in the prominent case of *operons* in prokaryotes [10]. Gene clusters are also prevalent in eukaryotes, even in animals, where the HOX gene cluster is without doubt the best studied representative. HOX genes are transcription factors that regulate the embryological development of the metazoan body plan [12].

Yet, the function of many genes is often barely understood or entirely unknown despite the increasing number of whole genome data that is becoming available. Hence, a popular approach in comparative genomics is to work this way backwards, starting with the investigation of conserved gene proximity in genomes of a reasonably phylogenetically diverse set of species. Here, the underlying assumption is made that accumulated genome rearrangements will have shuffled the genome sequences sufficiently so that natural selection becomes a plausible cause of conserved gene neighborhoods. By identifying homologous sets of genes that are consistently close to each other across several species, candidate gene clusters are identified that are then subject to more thorough functional analysis.

More recently, new technologies emerged, allowing the study of the spatial structure of genomes. *High-throughput chromosome conformation capture* (Hi-C), the most popular among such approaches, allows assessing the conformation of the chromatin structure in a cell sample through measuring the number of observed contacts between DNA regions [3]. The Hi-C method makes use of formaldehyde to covalently bond proteins and DNA strings which are located next to each other in the cell. After crosslinking, the cells are lysed and the DNA is digested by a restriction enzyme. Digested fragments bonded by the same protein are ligated. Sequencing the hybrid sequences reveals three-dimensional contacts between their genomic origins. The outcome of the experiment is a table, called *Hi-C map*, that records observed contacts. Each row and each column of the Hi-C map represents an equally sized segment of the genome sequence, and a count in each cell indicates how often sequences of the corresponding segments have been observed during the experiment. The size of these segments is known as *resolution*. It is a crucial parameter regarding the quality of the data. The higher the resolution of the chromatin structure is, the smaller is the segment size, but also the more data is needed to get significant results. An increasing number of Hi-C maps has recently been made publicly available (human and mouse [8], fruit fly [16]) and is used to answer numerous biological questions, starting from gene regulation and replication timing [8,13] to genome scaffolding and haplotyping [4,15].

Gene cluster discovery has sparked the development of various computational models for identifying sets of genes that exhibit close proximity. Such models typically rely on abstract data structures known as *gene order* sequences, which describe the succession of genes in chromosomes. In doing so, each element of a gene order sequence is the identifier of a gene's associated gene family. A popular method to find gene clusters is based on the identification of *common intervals* in these sequences, which are intervals with an identical set of elements (i.e. gene family identifiers), independent of the elements' order and multiplicity [7,14,18]. Since their first mentioning in [18], common intervals became the source for several generalizations [11,22], among others δ-*teams* [9]. δ-Teams are sets of elements where the distance between any two successors across all sequences is bounded by a given threshold $\delta \geq 0$. This flexible model not only facilitates the detection of gene clusters that are interspersed by unrelated inserted genes, but also allows the consideration of distance in terms of nucleotide base pairs between genes.

Even more than gene proximity on linear DNA sequences, the spatial conformation of chromosomes may provide a pivotal indicator for common regulation and/or associated function of sets of genes. Evidence of *spatial gene clusters* has been put forward already by Thévenin *et al.* [17] who studied spatial proximity within functional groups of genes in the human genome. In this work, we present the first spatial gene cluster model. Our model extends the δ-teams model from sequences to undirected weighted graphs, facilitating the detection of genes that are consistently spatially close in multiple species. In doing so, our method integrates Hi-C and genome sequence data into weighted undirected graphs, where vertices represent gene family identifiers of genes and weighted edges correspond to distances obtained from Hi-C data.

The remainder of this manuscript is organized as follows: In Sect. 2 we formally define δ-teams on graphs and present an algorithm for their discovery. We then extend our approach to finding δ-teams *with families*, i.e., the case where vertices across graphs are related through a common family membership, allowing multiple members of the same family to be part of the same graph. In Sect. 3 we show how δ-teams can be used to find candidate sets of spatial gene clusters using a combination of genome and Hi-C data of two or more species. We evaluate our approach using data from human and mouse. Section 4 concludes this manuscript and provides an outlook on future work.

An implementation of our method in the `Python` programming language is available from http://github.com/danydoerr/GraphTeams.

2   Discovering δ-Teams in Graphs with Shared Vertex Sets

In this section we discuss the general problem of identifying single linkage clusters in a collection of graphs, where the weakest link in a cluster does not exceed a given distance threshold δ. We call such clusters δ-teams to remain in line with previous literature which studied the equivalent problem on permutations and sequences.

To simplify presentation, we describe only the case of two input graphs G and H in detail. The general case can be trivially inferred. In fact, our implementation (see Sect. 3) supports comparison of two or more graphs.

We study undirected graphs $G = (V, E)$ with distance measure $d_G \colon V \times V \to \mathbb{R}_{>0}$. While subsequent definitions adhere to the general case, for all our purposes we assume edge-weighted graphs and use as distance measure the length of the shortest path between any two vertices, measured by the sum of the path's edge weights if such exists and ∞ otherwise. We use $E(G), V(G)$ to denote the edge and vertex set of a graph G, respectively. Since we will refer frequently to sets of vertices in one of several graphs, we will indicate the origin of a vertex set $S \subseteq V(G)$ of a graph G through subscript notation, i.e. S_G, whenever this information is relevant. We are interested in sets of vertices that are connected through paths on which the distance between two successive members is bounded by δ:

Definition 1 (δ-set). *Given a graph G with distance measure d_G and a threshold value $\delta \geq 0$, a vertex set $S \subseteq V(G)$ is a δ-set if for each pair of vertices $u, v \in S$ there exists a sequence $P = (u, \ldots, v) \subseteq S$ such that the distance $d_G(w, z)$ between any two consecutive vertices w and z of P is less than or equal to δ.*

The subsequent definitions establish relations between δ-sets across two graphs G and H with shared vertex set $V_\cap = V(G) \cap V(H)$. In doing so, we assume that there is a common non-empty set of vertices of the two graphs that is subject to subsequent analysis. Vertices that are unique to either of the two graphs are disregarded, yet may be relevant due to their involvement in paths between common vertices.

Definition 2 (δ-cluster). *Given two graphs G and H with distance measures d_G and d_H, respectively, and a threshold value $\delta \geq 0$, a vertex set $S \subseteq V_\cap$ is a δ-cluster if it is a δ-set in both G and H under distance measures d_G and d_H, respectively.*

Definition 3 (δ-team). *Given two graphs G and H, a δ-cluster S of G and H is a δ-team if it is maximal, i.e., there is no δ-cluster S' of G and H such that $S \subsetneq S'$.*

Example 1. The two graphs G and H depicted in Fig. 1(a) have three δ-teams: 1-team $\{d, f\}$; 2-team $\{c, d, f\}$, and 3-team $\{a, c, d, f, g\}$. The set $\{c, d, f, g\}$ exemplifies a non-maximal 3-cluster of G and H.

2.1   Finding δ-Teams by Decomposing Graphs with Divide-and-Conquer

Given the above definitions, the following computational problem naturally arises and is subject to this work:

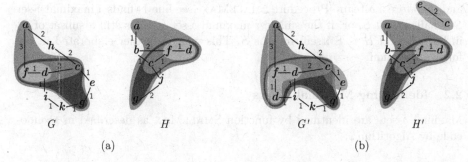

Fig. 1. Examples of δ-teams and δ-clusters in graphs without families (a) and with families (b). δ-teams and -clusters are highlighted by areas of shared color. Edge labels indicate weights. Vertices in (b) are represented by their family identifier.

Problem 1. Given two graphs G and H with distance measures d_G and d_H, respectively, and a threshold value $\delta \geq 0$, find all δ-teams of G and H.

The first observation that is key to addressing the problem at hand, is that two δ-teams cannot overlap. The following lemma, in which $\text{Teams}_\delta(S)$ denotes the set of δ-teams of vertex set S, is basis to all permutation-based (gene-) team algorithms and holds true for the here proposed generalization, too (the disjoint union of two sets is denoted by $\uplus$):

Lemma 1 *[2,9]. Given two graphs G and H with common vertex set $V_\cap$ and a threshold value $\delta \geq 0$, there exists a partition $\{V', V''\}$ of $V_\cap = V' \uplus V''$ such that $\text{Teams}_\delta(V_\cap) = \text{Teams}_\delta(V') \uplus \text{Teams}_\delta(V'')$.*

The lemma leads to a simple divide-and-conquer approach which has already been applied by He and Goldwasser [9] for the restricted case of sequential data. Here, we apply this lemma to general graphs. Algorithm DECOMPOSE divides the common vertex subset $S \subseteq V_\cap$ of graphs G and H into smaller subsets as long as S is not a δ-set in both graphs.

Algorithm 1. DECOMPOSE(S)

Input: graphs G, H; vertex subset $S \subseteq V_\cap$, $S \neq \emptyset$; threshold value $\delta \geq 0$
Output: all δ-teams of G and H within S
1: $S' \leftarrow$ SMALLMAX(S, S) // find a smaller maximal δ-set $S' \subseteq S$ of G or H
2: **if** $|S| = |S'|$ **then**
3:     **return** $\{S\}$
4: **else**
5:     **return** DECOMPOSE$(S') \cup$ DECOMPOSE$(S \setminus S')$
6: **end if**

Because Algorithm 1 proceeds from larger to smaller sets, a vertex set S, identified by the algorithm, that is a δ-set in both G and H is always maximal and therefore a δ-team. Procedure SMALLMAX (see line 1) finds a maximal δ-set S' smaller than S, or, if the smallest maximal δ-set (that is still a subset of S) in both G and H is S itself, returns S. This will be further elaborated in the following section.

2.2   Identifying Maximal δ-Sets

Maximal δ-sets are identified by function SMALLMAX as described in pseudo-code by Algorithm 2.

Algorithm 2. SMALLMAX(S_G, S_H)

Input: graphs G, H; vertex subsets $S_G \subseteq V(G)$ and $S_H \subseteq V(H)$; threshold value
      $\delta \geq 0$
Output: a maximal δ-set $S'_G \subseteq S_G$ or $S'_H \subseteq S_H$
  1: choose random vertices $u \in S_G$ and $v \in S_H$
  2: initialize sets $S'_G = \{u\}$, $S'_H = \{v\}$
  3: initialize Boolean variables p_G, p_H with **True**
  4: **while** $(p_G$ **or** $S'_G = S_G)$ **and** $(p_H$ **or** $S'_H = S_H)$ **and** $(p_G$ **or** $p_H)$ **do**
  5:     **for each** graph $X = G, H$ **do**
  6:         **if** $\exists\, s \in S_X \backslash S'_X$ s.t. $\exists\, s' \in S'_X$ with $d_X(s, s') \leq \delta$ **then**
  7:             add vertex s to set S'_X
  8:         **else**
  9:             $p_X \leftarrow$ **False**
 10:         **end if**
 11:     **end for**
 12: **end while**
 13: **if** $(\neg p_G$ **and** $S'_G \neq S_G)$ **then return** S'_G **else return** S'_H

Note that procedure SMALLMAX is drafted for a general setting that permits the discovery of different vertex sets in graphs G and H, respectively. In doing so, SMALLMAX can also be used in the case of finding δ-teams *with families* that is subject of Sect. 2.4. For now, the input sets S_G and S_H are identical.

SMALLMAX identifies a smaller maximal δ-set in either vertex set S_G or S_H. In each iteration (lines 4–12), the algorithm searches in each graph $X = G, H$ a vertex s of set S_X which has not been previously visited and that has distance at most δ from any already visited node. To this end, a list S'_X is maintained that keeps track of already visited vertices of set S_X. Boolean variables p_X indicate whether unvisited, yet reachable vertices in set $S_X \backslash S'_X$ could be found in graph X. The iteration is controlled by three different cases (line 4): If no unvisited node can be found, SMALLMAX has identified either a smaller δ-set of S_X or, if the traversal is exhausted, S_X itself. In the former case, the procedure stops and returns the visited subset S'_X of S_X. In the latter case, the algorithm continues the search for a smaller δ-set in the corresponding other vertex set S_Y, $Y = \{G, H\} \backslash X$, and will return such if found. Otherwise, the smallest maximal δ-sets in both S_G and S_H are the sets themselves. This also leads to a disruption of the while-loop (lines 4–12) and, by convention, the return of set S'_H $(= S_H)$.

Because SMALLMAX does not go further than distance δ from any already visited node of S'_X, it is clear that the returned vertex set is a δ-set. It is also maximal, because the algorithm does not stop prior to having found all vertices of S_X that can be reached from the starting node (which is also a member of S_X and S'_X).

The time complexity of algorithm DECOMPOSE depends on the number of its own recursive function calls. The decomposition of set S into sets S' and $S \backslash S'$ that is performed in line 5 of DECOMPOSE takes $O(|S|)$ time, but is overshadowed by the time complexity of SMALLMAX. For SMALLMAX, the most costly operation is the search for the next node s of $S_X \backslash S'_X$. This can be found through successive traversal of each graph using *breadth-first search* (BFS) outgoing from any arbitrary vertex of sets S_G and S_H, respectively. The BFS determines the running time of SMALLMAX and requires $O(|V(G)| + |E(G)| + |V(H)| + |E(H)|)$ time. In the worst case, DECOMPOSE needs $|V_\cap|$ iterations to decompose the initial, shared vertex set $V_\cap$.

This leads to an overall running time of $O(|V_\cap| \cdot (|V(G)| + |E(G)| + |V(H)| + |E(H)|))$ for Algorithm 1.

2.3 The Special Case of Shortest-Path Graphs

In the special case where each pair of vertices u, v of vertex set $V_\cap$ has a directly connecting edge whenever their distance is smaller or equal to δ, SMALLMAX takes $O(|V_\cap|)$ time in each iteration. This observation leads to an alternative approach for the general case that may in practice be faster for certain instances or applications: From the input graphs G, H two new graphs G' and H' are derived by computing shortest paths between all pairs of vertices in $V(G)$ and $V(H)$, respectively. In the new graph G' two vertices $u, v \in V(G) = V(G')$ are connected with an edge of weight 1 if their distance is smaller than δ and, similarly, for graph H'. Then, the enumeration of δ-teams of G and H is equivalent to computing 1-teams in G' and H'. Our implementation includes an option for the computation of δ-teams using this alternative approach. Shortest paths are obtained with Floyd-Warshall's algorithm which has a running time of $O(|V|^3)$ [5].

2.4 δ-Teams with Families

Family labels allow correspondences between vertices of the input graphs G and H that go beyond 1-to-1 assignments, which is the scenario best suitable for our application as further explained in Sect. 3. Given a graph $G = (V, E)$, let $\mathcal{F}: V \to F$ be a surjective mapping between vertices and families.

We extend the concepts of δ-cluster and δ-team to families as follows:

Definition 4 (δ-cluster with families). *Given two graphs G and H with distance measures d_G and d_H, respectively, a family mapping $\mathcal{F}$ and a threshold value $\delta \geq 0$, a pair of vertex sets (S_G, S_H) with $S_G \subseteq V(G)$ and $S_H \subseteq V(H)$, is a δ-cluster if (i) $\mathcal{F}(S_G) = \mathcal{F}(S_H)$ and (ii) S_G and S_H are δ-sets in G and H under distance measures d_G and d_H, respectively.*

Definition 5 (δ-team with families). *Given two graphs G and H, a δ-cluster (S_G, S_H) of G and H is a δ-team if it is maximal, i.e., there is no other δ-cluster (S'_G, S'_H) of G and H such that $S_G \subseteq S'_G$ and $S_H \subseteq S'_H$.*

Example 2. The two graphs G' and H' depicted in Fig. 1(b) have four δ-teams that are in the following represented by their family set: 1-team $\{d, f\}$; 2-teams $\{c, d, f\}$ and $\{c, e\}$, and 3-team $\{a, c, d, f, g\}$. The set $\{c, d, f, g\}$ exemplifies a non-maximal 3-cluster of G' and H'.

With the generalization to families, Lemma 1 is no longer applicable. However, Wang *et al.* [19] provide an adaptation which shows that the original divide-and-conquer approach can be trivially extended:

Lemma 2 *[19].* *Given two graphs G and H, a family mapping $\mathcal{F}$ and a threshold value $\delta \geq 0$, let $S_G \subseteq V(G)$, $S_H \subseteq V(H)$, s.t. $\mathcal{F}(S_G) = \mathcal{F}(S_H)$ and B be a maximal δ-set of S_G or S_H. W.l.o.g. let $B \subseteq S_G$, then $\mathrm{Teams}_\delta(S_G, S_H) = \mathrm{Teams}_\delta(B, S'_H) \cup \mathrm{Teams}_\delta(S_G \backslash B, S''_H)$, where $S'_H = \{v \in S_H \mid \mathcal{F}(v) \in \mathcal{F}(B)\}$ and $S''_H = \{v \in S_H \mid \mathcal{F}(v) \in \mathcal{F}(S_G \backslash B)\}$.*

The adaptations to algorithm DECOMPOSE are a straightforward implementation of Lemma 2 and are shown in Algorithm 3 (DECOMPOSEFAMILIES).

Algorithm 3. DECOMPOSEFAMILIES(S_G, S_H)

Input: Graphs G and H, mapping $\mathcal{F}$, vertex sets $S_G \subseteq V(G)$ and $S_H \subseteq V(H)$ such that $\mathcal{F}(S_G) = \mathcal{F}(S_H) \neq \emptyset$, distance measures d_G, d_H, and $\delta \geq 0$
Output: all δ-teams of G and H that are subsets of or equal to (S_G, S_H)
    // *find a maximal δ-set $S'_X \subseteq S_X$, where X is a placeholder for graph G or H*
1:   $S'_X \leftarrow$ SMALLMAX(S_G, S_H)
2:   **if** $S_X = S'_X$ **then**
3:     **return** $\{(S_G, S_H)\}$
4:   **else**
5:     $Y \leftarrow \{G, H\} \backslash X$
6:     $S'_Y \leftarrow \{v \in S_Y \mid \mathcal{F}(v) \in \mathcal{F}(S'_X)\}$
7:     $S''_Y \leftarrow \{v \in S_Y \mid \mathcal{F}(v) \in \mathcal{F}(S_X \backslash S'_X)\}$
8:     **return** DECOMPOSEFAMILIES(S'_X, S'_Y) $\cup$ DECOMPOSEFAMILIES($S_X \backslash S'_X, S''_Y$)
9:   **end if**

To efficiently retrieve vertices associated with families of $\mathcal{F}(S'_X)$ and $\mathcal{F}(S_X \backslash S'_X)$ (see lines 6 and 7 of Algorithm 3), we follow Wang *et al.* [19] and maintain a table of linked lists that maps family identifiers with its members in each respective graph. $\mathcal{F}(S'_X)$ can be built in $O(|S'_X|)$ time while $\mathcal{F}(S_X \backslash S'_X)$ needs $O(|S_X|)$ time. Afterwards, it is possible to build S'_Y and S''_Y in $O(|S_Y|)$ time. The runtime of SMALLMAX remains the same for Algorithm 3. Yet, because the input sets S_G and S_H can no longer be decomposed into disjoint sets, Algorithm 3 requires overall $O((|V(G)| + |E(G)|) \cdot (|V(H)| + |E(H)|))$ time and $O(|V(G)| + |E(G)| + |V(H)| + |E(H)|)$ space.

3   Application to Spatial Gene Cluster Discovery

We will now show how the discovery of δ-teams with families allows to find spatial gene clusters in genomic data of two or more species. We implemented Algorithm 3 in the Python programming language and provide an entirely automated Snakemake workflow for the identification of spatial gene clusters. Our workflow takes as input the fully assembled sequences of a collection of genomes as well as their corresponding Hi-C maps. It normalizes the Hi-C maps, establishes relationships between Hi-C segments and genes, and constructs weighted graphs that are then input to Algorithm 3. Further, our workflow allows the computation of a ranking scheme for gene cluster candidates based on shared functional associations of their members when provided with additional GO-annotations. Our approach is, to the best of our knowledge, the first of its kind that is capable of identifying spatial gene clusters. Our Snakemake workflow can be obtained from http://github.com/danydoerr/GraphTeams.

3.1   Constructing Graphs from Genome Sequences and Hi-C Data

For each genome, we construct an undirected weighted graph in which vertices correspond to genes that are labeled with the identifier of their associated gene family and in which weighted edges correspond to distances obtained from the contact counts of the genomes' respective Hi-C maps. Then, δ-teams (with families according to the genes' families) in the constructed graphs will correspond to spatial gene cluster candidates.

We first map the Hi-C data onto their chromosomal sequences. In doing so, we associate genes with segments of the Hi-C map. Consequently, contact counts between genes correspond to the contact counts of their associated segments. The value of a contact count does not represent a distance but a closeness score, hence a transformation is needed. We define the *distance* between two genes g_i, g_j associated with Hi-C map M as

$$d_M(g_i, g_j) = \max_{k,l}(M_{kl}) + 1 - M_{ij}. \tag{1}$$

Hi-C maps are symmetric matrices, therefore $d_M(g_i, g_j) = d_M(g_j, g_i)$. Whenever two gene pairs fall into the same Hi-C segment, their distance is estimated by their proximity on the DNA sequence. To this end, each base pair between the midpoints of two genes is scored with a relative contact count of C/r, where C is the average contact count between two adjacent segments in the Hi-C map, i.e., the mean of $M_{i,i+1}$ of Hi-C map M, and r is the resolution of the Hi-C map, i.e., the size of its segments. This estimator works well for our purposes because Hi-C data shows strong correlations with distances on the DNA sequence.

It is common that Hi-C maps contain large numbers of empty cells as a result of erroneous measurements and deliberate blanking of the contact counts around the centromere. We do not apply any correction to such cells except to those that correspond to adjacent segments, i.e., the $M_{i,i+1}$ cells. Here, we use the same estimator as described above for genes falling into the same cell of the Hi-C map.

Because we will compare distances obtained from different Hi-C maps, we must ensure that they all use the same scale. We do this by multiplying all distances of each Hi-C map M with a factor $c/(\max_{k,l}(M_{kl}))$ where c is the *average maximum contact count* across all Hi-C maps.

3.2   Quantifying Functional Associations of Gene Clusters Using Gene Ontology Annotations

We quantify functional associations between genes of a gene cluster candidate by testing against the *null* hypothesis that a gene in a gene cluster is functionally not more associated to one of its co-members than to any other genes in the genome. To this end, we make use of *Gene Ontology* (GO) [1] annotations and relate between gene functions by means of the gene ontology hierarchy that corresponds to the domain "Biological Processes". In doing so, we measure *GO-based functional dissimilarity* (GFD) [6] between pairs of GO-annotated genes. Given a directed acyclic graph $G = (V, E)$ corresponding to a GO-hierarchy, $r_G(g) = \{v \in V(G) \mid g$ associated with $v\}$ denotes the set of *GO terms*, i.e., vertices of the GO hierarchy G, with which gene g is associated. Further, $p_G(u, v)$ denotes the length of the shortest path between two vertices $u, v \in V$ measured in the number of separating nodes. The GFD between two GO-annotated genes g and g' is then defined as

$$gfd_G(g, g') = \min_{(u,v) \in r_G(g) \times r_G(g')} \left(\frac{p_G(u, v)}{depth_G(u) + depth_G(v)} \right), \qquad (2)$$

where $depth_G(w)$ is the length of the path from the root vertex of G to vertex w. This measure gives then rise to the *gene cluster penalty* defined for a gene set $C \subseteq G$ of a genome $\mathcal{G}$ as follows:

$$\phi_G(C, \mathcal{G}) = \sum_{g \in C} \left(\min_{g' \in C \setminus \{g\}} gfd(g, g') - \min_{g'' \in \mathcal{G} \setminus \{g\}} gfd(g, g'') \right). \qquad (3)$$

In our analysis, we rank gene clusters according to p-values empirically computed from sample pools of size of 10^7 which are drawn for each gene cluster size, respectively.

3.3   Finding Candidates of Spatial Gene Clusters in Human and Mouse

We used the approach described in Sect. 3.1 to find spatial gene cluster candidates in human and mouse. To this end, we queried the Ensemble Genome Browser (release 88) [23] to obtain information about orthologous genes of the human reference sequence GRCh38.p10 and the mouse reference sequence GRCm38.p5. The obtained data consists of 19,843 human genes that are

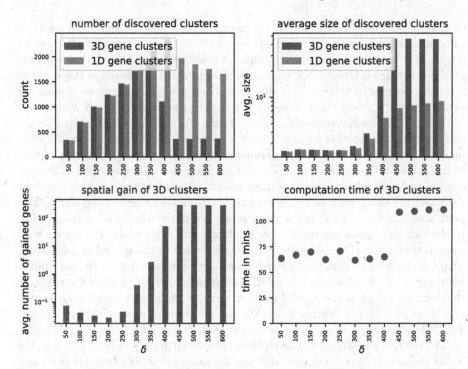

Fig. 2. Results of Algorithm 3 for different values of δ. Graphs are constructed from Hi-C datasets of human and mouse. The plots show for each threshold value δ, the number of discovered clusters (upper left) and their average sizes (upper right) in the spatial and sequential graphs, respectively, the average number of gained genes in the 3D gene clusters versus the 1D gene clusters (lower left), and the computation time for the 3D gene clusters (lower right).

orthologous to 20,647 mouse genes. The intra-chromosomal Hi-C maps of the human and mouse genomes that we use in this study were first published in Dixon *et al.* [8] and have a resolution of 40 kb.

The graphs for the human and mouse datasets were constructed as previously described. Subsequently, we used them to find δ-teams for different values of δ. All computations were performed on a Dell RX815 machine with 64 2.3 GHz AMD Opteron processors and 512 GB of shared memory. The running times for computing all δ-teams for each value of δ are shown in the bottom right plot of Fig. 2 and range from 62 min for $\delta = 200$ to 111 min for $\delta = 600$. The plot indicates a sharp increase of running time for $\delta > 400$ that correlates with the increase of the size of identified δ-teams in our dataset.

Apart from the performance of our algorithm, we also wanted to investigate how suitable spatial data is to improve the search for gene clusters. Next to the graphs that were generated as described in Sect. 3.1 and which we will further call *spatial graphs*, we constructed a second type of graphs, called *sequential graphs*. In the latter, distances between genes are limited to those that are adjacent in their respective genome sequences. Connecting these adjacent genes, we used the same distances as in the former graph so that distances between both types of graphs are comparable. Since our algorithm is a direct generalization of previous methods acting on linear DNA sequences, we can generate results of these methods using sequential graphs. We call δ-teams that are found in spatial graphs *3D gene clusters*, whereas those in sequential graphs are called *1D gene clusters*.

Figure 2 shows the results for both graphs. In the plot on the top left, we can see that the number of gene clusters grows for both types of graphs with increasing values of δ while the number of 3D gene clusters is slightly higher than that of 1D gene clusters. This changes after $\delta = 350$ when more 3D gene clusters are merged than new instances are found, leading to a rapid decrease in their number along with an increase in their size (see plot on the top right). The peak associated with this phenomenon is delayed in the sequential graphs, owing to the fact that genes are there more stretched out. This is also the reason why we find that some 1D gene clusters are much denser in the spatial graphs. More surprisingly, we also find gene clusters that can only be found in spatial graphs for a given threshold value δ. We call the average amount of genes in a cluster that can be found in the spatial graphs, but not in the sequential ones *spatial gain* (see plot at the bottom left). We see an increase in spatial gain around $\delta = 250$ until a saturation seems to be reached at $\delta = 450$.

We further investigated gene clusters discovered with $\delta = 350$, which strike a fair balance between number and size as can be readily observed from our previous analysis. The datasets of both, 3D and 1D gene clusters, were used to evaluate functional associations between gene cluster members. To this end, GO-annotations of the human genome were obtained from [1] to compute gene cluster penalties and to rank gene clusters according to their empirical p-value as described in Sect. 3.2. In the obtained gene ontology dataset, 15,737 out of 19,843 human genes were associated with one or more GO-terms. Because the analysis is restricted to those genes with annotated GO-terms, only 1,559 out of 1,961 3D gene clusters and 1,669 out of 2,118 1D gene clusters could be further investigated. 18.54% of the 3D gene clusters and 18.33% of the 1D gene clusters exhibited a significant empirical p-value for $p < 0.05$. Overall, significant 3D gene clusters tend to include more (annotated) genes (total: 930) than their 1D counterparts (total: 886). Table 1 lists the top 20 3D gene clusters that are either not found in the set of significant 1D gene clusters, or only partially found, or broken into two or more sub-clusters. We can see that many of them are already known from the literature. E.g., we find four clusters of *olfactory receptor* (OR) genes on different chromosomes, the *taste receptor type 2* (TAS2R) gene cluster and the HOXC gene cluster. The latter is one of three clusters among the top 20 but can be found in the 1D results only as a composition of sub-clusters. Therefore, these genes seem to be even closer together in 3D than on the DNA strand. The same is true for other clusters, such as that of the *testis-specific protein Y-encoded* (TSPY) and *superfamily Ig belonging lectins* (SIGLEC) which were not even partially detected in the 1D graphs.

Table 1. Top 20 3D gene clusters with smallest p-value. Clusters that can be found as split sub-clusters in the 1D results are marked by an asterisk. Those completely absent in the 1D results are marked by a plus.

Name	Genes	Penalty	p-Value
HOXC*	HOTAIR_2, HOTAIR_3, HOXC10, HOXC11, HOXC12, HOXC13, HOXC4, HOXC5, HOXC6, HOXC8, HOXC9	0.006	$1 \cdot 10^{-7}$
OR	OR5AP2, OR5AR1, OR5M1, OR5M10, OR5M11, OR5M3, OR5M8, OR5M9, OR5R1, OR8K1, OR8U1, OR9G1, OR9G	0	$1 \cdot 10^{-7}$
IGHV*	IGHV3-11, IGHV3-13, IGHV3-20, IGHV3-21, IGHV3-23, IGHV3-30, IGHV3-33, IGHV3-35, IGHV3-64D, IGHV3-7	0	$1 \cdot 10^{-7}$
KRTAP*	KRTAP13-1, KRTAP13-2, KRTAP13-3, KRTAP13-4, KRTAP15-1, KRTAP24-1, KRTAP26-1, KRTAP27-1	0	$1 \cdot 10^{-7}$
TAS2R	TAS2R14, TAS2R19, TAS2R20, TAS2R31, TAS2R46, TAS2R50	0	$3.70 \cdot 10^{-6}$
OR	OR2A12, OR2A14, OR2A25, OR2A5	0	$9.09 \cdot 10^{-5}$
ZSCAN4	NKAPL, ZKSCAN3, ZKSCAN4, ZSCAN26	0.006	0.00015
TRAV	TRAV12-1, TRAV12-2, TRAV12-3, TRAV13-1, TRAV13-2, TRAV17, TRAV18, TRAV19, TRAV22, TRAV23DV6, TRAV5, TRAV8-1, TRAV8-3, TRAV9-2	0	0.00037
OR	OR5AC1, OR5H1, OR5H14	0	0.00037
IGHV+	IGHV1-18, IGHV1-24, IGHV1-3	0	0.00037
BTN3+	BTN3A1, BTN3A2, BTN3A3	0	0.00037
(Unnamed)	GTF2A1L, STON1, STON1-GTF2A1L	0	0.00037
CYP3A	CYP3A4, CYP3A43, CYP3A5, CYP3A7, CYP3A7-CYP3A51P	0.028	0.00037
(Unnamed)	ADGRE1, C3, CD70, GPR108, TNFSF14, TRIP10, VAV1	0.057	0.00047
ZNF	CCDC106, FIZ1, U2AF2, ZNF524, ZNF580, ZNF784, ZNF865	0.097	0.00110
OR	OR8B12, OR8B4, OR8B8	0.012	0.00376
KIR	KIR2DL1, KIR2DL3, KIR2DL4, KIR2DS4, KIR3DL1, KIR3DL2, KIR3DL3	0.179	0.00243
MMP	MMP12, MMP13, MMP3	0.035	0.00486
TSPY+	TSPYL1, TSPYL4	0	0.00504
SIGLEC+	SIGLEC12, SIGLEC8	0	0.00504

4 Discussion and Outlook

The enumeration of common intervals in sequences has been subject to various extensions including δ-teams. Here, we described a generalization of δ-teams from sequences to graphs. We presented a novel algorithm for the enumeration of δ-teams that, when trivially extended to k graphs $G_i = (V_i, E_i)$, for $i = 1, \ldots, k$, will run in $O(\sum_i |V_i| + |V_\cap| \cdot \sum_i (|V_i| + |E_i|))$ time and $O(\sum_i (|V_i| + |E_i|))$ space, where $V_\cap = V_1 \cap \cdots \cap V_k$. Our algorithm beats the naive approach that requires $O(\sum_i |V_i|^3)$ time and $O(\sum_i |V_i|^2)$ space by computing all-pairs-shortest-paths and then using a standard single linkage clustering algorithm to enumerate δ-teams. Further, we provide an algorithm for the computation of δ-teams that, when trivially extended to k graphs *with families*, will run in $O(k \cdot \prod_i (|V_i| + |E_i|))$ time and $O(k \cdot \sum_i (|V_i| + |E_i|))$ space.

In comparison, the best algorithm for the enumeration of δ-teams in k permutations of size n runs in $O(k \cdot n \cdot \log N)$ time, where N denotes the number of reported δ-teams [20]. The best algorithm that solves the corresponding family-based problem for k sequences of lengths $n_1, \ldots, n_k$ runs in $O(k \cdot C \cdot \log(n_1 \cdots n_k))$ time, where C is a factor accounting for the number of possible 1:1 assignments between family members across the k graphs [19]. The differences in running time between the permutation-, sequence- and our graph-based algorithms reflect the fact that the latter solve much harder problems. Nevertheless, further studies may lead to improved algorithms. It seems possible that the problem of finding δ-teams in graphs *without families* could be solved faster with the help of a guide tree that allows to find a maximal δ-set by traversing each graph in fewer steps than required by an exhaustive graph traversal. Alternatively, a randomized variant of our algorithm could assert a better expected running time.

The presented algorithmic work could also be extended into another direction, by allowing the direct computation of the *single-linkage hierarchy*. This makes the gene cluster analysis no longer dependent on a fixed δ, but will provide all possible δ-clusters through a single computation. This idea has also been applied for δ-teams in sequences, where the hierarchy is called *gene team tree* [21,24].

By identifying δ-teams *with families*, we provide a flexible model that is well suitable to capture the complexity of biological datasets such as those at hand. Our presented algorithm and our implementation are fast enough to conveniently process large graphs as demonstrated in the evaluation of this study. The identification of all δ-clusters in the studied Hi-C dataset of human and mouse took between 62 and 111 min on state-of-the-art hardware.

Finally, we evaluated functional associations between members of 3D and 1D gene cluster candidates, respectively. Our experimental evaluation provides further evidence for the existence of spatial gene clusters, that is, sets of functionally associated genes whose members are closer to each other in the 3D space than on the chromosomal sequence.

Acknowledgements. We are very grateful to Krister Swenson for kindly providing the Hi-C data used in this study and for his many valuable suggestions. We wish to thank Pedro Feijão for many fruitful discussions in the beginning of this project. This work was partially supported by DFG GRK 1906/1.

References

1. Ashburner, M., Ball, C.A., Blake, J.A., Botstein, D., Butler, H., Cherry, J.M., Davis, A.P., Dolinski, K., Dwight, S.S., Eppig, J.T., Harris, M.A., Hill, D.P., Issel-Tarver, L., Kasarskis, A., Lewis, S., Matese, J.C., Richardson, J.E., Ringwald, M., Rubin, G.M., Sherlock, G.: Gene ontology: tool for the unification of biology. Nat. Genet. **25**(1), 25–29 (2000)
2. Beal, M., Bergeron, A., Corteel, S., Raffinot, M.: An algorithmic view of gene teams. Theoret. Comput. Sci. **320**(2–3), 395–418 (2004)

3. Belton, J.M., McCord, R.P., Gibcus, J.H., Naumova, N., Zhan, Y., Dekker, J.: Hi-C: a comprehensive technique to capture the conformation of genomes. Methods **58**(3), 268–276 (2012)
4. Burton, J.N., Adey, A., Patwardhan, R.P., Qiu, R., Kitzman, J.O., Shendure, J.: Chromosome-scale scaffolding of de novo genome assemblies based on chromatin interactions. Nat. Biotechnol. **31**(12), 1119–1125 (2013)
5. Cormen, T.H., Leiserson, C.E., Rivest, R.L.: Introduction to Algorithms. MIT Press, Cambridge (1990)
6. Díaz-Díaz, N., Aguilar-Ruiz, J.S.: Go-based functional dissimilarity of gene sets. BMC Bioinform. **12**(1), 360 (2011)
7. Didier, G., Schmidt, T., Stoye, J., Tsur, D.: Character sets of strings. J. Discret. Algorithms **5**(2), 330–340 (2006)
8. Dixon, J.R., Selvaraj, S., Yue, F., Kim, A., Li, Y., Shen, Y., Hu, M., Liu, J.S., Ren, B.: Topological domains in mammalian genomes identified by analysis of chromatin interactions. Nature **485**(7398), 376–380 (2012)
9. He, X., Goldwasser, M.H.: Identifying conserved gene clusters in the presence of homology families. J. Comput. Biol. **12**(6), 638–656 (2005)
10. Jacob, F., Perrin, D., Sanchez, C., Monod, J.: Operon: a group of genes with the expression coordinated by an operator. C. R. Hebd. Seances Acad. Sci. **250**, 1727–1729 (1960)
11. Jahn, K.: Efficient computation of approximate gene clusters based on reference occurrences. J. Comput. Biol. **18**(9), 1255–1274 (2011)
12. Larroux, C., Fahey, B., Degnan, S.M., Adamski, M., Rokhsar, D.S., Degnan, B.M.: The NK homeobox gene cluster predates the origin of Hox genes. Curr. Biol. **17**(8), 706–710 (2007)
13. Ryba, T., Hiratani, I., Lu, J., Itoh, M., Kulik, M., Zhang, J., Schulz, T.C., Robins, A.J., Dalton, S., Gilbert, D.M.: Evolutionarily conserved replication timing profiles predict long-range chromatin interactions and distinguish closely related cell types. Genome Res. **20**(6), 761–770 (2010)
14. Schmidt, T., Stoye, J.: Gecko and GhostFam: rigorous and efficient gene cluster detection in prokaryotic genomes. Methods Mol. Biol. **396**, 165–182 (2007). (Chapter 12)
15. Selvaraj, S., Dixon, J.R., Bansal, V., Ren, B.: Whole-genome haplotype reconstruction using proximity-ligation and shotgun sequencing. Nat. Biotechnol. **31**(12), 1111–1118 (2013)
16. Sexton, T., Yaffe, E., Kenigsberg, E., Bantignies, F., Leblanc, B., Hoichman, M., Parrinello, H., Tanay, A., Cavalli, G.: Three-dimensional folding and functional organization principles of the Drosophila genome. Cell **148**(3), 458–472 (2012)
17. Thévenin, A., Ein-Dor, L., Ozery-Flato, M., Shamir, R.: Functional gene groups are concentrated within chromosomes, among chromosomes and in the nuclear space of the human genome. Nucleic Acids Res. **42**(15), 9854–9861 (2014)
18. Uno, T., Yagiura, M.: Fast algorithms to enumerate all common intervals of two permutations. Algorithmica **26**(2), 290–309 (2000)
19. Wang, B.F., Kuo, C.C., Liu, S.J., Lin, C.H.: A new efficient algorithm for the gene-team problem on general sequences. IEEE/ACM Trans. Comput. Biol. Bioinform. **9**(2), 330–344 (2012)
20. Wang, B.F., Lin, C.H.: Improved algorithms for finding gene teams and constructing gene team trees. IEEE/ACM Trans. Comput. Biol. Bioinform. **8**(5), 1258–1272 (2010)
21. Wang, B.F., Lin, C.H., Yang, I.T.: Constructing a gene team tree in almost $O(n \lg n)$ time. IEEE/ACM Trans. Comput. Biol. Bioinform. **11**(1), 142–153 (2014)

22. Winter, S., Jahn, K., Wehner, S., Kuchenbecker, L., Marz, M., Stoye, J., Böcker, S.: Finding approximate gene clusters with Gecko 3. Nucleic Acids Res. **44**(20), 9600–9610 (2016)
23. Yates, A., Akanni, W., Amode, M.R., Barrell, D., Billis, K., Carvalho-Silva, D., Cummins, C., Clapham, P., Fitzgerald, S., Gil, L., Girn, C.G., Gordon, L., Hourlier, T., Hunt, S.E., Janacek, S.H., Johnson, N., Juettemann, T., Keenan, S., Lavidas, I., Martin, F.J., Maurel, T., McLaren, W., Murphy, D.N., Nag, R., Nuhn, M., Parker, A., Patricio, M., Pignatelli, M., Rahtz, M., Riat, H.S., Sheppard, D., Taylor, K., Thormann, A., Vullo, A., Wilder, S.P., Zadissa, A., Birney, E., Harrow, J., Muffato, M., Perry, E., Ruffier, M., Spudich, G., Trevanion, S.J., Cunningham, F., Aken, B.L., Zerbino, D.R., Flicek, P.: Ensembl 2016. Nucleic Acids Res. **44**(D1), D710 (2016)
24. Zhang, M., Leong, H.W.: Gene team tree - a hierarchical representation of gene teams for all gap lengths. J. Comput. Biol. **16**(10), 1383–1398 (2009)

Inferring Local Genealogies on Closely Related Genomes

Ryan A. Leo Elworth and Luay Nakhleh[(⊠)]

Department of Computer Science, Rice University,
6100 Main Street, Houston, TX 77005, USA
{ral8,nakhleh}@rice.edu

Abstract. The relationship between the evolution of a set of genomes and of individual loci therein could be very complex. For example, in eukaryotic species, meiotic recombination combined with effects of random genetic drift result in loci whose genealogies differ from each other as well as from the phylogeny of the species or populations—a phenomenon known as incomplete lineage sorting, or ILS. The most common practice for inferring local genealogies of individual loci is to slide a fixed-width window across an alignment of the genomes, and infer a phylogenetic tree from the sequence alignment of each window. However, at the evolutionary scale where ILS is extensive, it is often the case that the phylogenetic signal within each window is too low to infer an accurate local genealogy. In this paper, we propose a hidden Markov model (HMM) based method for inferring local genealogies conditional on a known species tree. The method borrows ideas from the work on coalescent HMMs, yet approximates the model parameterization to focus on computationally efficient inference of local genealogies, rather than on obtaining detailed model parameters. We also show how the method is extended to cases that involve hybridization in addition to recombination and ILS. We demonstrate the performance of our method on synthetic data and one empirical data set, and compare it to the sliding-window approach that is, arguably, the most commonly used technique for inferring local genealogies.

1 Introduction

The coalescent is a stochastic process that considers a sample of alleles from a single population and its genealogical history under certain assumptions such as those made by the Wright-Fisher (W-F) model [21]. Under the coalescent, the genealogy of a sample takes the shape of a tree that links all extant samples to their most recent common ancestor (MRCA); Fig. 1(a). This model was extended to model multiple populations related by a tree structure, giving rise to the multispecies coalescent [6].

Part of this research was conducted while RALE was funded by a training fellowship from the National Library of Medicine (Award T15LM007093; PD Lydia E. Kavraki). Furthermore, research was funded in part by NSF grants CCF-1541979 and DMS-1547433.

J. Meidanis and L. Nakhleh (Eds.): RECOMB CG 2017, LNBI 10562, pp. 213–231, 2017.
DOI: 10.1007/978-3-319-67979-2_12

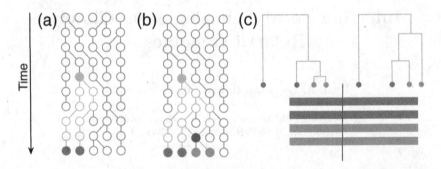

Fig. 1. The coalescent and recombination. (a) A single population of five individuals over 10 generations. The MRCA (top solid circle) of a sample of two extant individuals (two solid circles at the bottom) is shown. (b) The coalescent with recombination. The genealogy of a sample of four extant individuals (four solid circles at the bottom) from their MRCA (top solid circle) is shown. The recombination node (solid circle in the second layer from the bottom) results in an ancestral recombination graph (ARG) for the genealogy. (c) The multispecies coalescent with recombination viewed as a process along the genomes: The local genealogy changes as a recombination breakpoint (the vertical line) is encountered along the multiple sequence alignment (the four horizontal bars).

When recombination occurs, the genealogy takes the shape of a rooted, directed, acyclic graph known as an ancestral recombination graph, or ARG; Fig. 1(b). In this case, "walking" across the genome alignment, the local genealogy observed at each site could change as a recombination breakpoint is encountered; Fig. 1(c). Therefore, while the genomes evolve within the branches of a species tree (barring reticulation), individual genomic regions, or loci, across the genomes could have local genealogies that differ in shape from each other and from that of the species tree. Elucidating the species tree itself as well as the individual local genealogies is of great interest [13]. Sliding a fixed-size window across the genome alignment and building a tree on the sequence alignment pertaining to each window is the most common practice. For example, this is the approach employed in analyzing the genomes of *Staphylococcus aureus* [38], butterflies [42], and mosquitos [12], to demonstrate the extent of ILS. However, for genomes that are evolutionarily close enough for incomplete lineage sorting (ILS) to occur, the signal within a window could be problematic: for small window sizes, the signal could be weak to infer a tree with well supported branches, and for large window sizes, recombination would be extensive within the window rendering the assumption of a treelike genealogy incorrect. For example, in the work of [12], the window size was set to 50 kb, and, in a more recent study of sparrow genomes [10], the window size was even set to 100 kb. These are window sizes large enough to contain extensive recombination breakpoints within each window. There has already been debate in the community as to the size of genomic regions that would be recombination-free (or almost recombination-free) and could truly have a single underlying evolutionary tree [9,36]. In particular, it is claimed that a recombination-free locus is about 12bp or shorter.

In this work, we present a hidden Markov model (HMM) based method for inferring local genealogies from genomic data conditional on the knowledge of a species phylogeny. Wiuf and Hein [39] observed that to sample an ARG and its parameters from genomic sequences, a process can be followed that moves along the sequences and modifies the local genealogy as recombination break-points are encountered; This observation allowed for introducing the *sequentially Markov coalescent* [25] and *coalescent hidden Markov model* (coalescent HMMs) [16]. These models are very detailed and their main utility is to obtain accurate estimates of evolutionary parameters, such as ancestral population sizes and recombination rates (in most applications of these models, the local genealogies are treated as nuisance parameters). Here, we make use of the observation of [39] but depart from the detailed and computationally intensive work of [16,25], as we focus on inferring local genealogies, rather than on estimating evolutionary parameters. This departure allows us to make the computation faster for obtaining good estimates of the local genealogies. In particular, we "integrate out" the branch lengths of local genealogies in computing emission probabilities and use the Normalized Rooted Branch Score, or NRBS, distance [15] to approximate the transition probabilities. Running standard, efficient algorithms on the resulting HMM and the genomic sequence data provides estimates of local genealogies along with their posterior support. We demonstrate the utility of our model on both simulated and biological data sets and discuss its limitations, in addition to comparing its performance against the traditional window-sliding approach. Furthermore, we show how our model can be extended, with a slight modification, to account for cases of gene flow and show results of corresponding simulations with reticulate evolution.

Other earlier methods exist for elucidating local genealogies; e.g., [2,4,22, 26,29,34,40]. All this work notwithstanding, high profile analyses continue to employ the simple sliding-window approach [10,12,42]. The focus of our work here is to highlight the performance of a sliding-window approach on data sets that consist of or include closely related genomes, and demonstrate the gains one can obtain by incorporating spatial dependencies along the genome in improving the accuracy of obtained local genealogies.

2 Model and Methods

Let S be a set of aligned genomes on taxa X and S_i be the ith site in the alignment. Given a species tree Ψ on X, we denote by $H(\Psi)$ the set of all possible coalescent histories [5]; Fig. 2. Every S_i has evolved down a local coalescent, or gene, history in $H(\Psi)$. We define a hidden Markov model (HMM) $\mathcal{M}$ whose states are $q_1, q_2, \ldots, q_{|H(\Psi)|}$, such that state q_i corresponds to coalescent history $g_i \in H(\Psi)$; see Fig. 2. Each internal node of the species tree Ψ has a divergence time (in years) associated with it, such that τ_v is the divergence time of node v (all the leaves are assumed to be at time 0). Furthermore, with every branch

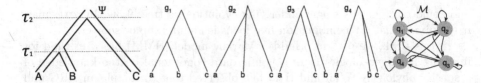

Fig. 2. A species tree Ψ on three taxa A, B, and C, and the four possible coalescent (or, gene) histories on three taxa g_1, g_2, g_3, and g_4. The HMM $\mathcal{M}$ has four states that correspond to the four coalescent histories. Divergence time τ_1 is when species A and B diverged from their MRCA and divergence time τ_2 is when species C and the MRCA of A and B diverged from their MRCA. While the coalescence times of both nodes in each of g_2, g_3, and g_4 must be larger than τ_2, the coalescence time of a and b in g_1 must be between τ_1 and τ_2.

e of the species tree we associate two quantities, N_e and g_e, which correspond to the population size and generation time for the population corresponding to that branch.

The joint probability of a sequence H of hidden states and a sequence alignment $\mathcal{S}$, both of length n, is

$$P(\mathcal{S}, H) = P(H_1) \prod_{i=1}^{n} [P(\mathcal{S}_i|H_i) \cdot P(H_i|H_{i-1})]. \tag{1}$$

The likelihood of the model $\mathcal{M}$ is

$$L(\mathcal{M}) = \sum_H P(\mathcal{S}, H) \tag{2}$$

which can be efficiently computed by the forward (or backward) algorithm [8]. Furthermore, the local genealogy and its support for each site are efficiently computed using the Viterbi, forward, and backward algorithms [8] once the model is parameterized, e.g., to maximize the likelihood. For the initial probability, the method of [5] can be used to compute $P(H_1 = q_i) = P(g_i|\Psi)$. In our analyses below, we used a uniform distribution for the initial probability given the very small number of taxa we use.

We now describe how we derive the other two terms in the equation for $P(\mathcal{S}, H)$: the emission probability $P(\mathcal{S}_i|H_i)$ and the transition probability $P(H_i|H_{i-1})$.

2.1 Emission Probabilities

The HMM emits an observation $\mathcal{S}_k$ from state q_i with probability $p(\mathcal{S}_k|q_i, \Psi, \theta)$, where θ are the parameters of the assumed substitution model. In this work, we assume the Jukes-Cantor model [20] of sequence evolution, and θ consists only of μ, the per individual substitution rate per site per generation. The role that the species tree Ψ plays in determining the emission probabilities is in that

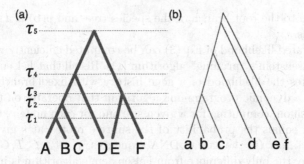

Fig. 3. (a) A species tree Ψ with times at its internal nodes in years. All the leaves are assumed to have time 0. (b) A coalescent history h, where ab coalesce in AB, abc coalesce in ABC, $abcd$ coalesce in $ABCD$, $abcde$ and $abcdef$ coalesce above the root.

it puts constraints on the ranges over which to integrate the branch lengths of the coalescent history. Let us consider species tree Ψ in Fig. 3(a) and coalescent history g in Fig. 3(b).

Given the times associated with the nodes of the tree Ψ, the lengths, in years, of the branches of history h satisfy:

- $\tau_1 \leq \ell_a, \ell_b \leq \tau_2$;
- $0 \leq \ell_{ab} \leq (\tau_4 - \tau_1)$;
- $\tau_2 \leq \ell_c \leq \tau_4$;
- $0 \leq \ell_{abc} \leq (\tau_4 - \tau_2)$;
- $\tau_4 \leq \ell_d \leq \tau_5$;
- $0 \leq \ell_{abcd} < \infty$;
- $\tau_5 \leq \ell_e < \infty$;
- $0 \leq \ell_{abcde} < \infty$; and,
- $\tau_5 \leq \ell_f < \infty$.

To turn these bounds into units of expected numbers of mutations, each bound for branch x (e.g., $x = abcd$ for the branch incoming into the MRCA of a, b, c, d in h) is multiplied by θ_x/g_x, where $\theta_x = 2N_x\mu$, where N_x is the population size associated with branch x, and g_x is the generation time (number of years per generation) for branch x.

Let ζ be a vector of upper and lower bounds on the lengths (in expected number of mutations per site per generation) of coalescent history g's branches. For every branch e in g, these bounds are derived from the constraints of the mapping between the nodes of g and the branches of the species tree as discussed above for the example in Fig. 3. Then, we have

$$p(S_k|q_i, \Psi, \theta) = p(S_k|g_i, \Psi, \theta) = \int_{\mathbf{b}} p(S_k|g_i, \zeta, \theta)p(\mathbf{b})d\mathbf{b} \qquad (3)$$

where the definite integral is taken over all branch lengths of g_i, where for branch e, the integration is over $[\zeta_{lower}, \zeta_{upper}]$, or $[\zeta_{lower}, \infty)$ in the case of gene history nodes mapped to the root branch of the species tree, and $p(\mathbf{b})$ is a prior on the branch lengths.

The integrated likelihood of Eq. (3) can be computed efficiently with a modification to Felsenstein's "pruning" algorithm [11]. Recall that Felsenstein's algorithm computes the likelihood of a gene history with fixed branch lengths. It does so using a dynamic programming algorithm that operates on the tree in a bottom-up fashion, computing for a given site and for each node v, the quantity $C(x, v)$ which equals the probability of the subtree rooted at v given that v is labeled with state x (in the case of DNA sequences, $x \in \{A, C, T, G\}$).

In our case, the only difference from Felsenstein's algorithm is in computing $C(x, v)$ for internal nodes. If v is a node whose children are u and w, then we have

$$
C(x, v) = \left(\int_{\zeta_{vu_{lower}}}^{\zeta_{vu_{upper}}} \left(\sum_y C(y, u) \cdot P_{x \to y}(b_{vu}) \right) p(b_{vu}) db_{vu} \right)
$$
$$
\cdot \left(\int_{\zeta_{vw_{lower}}}^{\zeta_{vw_{upper}}} \left(\sum_y C(y, w) \cdot P_{x \to y}(b_{vw}) \right) p(b_{vw}) db_{vw} \right) \quad (4)
$$

where $P_{x \to y}(b_{vu})$ is the probability of state x changing to state y over time that is given by the length of branch (v, u), b_{vu}. This is equal to

$$
C(x, v) = \left(\sum_y C(y, u) \cdot \left(\int_{\zeta_{vu_{lower}}}^{\zeta_{vu_{upper}}} \left(P_{x \to y}(b_{vu}) \right) p(b_{vu}) db_{vu} \right) \right)
$$
$$
\cdot \left(\sum_y C(y, w) \cdot \left(\int_{\zeta_{vw_{lower}}}^{\zeta_{vw_{upper}}} \left(P_{x \to y}(b_{vw}) \right) p(b_{vw}) db_{vw} \right) \right).
$$
$$(5)$$

Assuming an $Exp(1)$ prior on branch lengths, we have $p(b_{vu}) = e^{-b_{vu}}$. As pointed out in [1], extremely long branch lengths are not realistic and a proper branch length prior puts decreasing density on higher branch lengths as is the case with our $Exp(1)$ prior. Each of the integrations in Eq. (5) can be computed analytically under the JC model as

$$
\int_{\zeta_{e_{lower}}}^{\zeta_{e_{upper}}} (P_{x \to y}(b_e)) \, p(b_e) db_e = \int_{\zeta_{e_{lower}}}^{\zeta_{e_{upper}}} \frac{1}{4}(1 + 3e^{-4b_e/3}) e^{-b_e} db_e = -\frac{1}{4} e^{-\zeta_{e_{upper}}}
$$
$$
+ \frac{1}{4} e^{-\zeta_{e_{lower}}} - \frac{9}{28} e^{-7\zeta_{e_{upper}}/3} + \frac{9}{28} e^{-7\zeta_{e_{lower}}/3} \quad (6)
$$

for the case of $x = y$ and

$$
\int_{\zeta_{e_{lower}}}^{\zeta_{e_{upper}}} (P_{x \to y}(b_e)) \, p(b_e) db_e = \int_{\zeta_{e_{lower}}}^{\zeta_{e_{upper}}} \frac{1}{4}(1 - e^{-4b_e/3}) e^{-b_e} db_e = -\frac{1}{4} e^{-\zeta_{e_{upper}}}
$$
$$
+ \frac{1}{4} e^{-\zeta_{e_{lower}}} + \frac{3}{28} e^{-7\zeta_{e_{upper}}/3} - \frac{3}{28} e^{-7\zeta_{e_{lower}}/3} \quad (7)
$$

for the case of $x \neq y$.

2.2 Transition Probabilities

The length of a non-recombining tract before encountering a recombination breakpoint is exponentially distributed with intensity equal to the total branch length of the gene history (in units of scaled recombination distance, or centi-Morgans per megabase pair, cM/Mb) [17]. With this in mind, we approximate the transition probabilities by

$$
p(q_i, q_{i'}) = \begin{cases} 1 - \frac{1}{L*+1} & i = i' \\ \frac{log(NRBS(g_i,g_{i'})+e)}{\sum_{i,i'} log(NRBS((g_i,g_{i'}+e)))} \cdot \frac{1}{L*+1} & i \neq i' \end{cases} \tag{8}
$$

where L* is the expected tract length for the state we are currently in, or the reciprocal of the total branch length of the gene history in units of scaled recombination distance for the current state, i.e., $L* = \frac{1}{\sum_i b_i \cdot \lambda}$, where the sum is taken over all branch lengths of the genealogy. Here, we introduce the parameter λ to scale the branches of our trees to units of scaled recombination distance, which is to be learned from the data. Given that we are using integrated coalescent histories, our gene histories do not have specific branch lengths with which to derive L*. To obtain an estimate of L*, which requires specific branch lengths, we make another approximation by mapping our gene history nodes to the mid-point of the species tree branches that they coalesce inside of. This allows us to derive branch lengths which we convert to scaled recombination distance in order to finally derive L*, with gene history nodes coalescing above the species tree root being set to coalescing at the root node of the species tree.

For transitions between two different states, we make the simplifying assumption that when crossing a recombination breakpoint, the local genealogy changes to one that is very similar to the current one. We make use of the log scaled Normalized Rooted Branch Score, or NRBS, distance of [15]. The NRBS is the adaptation of the "branch score" given by [23] but for rooted trees. It gives a measure of tree similarity between two rooted trees with branch lengths. The transition probability between two different states is proportional to this distance and thus proportional to the similarity of the local genealogies in question.

2.3 Gene Flow: The Species Network Case

When the evolutionary history of the genomes under analysis involves gene flow (in addition to recombination), the species tree can be replaced by a phylogenetic network [24]. In this case, one can view every site in the genomic alignment as evolving down a local genealogy, but such a genealogy evolving within the branches of one of the parental trees inside the network [43].

One can view the phylogenetic network as a mixture of parental trees and each local genealogy evolves within one of the parental trees defined by the network [43].

Let $\Psi = \{\psi_1, \ldots, \psi_m\}$ and $H(\psi) = \{g_1, \ldots, g_r\}$ be the sets of all parental trees defined by the given phylogenetic network Ψ and all coalescent histories possible under a particular parental tree ψ, respectively. Then, the set of states of the HMM $\mathcal{M}$ in this case is $Q = \{q_{i,j} : 1 \leq i \leq m, 1 \leq j \leq r\}$, where state $q_{i,j}$ corresponds to parental tree ψ_i and coalescent history g_j. The emission probabilities in this case remain unchanged from the previous section (the constraints on the branch lengths of the coalescent history g_j in state $q_{i,j}$ come from the parental tree ψ_i). The initial probability is now calculated using the method of [41].

The transition probabilities, however, are modified slightly from the case of no gene flow. We assume that the HMM remains in a state with the same parental species tree with probability α and transitions to a state with a different parental species tree with probability $1 - \alpha$ (normalized by the number of parental species trees minus 1). Multiplying α by the functions described in the no gene flow case is sufficient to handle the case of switching gene trees while remaining in the same parental tree:

$$p(q_{i,j}, q_{i',j'}) = \frac{h(q_{i,j}, q_{i',j'})}{\sum_{u,v} h(q_{i,j}, q_{u,v})}, \tag{9}$$

where u and v iterate over all parental trees and local genealogies, respectively, and

$$h(q_{i,j}, q_{i',j'}) = \begin{cases} 1 - \frac{1}{L*+1} & i = i', j = j' \\ \alpha \cdot \frac{log(NRBS(g_i,g_{i'})+e)}{\sum_{i,i'} log(NRBS((g_i,g_{i'}+e)))} \cdot \frac{1}{L*+1} & i = i', j \neq j' \\ \frac{1-\alpha}{m-1} \cdot \frac{log(NRBS(g_i,g_{i'})+e)}{\sum_{i,i'} log(NRBS((g_i,g_{i'}+e)))} \cdot \frac{1}{L*+1} & i \neq i' \end{cases}. \tag{10}$$

The value of α is to be learned during the training of the HMM. With this approximation in our model, the HMM switching between states of different parental trees denotes entering or exiting an introgressed region.

2.4   Learning the Model Parameters

To sum it up, under our formulation, the model parameters that need to be learned are the times τ associated with the species tree's or network's nodes, the population sizes N and generation times g associated with the species tree's branches, the DNA substitution model parameters θ (μ for this study as we only consider the Jukes-Cantor model of evolution), λ, and, in the case of a phylogenetic network, α. In this work, we assume known values for N and g instead of learning them from the data. For parameter inference, we rely on hill-climbing heuristics to obtain the parameter settings that maximize the likelihood of our model, $L(\mathcal{M})$. For the experiments conducted in this study, the parameters were learned using the BOBYQA multivariate optimizer based on [31], part of the Apache Commons Math library in Java.

When *a priori* knowledge of the model parameters exists, that knowledge could be used instead of learning the corresponding parameters. Furthermore, to the best of our knowledge, no work currently exists on investigating the identifiability of parameters in coalescent HMMs. Such an investigation is worth pursuing in future research.

3   Experimental Evaluation

For various simulated and biological data sets, we trained the model and analyzed
the genomes using the Viterbi algorithm and/or the posterior decoding [32]. In
the simulated data sets, the true topology of the gene history under which each
site evolved is known. Therefore, in this case we are able to compare the true
topology of the gene histories to the ones our model is confident about based
on a combination of the forward and backward algorithms that compute the
posterior decoding [32].

To obtain the labeling of sites with local genealogies using our method, we
focused on the posterior decoding given by

$$P(H_i = g_k | \mathcal{S})$$

where H_i is the hidden state for site i in the alignment $\mathcal{S}$, and g_k is any of the
possible gene histories on the three taxa. This quantity is efficiently computable
using a combination of the forward and backward algorithms [32].

We also compare the results of our method against a commonly used method
of obtaining local genealogies where a fixed window is passed across a genomic
alignment and maximum likelihood or bayesian phylogenetic inference tools are
used to infer a local genealogy for the window. In [3,12], maximum likelihood
trees are built for every sequential non overlapping 50 kbp of their alignments.
We do the same in our analyses, comparing our results to the local genealogies
reported by RAxML [37] for sequential, non overlapping 50 kbp windows in our
simulated alignments.

3.1   Results on Simulated Data with No Gene Flow

Three-taxon genome alignments were generated by first generating gene histories with recombination using ms [18] followed by the generation of nucleotide
sequence alignments of length 500,000 using seq-gen [33] under the Jukes-Cantor
model of evolution [20]. The specified species tree provides the population structure for ms, and the various parameters were set as we describe below. The
program ms outputs a collection of gene genealogies, each corresponding to a
genomic locus (specified by the start and end locations on the genome). The
program seq-gen was then used to simulate sequence data down each genealogy (the length of the sequence alignment of each genealogy is specified by the
genomic coordinates produced by ms). Finally, the sequence alignments of the
loci are concatenated (while preserving the order produced by ms), and those are
fed to the method for inferring local genealogies.

As in [16], our simulations consisted of three populations where the parameters used approximate the human/chimp/gorilla evolutionary scenario. Here, we
label the human, chimp, and gorilla taxa as A, B, and C, respectively, and refer
back to the parameters from Fig. 2. We set τ_1 to four million years and τ_2 to
five and a half million years ago. We also use the same population parameters as
[16] with N_A, N_B, and N_C being 30,000 versus the N_{AC} and N_{ABC} population

sizes of 40,000. All generation times were set to be 25 (years per generation) and a recombination rate of r = 0.0075 was used, again following [16]. We set the mutation rate to $3.4 \cdot 10^{-8}$ from the estimates of human mutation rates in [27]. After generating the 500 kbp simulated data set, we heuristically obtained maximum likelihood estimates of the model parameters by using 160 runs of optimization allowing for up to 3,000 iterations per run.

Let K be a $1 \times 500,000$ vector indicating for each site in the alignment the true gene genealogy under which that site evolved (this is known in simulations). Let M^t be a $1 \times 500,000$ vector indicating for each site i in the alignment the gene genealogy g_s that satisfies $P(H_i = g_s|\mathcal{S}) > t$. Notice that when $t \leq 0.5$, more than one gene tree could satisfy the condition for site i. For each of the three possible gene trees g_s ($s = 1, 2, 3$), we define the true positives rate (TPR) and false positives rate (FPR) as

$$TPR(s,t) = \frac{|\{i : 1 \leq i \leq 500,000, K[i] = M^t[i] = g_s\}|}{|\{i : 1 \leq i \leq 500,000, K[i] = g_s\}|}$$

and

$$FPR(s,t) = \frac{|\{i : 1 \leq i \leq 500,000, K[i] \neq g_s, M^t[i] = g_s\}|}{|\{i : 1 \leq i \leq 500,000, K[i] \neq g_s\}|}.$$

Finally, for each of the three gene trees, g_s ($s = 1, 2, 3$), we plotted the receiver operating characteristic (ROC) curve, where t serves as the discrimination threshold. The three ROC curves for the three topologies for this simulation are shown in Fig. 4.

The results show a good performance of the method, despite its heavily simplifying approximations. To zoom in on the performance, we report on the TPR and FPR values for a few select thresholds in Table 1.

Clearly, as the discrimination threshold value increases, the true positive ratios decrease, but the false positives ratio also improves. Interestingly, the method performs consistently better in terms of correctly identifying the local genealogy g_1, whose topology is identical to that of the species tree, than either of the other two genealogies.

Alongside the analyses of our method, we split the alignment into non overlapping windows and generated maximum likelihood (ML) trees. For this, we added a fourth, outgroup taxon diverging 18 million years ago (corresponding to orangutans), also following [16], so as to enable inference of rooted three-taxon trees (the outgroup was used to root the ML trees). We used window sizes of 1 kbp, 10 kbp, and 50 kbp. In the case of window size 50 kbp, all windows for this no-gene flow case yielded the same ((a,b),c) topology in agreement with the divergence pattern of the overall population history and regardless of the true local genealogy. In the case of window size 10 kbp, only two of three possible topologies were estimated. The TPR and FPR of the window-sliding method for all three window sizes are shown in Table 1. Clearly, in these simulations, a window size of 1 kbp is the best choice, and is comparable in performance to our method with a discrimination threshold between 0.08 and 0.1. Nevertheless,

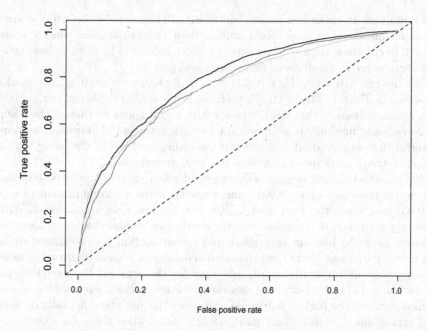

Fig. 4. ROC curves of the three gene topologies based on the posterior decoding support calculated by our method: $(((a, b), c)$ in black, $((a, c), b)$ in dark grey and $((b, c), a)$ in light grey. We used the ROC commands from the clinfun R package to generate this plot [35].

Table 1. True positive and false positive rates of our method for various posterior support thresholds, ARGweaver, and the window sliding method. The data was simulated with no gene flow, and with recombination rate $r = 0.0075$.

Discrimination threshold (t)	g_1 TPR	g_1 FPR	g_2 TPR	g_2 FPR	g_3 TPR	g_3 FPR
0.001	0.92	0.61	0.77	0.41	0.75	0.38
0.01	0.87	0.49	0.69	0.30	0.66	0.28
0.05	0.83	0.43	0.63	0.24	0.61	0.23
0.08	0.82	0.42	0.61	0.23	0.59	0.22
0.1	0.82	0.41	0.60	0.22	0.58	0.21
0.5	0.72	0.30	0.49	0.15	0.49	0.14
0.9	0.62	0.21	0.37	0.10	0.42	0.08
0.99	0.53	0.15	0.31	0.064	0.34	0.046
0.999	0.41	0.08	0.23	0.040	0.24	0.025
0.999999	0.14	0.011	0.078	0.0076	0.092	0.0067
N/A 1 kbp Windows	0.83	0.39	0.43	0.08	0.54	0.10
N/A 10 kbp Windows	0.97	0.92	0.083	0.034	0	0
N/A 50 kbp Windows	1	1	0	0	0	0
N/A ArgWeaver	0.84	0.33	0.47	0.10	0.53	0.09

it is important to make two points here. First, 10 kbp and 50 kbp window sizes are more commonly used in recent studies than 1 kbp or smaller window sizes. Second, in empirical data sets, we expect 1 kbp windows to have much less signal than we observe in simulations under our condition.

We also ran ARGweaver [34] (using its default parameter settings); the results are shown in Table 1. The method's performance is similar to that of our method with a discrimination threshold between 0.08 and 0.1 and to that of the 1 kbp window-sliding method. It is important to note here that ARGweaver is more powerful than our method and the window-sliding method in the sense that it computes many more quantities than the local genealogies.

We observed similar results of our method when varying the recombination rate as for the cases of $r = 0.0019$ and $r = 0.03$. With a recombination rate of $r = 0.03$, four times the rate used in [16], the method's accuracy drops. With extremely high rates of recombination the contiguous regions sharing an identical topology begin to become very short and recombination breakpoints start to become very frequent along the genome. As contiguous regions sharing the same topology become especially small, it is easy for there to not be enough signal for the learned HMM to switch states every few nucleotides, even with a higher learned λ value. Too high of a λ would also cause the areas that actually do have long contiguous regions to very easily switch states when there are alternative mutations by chance.

To better understand the behavior of methods for inference of local genealogies, particularly on closely related genomes, it is important to zoom in on the individual regions and assess the signal there. After all, the model local genealogy could be a fully resolved trees, but the true local genealogy would be completely unresolved if the the sequences have no substitutions. We zoomed in on a 1 kbp region of our simulated alignment from positions 14,501 to 15,500 in Fig. 5 and on a 1 kbp region from positions 58,001 to 59,000 in Fig. 6, where we see poor and excellent performance of the method, respectively.

In Fig. 5 we see three recombination breakpoints that alternate the true local genealogy, all of which are mislabeled by our method. The signal from the variable sites, however, is completely deceptive, and this figure highlights a pattern that would plague methods for inferring local genealogies. The first two regions (blue and green), are accompanied almost exclusively with variable sites which give signal for a gene history with topology ((b,c),a). We then see a very dense burst of variable sites supporting a gene history of ((a,c),b) whereas the true genealogy is the blue topology representing the ((a,b),c) gene history.

Conversely, in Fig. 6 we see our method working nearly perfectly in recovering the true gene history topologies as output by the simulation. We see two large blue and green regions recovered nearly exactly in both the Viterbi labeling and posterior decoding. Here, we see that the signal in the informative variable sites strongly supports the inferences as in the case of the extremely dense blue and green variable sites under the corresponding topology regions. Our method is even able to detect the recombination breakpoint occurring between these two regions almost exactly, within a few base pairs from the actual breakpoint.

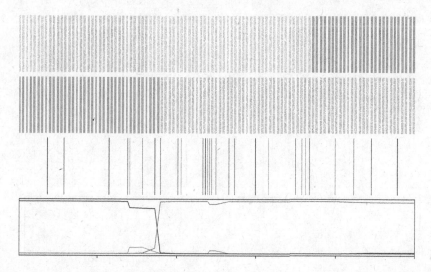

Fig. 5. Topology labels from alignment position 14,501 to 15,500 with recombination rate of r = 0.0075. (Top panel) True topologies as output by ms. (Second panel from top) Viterbi labelings from the trained HMM. (Third panel from top) Informative single nucleotide polymorphisms (sites where a and b have the same nucleotide and c is a different nucleotide shown in blue, etc. Invariant sites shown in white). (Bottom panel) Posterior support from the posterior decoding ($((a,b),c)$ in blue, $((a,c),b)$ in green and $((b,c),a)$ in red. (Color figure online)

It has been long recognized that recombinations could be undetectable due to lack of signal [19,30]. This could arise, for example, if recombination results in the crossover between two identical sequences. This issue makes it hard, if not impossible, for methods to correctly delineate recombination breakpoints and, consequently, local genealogies.

3.2   Results on Simulated Data with Gene Flow

In this case, we simulated data as above, with recombination rates $r = 0.0075$ and a migration event with migration rate $M = 0.4$ (the species phylogeny was a network obtained by adding a horizontal edge from C to B to the species tree in Fig. 2).

When we assume the evolutionary history of our genome alignment is a network, we are interested not only in the accuracy of the local gene genealogies, but also of the areas that were gained across species boundaries through introgression (in our model, these would be regions labeled by a different parental tree). In analyzing the performance of our method, in addition to posterior support for the true gene tree topologies as shown before, we are able to look at whether the method is able to correctly label regions gained from migration or introgression. For our simulated data sets, we assessed this by analyzing cases where the true simulated gene genealogy's branch lengths allow B and C to coalesce before they would otherwise be allowed to. Unfortunately, the ms software

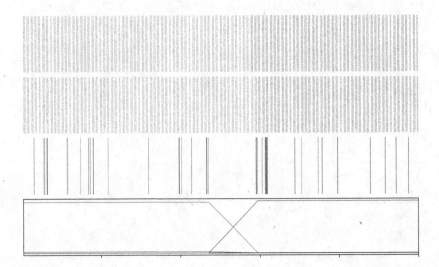

Fig. 6. Topology labels from alignment position 58,001 to 59,000 with recombination rate of $r = 0.0075$. (Top panel) True topologies as output by ms. (Second panel from top) Viterbi labels from the trained HMM. (Third panel from top) Informative single nucleotide polymorphisms (sites where a and b have the same nucleotide and c is a different nucleotide shown in blue, etc. Invariant sites shown in white). (Bottom panel) Posterior support from the posterior decoding $(((a, b), c)$ in blue, $((a, c), b)$ in green and $((b, c), a)$ in red. (Color figure online)

package does not support exact annotation of which gene trees are of introgressive origin, and modifying it to achieve this is not a simple task. Thus, our annotation of introgressed regions represents a lower bound on the true number of introgressed loci.

In this case, we focused on the performance of the method in terms of whether it elucidates that a site has been introgressed or not (that is, a binary classification problem). The ROC curve of the method is shown in Fig. 7. As above, the posterior decoding value was used as the discrimination threshold.

Once again, the method has good performance despite its simplistic approximations of the coalescent with recombination and gene flow. For example, it can achieve a TPR of about 0.8 and an FPR of about 0.3.

While the 50 kbp windows all yielded the same topology in the no gene flow case, we did see two windows with alternate topologies indicating introgressed tracts in this simulation.

3.3   Results on a Biological Data Set

In [16], regions of the human/chimp/gorilla genome alignment were analyzed and posterior support values were shown across the genome. We reevaluated a 32.6 kbp region with sufficient length and good patterns of large mixed gene genealogies to reanalyze with our new method. While in [16], the CoalHMM's

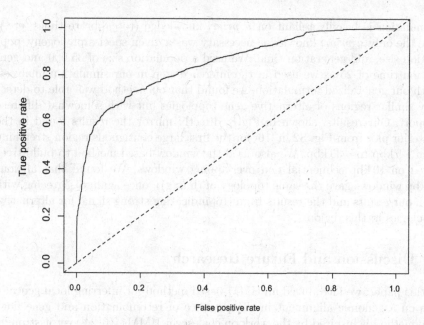

Fig. 7. ROC curve for labeling introgressed regions by our method based on varying posterior support as the discrimination threshold. We used the ROC commands from the clinfun R package to generate this plot [35].

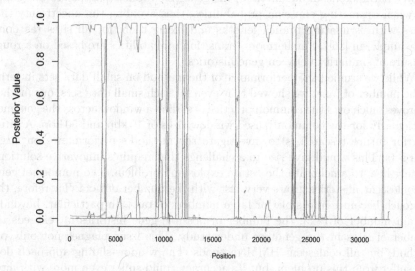

Fig. 8. Posterior support for a 32-kbp stretch of DNA obtained from human/chimp/gorilla genomes and analyzed in [16]. Posterior support for the three gene history topologies ((H,C),G), (H,(C,G)), and ((H,G),C) are in blue, green, and red, respectively. (Color figure online)

formulation is heavily reliant on *a priori* knowledge (to calibrate times, etc.), here the only *a priori* knowledge necessary was a given species phylogeny, population size, and generation time. We used a population size of 35,000 and generation time of 25 as we used in the inference step in our simulation analyses. With our generalized formulation, we found that our method was able to detect very similar regions of alternative gene topologies but with somewhat different support. Our results, shown in Fig. 8, directly mirror the results found in the posterior plot from Fig. S2 in [16] for the first large contiguous region stretching from ∼5 kbp to ∼35 kbp. We also used the window-based method to build trees based on 10 kbp sequential non overlapping windows. We found that all four 10 kbp windows gave the same topology of (h,(c,g)), once again disagreeing with both our results and the results from [16] indicating strong signal for alternative topologies in this region.

4   Discussion and Future Research

In this paper, we introduced an HMM-based method for inferring local genealogies on a genomic alignment in the presence of recombination and gene flow. The method is inspired by the work on coalescent HMMs [16,24], yet it strongly approximates the emission and transition probabilities since the focus here is on the local genealogies themselves, rather than accurate estimate of the evolutionary parameters. In our method, the gene tree branch lengths are integrated out when estimating emission probabilities, thus avoiding the uncertainty that comes with estimating branch lengths and fixing them for all sites (as done, for example, in [16]). Furthermore, transition probabilities are based on a rough measure of similarity between gene histories.

While we studied the performance of the method on small data sets in terms of the number of taxa, we showed that even for such small data sets, our method improves much over the common practice of sliding a window across the genomes, particularly for the commonly used window sizes of 10 kbp and 50 kbp. A direction for future research is to investigate the method's performance on larger data sets. This would give rise to a challenge that requires innovative solutions, namely how to ameliorate the "state explosion" problem. The number of gene, or coalescent, histories grows very fast with the number of taxa. Therefore, this approach becomes infeasible for large numbers of taxa. In particular, beyond a certain number of taxa, the number of possible gene histories far exceeds the number of sites in the genomes under study. This issue plagues not only our method, but all coalescent HMM methods. The window-sliding approach does not suffer from this problem, but its accuracy could suffer even more with more taxa. A marriage of the two approaches could provide one solution as follows. Sliding a window across the genomes yields a set of plausible local genealogies. However, this set could miss some of the potential local genealogies. Therefore, the set of trees identified by the window-sliding approach could be enriched. One way to enrich the set is to add, for example, all 1-NNI (nearest neighbor interchange) neighbors of all trees. Finally, this enriched set of local genealogies constitute the main states of the HMM. We will explore this aspect of the method.

As Hein *et al.* [14] noted, "local trees cannot be reduced to coalescent topologies or unrooted tree topologies, because trees with long branches will on average encounter a recombination breakpoint sooner than trees with short branches." Nevertheless, introducing states for every possible set of branch lengths is not possible. A strong approximation made by our method is that of the total length of a local genealogy (the b_i values in calculating the quantity $L*$). This approximation could be problematic in certain cases of larger data sets. A happy medium between integrating out branch lengths and using fixed ones must be found for all these approaches to scale. Still, an important message of this study is that sliding a window across the genomes to quantify incongruence is not the solution; after all, it could very well be the case that in cases of extensive incongruence the phylogenetic signal within a window is too small to infer accurate local genealogies. The Markov dependence on the neighboring genealogies in HMM-based approaches help ameliorate this problem as we demonstrated. However, it is important to acknowledge that at such low evolutionary scales, all these methods will run into problems of parameter (lack of) identifiability due to low signal.

In addition to our study being limited in the number of taxa, there are a number of simplifications to the evolutionary process that are assumed. Further research is required to expand the method to handle cases of gaps in genomic alignments, varied recombination and mutation rates across the genome (as in the case of recombination hotspots) and in expanding the mathematics of the integrated branch length emission probabilities to go beyond the Jukes-Cantor model. Additionally, augmenting the model to handle larger data sets in terms of the number of taxa will allow for more studies of existing empirical data sets.

Last but not least, we conducted a simple comparison to the ARGweaver method here. In future research, we will conduct a more through study of the various methods that have been devised for inferring local genealogies. In the case of gene flow, we will compare the performance of our method to others that identify introgression in sequence alignments, such as [7,28].

References

1. Alfaro, M.E., Holder, M.T.: The posterior and the prior in bayesian phylogenetics. Annu. Rev. Ecol. Evol. Syst. **37**, 19–42 (2006)
2. Boussau, B., Guéguen, L., Gouy, M.: A mixture model and a hidden Markov model to simultaneously detect recombination breakpoints and reconstruct phylogenies. Evol. Bioinf. Online **5**, 67 (2009)
3. Heliconius Genome Consortium, et al.: Butterfly genome reveals promiscuous exchange of mimicry adaptations among species. Nature **487**(7405), 94–98 (2012)
4. de Oliveira Martins, L., Leal, E., Kishino, H., Kishino, H.: Phylogenetic detection of recombination with a Bayesian prior on the distance between trees. PLoS One **3**(7), e2651 (2008)

5. Degnan, J.H., Salter, L.A.: Gene tree distributions under the coalescent process. Evolution **59**(1), 24–37 (2005)
6. Degnan, J., Rosenberg, N.: Gene tree discordance, phylogenetic inference and the multispecies coalescent. Trends Ecol. Evol. **24**(6), 332–340 (2009)
7. Durand, E.Y., Patterson, N., Reich, D., Slatkin, M.: Testing for ancient admixture between closely related populations. Mol. Biol. Evol. **28**(8), 2239–2252 (2011)
8. Durbin, R., Eddy, S., Krogh, A., Mitchison, G.: Biological Sequence Analysis: Probabilistic Models of Proteins and Nucleic Acids. Cambridge University Press, Cambridge (1998)
9. Edwards, S.V., Xi, Z., Janke, A., Faircloth, B.C., McCormack, J.E., Glenn, T.C., Zhong, B., Wu, S., Lemmon, E.M., Lemmon, A.R., Leache, A.D., Liu, L., David, C.C.: Implementing and testing the multispecies coalescent model: a valuable paradigm for phylogenomics. Mol. Phylogenet. Evol. **94**, 447–462 (2016)
10. Elgvin, T.O., Trier, C.N., Tørresen, O.K., Hagen, I.J., Lien, S., Nederbragt, A.J., Ravinet, M., Jensen, H., Sætre, G.-P.: The genomic mosaicism of hybrid speciation. Sci. Adv. **3**(6), e1602996 (2017)
11. Felsenstein, J.: Evolutionary trees from DNA sequences: a maximum likelihood approach. J. Mol. Evol. **17**(6), 368–376 (1981)
12. Fontaine, M.C., Pease, J.B., Steele, A., Waterhouse, R.M., Neafsey, D.E., Sharakhov, I.V., Jiang, X., Hall, A.B., Catteruccia, F., Kakani, E., Mitchell, S.N., Wu, Y.-C., Smith, H.A., Love, R.R., Lawniczak, M.K., Slotman, M.A., Emrich, S.J., Hahn, M.W., Besansky, N.J.: Extensive introgression in a malaria vector species complex revealed by phylogenomics. Science **347**(6217), 1258524 (2015)
13. Hahn, M.W., Nakhleh, L.: Irrational exuberance for resolved species trees. Evolution **70**(1), 7–17 (2016)
14. Hein, J., Schierup, M.H., Wiuf, C.: Gene Genealogies, Variation and Evolution. Oxford University Press, Oxford (2005)
15. Heled, J., Drummond, A.J.: Bayesian inference of species trees from multilocus data. Mol. Biol. Evol. **27**(3), 570–580 (2010)
16. Hobolth, A., Christensen, O., Mailund, T., Schierup, M.: Genomic relationships and speciation times of human, chimpanzee, and gorilla from a coalescent hidden Markov model. PLoS Genet. **3**(2), e7 (2007). doi:10.1371/journal.pgen.0030007
17. Hudson, R.R.: Gene genealogies and the coalescent process. Oxford Surv. Evol. Biol. **7**(1), 44 (1990)
18. Hudson, R.R.: Generating samples under a wright-fisher neutral model of genetic variation. Bioinformatics **18**(2), 337–338 (2002)
19. Hudson, R.R., Kaplan, N.L.: Statistical properties of the number of recombination events in the history of a sample of DNA sequences. Genetics **111**(1), 147–164 (1985)
20. Jukes, T., Cantor, C.: Evolution of protein molecules. In: Munro, H. (ed.) Mammalian Protein Metabolism, pp. 21–132. Academic Press, NY (1969)
21. Kingman, J.F.C.: The coalescent. Stochast. Processes Appl. **13**, 235–248 (1982)
22. Kosakovsky Pond, S.L., Posada, D., Gravenor, M.B., Woelk, C.H., Frost, S.D.: Automated phylogenetic detection of recombination using a genetic algorithm. Mol. Biol. Evol. **23**(10), 1891–1901 (2006)
23. Kuhner, M.K., Felsenstein, J.: A simulation comparison of phylogeny algorithms under equal and unequal evolutionary rates. Mol. Biol. Evol. **11**, 459–468 (1994)
24. Liu, K., Dai, J., Truong, K., Song, Y., Kohn, M.H., Nakhleh, L.: An HMM-based comparative genomic framework for detecting introgression in eukaryotes. PLoS Comput. Biol. **10**(6), e1003649 (2014)

25. McVean, G.A., Cardin, N.J.: Approximating the coalescent with recombination. Philos. Trans. R. Soc. London B: Biol. Sci. **360**(1459), 1387–1393 (2005)

26. Minichiello, M.J., Durbin, R.: Mapping trait loci by use of inferred ancestral recombination graphs. Am. J. Hum. Genet. **79**, 910–922 (2006)

27. Nachman, M.W., Crowell, S.L.: Estimate of the mutation rate per nucleotide in humans. Genetics **156**(1), 297–304 (2000)

28. Pease, J.B., Hahn, M.W.: Detection and polarization of introgression in a five-taxon phylogeny. Syst. Biol. **64**(4), 651–662 (2015)

29. Pond, S.L.K., Posada, D., Stawiski, E., Chappey, C., Poon, A.F., Hughes, G., Fearnhill, E., Gravenor, M.B., Brown, A.J.L., Frost, S.D.: An evolutionary model-based algorithm for accurate phylogenetic breakpoint mapping and subtype prediction in HIV-1. PLoS Comput. Biol. **5**(11), e1000581 (2009)

30. Posada, D., Crandall, K., Holmes, E.: Recombination in evolutionary genomics. Annu. Rev. Genet. **36**, 75–97 (2002)

31. Powell, M.J.: The bobyqa algorithm for bound constrained optimization without derivatives. Cambridge NA Report NA2009/06. University of Cambridge, Cambridge (2009)

32. Rabiner, L.: A tutorial on hidden Markov models and selected applications in speech recognition. Proc. IEEE **2**(2), 257–286 (1989)

33. Rambaut, A., Grass, N.C.: Seq-Gen: an application for the Monte Carlo simulation of DNA sequence evolution along phylogenetic trees. Comput. Appl. Biosci.: CABIOS **13**(3), 235–238 (1997)

34. Rasmussen, M.D., Hubisz, M.J., Gronau, I., Siepel, A.: Genome-wide inference of ancestral recombination graphs. PLoS Genet. **10**(5), e1004342 (2014)

35. Seshan, V.E.: clinfun: Clinical trial design and data analysis functions. R Package Version, **1**(6) (2014)

36. Springer, M.S., Gatesy, J.: The gene tree delusion. Mol. Phylogenet. Evol. **94**, 1–33 (2016)

37. Stamatakis, A.: RAxML-VI-HPC: maximum likelihood-based phylogenetic analyses with thousands of taxa and mixed models. Bioinformatics **22**(21), 2688–2690 (2006)

38. Takuno, S., Kado, T., Sugino, R., Nakhleh, L., Innan, H.: Population genomics in bacteria: a case study of staphylococcus aureus. Mol. Biol. Evol. **29**(2), 797–809 (2012)

39. Wiuf, C., Hein, J.: Recombination as a point process along sequences. Theor. Popul. Biol. **55**, 248–259 (1999)

40. Wu, Y.: New methods for inference of local tree topologies with recombinant snp sequences in populations. IEEE/ACM Trans. Comput. Biol. Bioinf. (TCBB) **8**(1), 182–193 (2011)

41. Yu, Y., Degnan, J.H., Nakhleh, L.: The probability of a gene tree topology within a phylogenetic network with applications to hybridization detection. PLoS Genet. **8**(4), e1002660 (2012)

42. Zhang, W., Dasmahapatra, K.K., Mallet, J., Moreira, G.R., Kronforst, M.R.: Genome-wide introgression among distantly related heliconius butterfly species. Genome Biol. **17**, 25 (2016)

43. Zhu, J., Yu, Y., Nakhleh, L.: In the light of deep coalescence: Revisiting trees within networks. BMC Genom. **17**(14), 271 (2016)

Enhancing Searches for Optimal Trees Using SIESTA

Pranjal Vachaspati and Tandy Warnow[⊠]

Department of Computer Science, University of Illinois, Urbana, IL 61801, USA
vachasp2@illinois.edu

Abstract. Many supertree estimation and multi-locus species tree estimation methods compute trees by combining trees on subsets of the species set based on some NP-hard optimization criterion. A recent approach to computing large trees has been to constrain the search space by defining a set of "allowed bipartitions", and then use dynamic programming to find provably optimal solutions in polynomial time. Several phylogenomic estimation methods, such as ASTRAL, the MDC algorithm in PhyloNet, and FastRFS, use this approach. We present SIESTA, a method that allows the dynamic programming method to return a data structure that compactly represents all the optimal trees in the search space. As a result, SIESTA provides multiple capabilities, including: (1) counting the number of optimal trees, (2) calculating consensus trees, (3) generating a random optimal tree, and (4) annotating branches in a given optimal tree by the proportion of optimal trees it appears in. SIESTA is available in open source form on github at https://github.com/pranjalv123/SIESTA.

1 Introduction

Phylogeny estimation is generally approached as a statistical estimation problem, and finding the best tree for a given dataset is typically based on methods that are computationally very intensive; for example, maximum likelihood phylogeny estimation is NP-hard [19] and Bayesian MCMC methods require a long time to converge. For this reason, among others, the calculation of very large phylogenies is often enabled by divide-and-conquer methods that use "supertree methods" to combine smaller trees into larger trees. A more common use of supertree methods is to combine trees computed by independent research groups on different datasets into a single tree on a large dataset. Supertree methods are very popular and an area of active research in the computational phylogenetics community [3].

Species tree estimation, even for small numbers of species, is also difficult because of multiple processes that create differences in the evolutionary history across the genome; examples of such processes include incomplete lineage sorting (ILS), gene duplication and loss (GDL), and horizontal gene transfer (HGT) [12]. Species tree estimation is therefore performed using multiple loci from throughout the genomes of the different organisms, and is referred to as

© Springer International Publishing AG 2017
J. Meidanis and L. Nakhleh (Eds.): RECOMB CG 2017, LNBI 10562, pp. 232–255, 2017.
DOI: 10.1007/978-3-319-67979-2_13

"phylogenomics". One of the standard approaches for species tree estimation is to compute gene trees (i.e., trees on different genomic regions) and then combine the trees together into a species tree under statistical models of evolution, such as the multi-species coalescent (which models ILS), that allow for gene tree heterogeneity. Examples of methods that construct species trees by combining gene trees and that are statistically consistent under the multi-species coalescent model include ASTRAL [15,16], GLASS [17], the population tree in BUCKy [8], MP-EST [10], NJst [9], and a modification of NJst called ASTRID [29].

This approach, called "summary methods", shares algorithmic features in common with supertree methods in that both construct trees on the set of species by combining trees on subsets of the species set; the difference between the two types of methods is that in the supertree context, the assumption is that the heterogeneity observed between these "source trees" is due only to estimation error, while in the phylogenomic context the assumption is that source trees can differ from the species tree due to a combination of estimation error and true heterogeneity resulting from ILS, GDL, HGT, or some other causes.

Summary methods and supertree methods are often based on attempts to solve NP-hard problems, and so typically use heuristics (a combination of hill-climbing and randomization) to search for optimal trees. While these heuristics can be highly effective on small datasets, they are often very slow and there are no guarantees about the solutions they find.

An alternative approach to the use of heuristic searches is constrained exact optimization, whereby the solution space is first constrained using the input source trees, and then an exact solution to the optimization problem is found within that constrained space. This approach can lead to polynomial time methods (where the running time depends on the size of the constraint space as well as on the input) that can have outstanding accuracy. The first use of this approach was presented in Hallet and Lagergren [7], which provided a method to find a species tree minimizing the duplication-loss reconciliation cost given a set of estimated gene trees. Since then, many other constrained exact optimization methods have been developed in phylogenomics for different purposes, including species tree estimation from sets of gene trees under gene duplication and loss models or under the multi-species coalescent model, or improving gene trees given a species tree [2,4,15,16,26,27,30,31].

Most of these approaches constrain the search space using a set of "allowed bipartitions", which we define here. Each edge e in an unrooted tree T on a set S of species defines a bipartition π_e of S (also called a "split"), obtained by deleting e but not its endpoints from T; hence, every tree T can be defined by its set of bipartitions $C(T) = \{\pi_e : e \in E(T)\}$. The constraints imposed by these algorithms are obtained by specifying a set X of allowed bipartitions so that the returned tree T must satisfy that $C(T) \subseteq X$. The set X is used to define a set of "allowed clades" (comprised of the halves of the bipartitions, plus the full set of species), and dynamic programming is then used on the set of allowed clades to find an optimal solution to the optimization problem. The set X has an impact on the empirical performance, but even simple ways of defining X can result in very good accuracy and provide guarantees of statistical consistency under statistical models of evolution [16,30].

The constrained exact optimization approach has multiple advantages over heuristic search techniques. From an empirical perspective, the dynamic programming approach is frequently faster, and if the constraint space is selected well it is often more accurate than alternative approaches that typically use heuristic searches for optimal solutions. From a theoretical perspective, the ability to provably find an optimal solution within the constraint space is often sufficient to prove statistical consistency under a statistical model of evolution (e.g., under the multi-species coalescent model); hence, many of the methods that use constrained exact optimization can be proven statistically consistent, even for very simple ways of defining the constraint set.

These constrained exact optimization methods typically have excellent accuracy in terms of scores for the optimization problems they address (established on both biological and simulated datasets) and topological accuracy of the trees they compute (as established using simulated datasets). A basic limitation of these methods, however, is that they return a single optimal tree, even though there can be multiple optima on some inputs. This limitation reduces the utility of the methods.

We present SIESTA (Summarizing Implicit Exact Species Trees Accurately), an algorithmic tool that can be used to enhance these dynamic programming methods for finding optimal trees. The input to SIESTA is the set $\mathcal{T}$ of source trees, the constraint set X of allowed bipartitions, and a scoring function w that assigns scores to tripartitions of the taxon set (and which is derived from the optimization function F that assigns scores to trees and the set $\mathcal{T}$, as we show later); SIESTA returns a data structure $\mathcal{I}$ that represents the set $\mathcal{T}^*$ of trees that optimize the function F subject to the constraint that every bipartition in every tree in $\mathcal{T}^*$ is in X. This data structure $\mathcal{I}$ enables the user to explore the set of optimal trees in various ways. In this study, we use SIESTA to compute consensus trees and the maximum clade credibility (MCC) tree, to count the number of optimal trees, and to report the frequency of each bipartition in the set of optimal trees. We explore the impact of using SIESTA with two methods that use the dynamic programming method for constrained exact optimization: the supertree method FastRFS [30] and the species tree estimation method ASTRAL [16], which addresses gene tree heterogeneity due to ILS.

The remainder of the paper is organized as follows. The performance study is described in Sect. 2, and the results of that study are presented in Sect. 3. We discuss the trends observed in our experiment, and the impact of using SIESTA in supertree estimation and multi-locus species tree estimation, in Sect. 4. The conclusions are presented in Sect. 5. Details of SIESTA's algorithm design and running time analysis are provided in Sect. 6. The simulated datasets analyzed in this paper are available on FigShare at [28].

2 Experiments

2.1 Overview

We tested SIESTA in conjunction with the supertree method FastRFS (using the enhanced version) and the coalescent-based species tree estimation method ASTRAL on 2 biological and 1765 simulated datasets. For each dataset we examined, we used SIESTA to compute the set of optimal solutions, and then computed consensus trees for these trees. We computed the strict consensus tree, which is the unique tree whose bipartitions appear in every optimal tree. We report the average of the FN and FP error rates.

2.2 Methods

Standard Methods. We report results for FastRFS v2.0 (using the enhanced variant, as described in [30]) on the supertree datasets, since this technique gave the best performance, and improved on MRL [18], a leading supertree method, as well as on ASTRAL-II. We also report results for ASTRAL-II (ASTRAL v4.11.1) on the phylogenomic datasets using default settings. We used RAxML v8.2.4 [23] to estimate gene trees (using options -m GTRGAMMA -p 12345) and to run MRL within FastRFS (using options -m BINGAMMA -p 12345).

ASTRAL-SIESTA. The use of SIESTA to process ASTRAL trees is called ASTRAL-SIESTA: this is the algorithm that computes the data structure for the optimal trees computed by ASTRAL, and then returns the strict consensus of the optimal trees as well as the Maximum Clade Credibility (MCC) tree. The output of ASTRAL-SIESTA can also be used for other explorations of the set of optimal trees, including annotating edges in a given candidate species tree with branch support based on the frequency of the edge appearing in the set of optimal trees. ASTRAL-SIESTA uses ASTRAL v4.11.1 (with the -q option) to compute the Maximum Clade Credibility (MCC) tree, which is based on the ASTRAL-II posterior support values [21].

FastRFS-SIESTA. The use of SIESTA to process FastRFS trees is called FastRFS-SIESTA: this is the algorithm that computes the data structure for the optimal trees computed by FastRFS, and then returns the strict consensus of the optimal trees. The output of FastRFS-SIESTA can also be used for other explorations of the set of optimal trees, including annotating edges in a given candidate supertree with branch support based on the frequency of each edge appearing in the set of optimal trees. FastRFS-SIESTA uses ASTRAL v4.7.8 to compute the set X of allowed bipartitions (using the option -k searchspace_norun).

Simulated Supertree Datasets. The simulated supertree datasets were originally provided in [25], and have been used to explore the accuracy of several supertrees methods [18,30]. We explore the results on the datasets with 1000 taxa, which are the hardest datasets in this collection; results on 100 and 500 taxa are shown in the supplement. Each replicate contains one "scaffold" tree and several clade-based trees. The scaffold tree is based on a random sample of the species, and contains 20%, 50%, 75%, or 100% of the taxa sampled uniformly at random from the leaves of the tree. The clade-based trees are based on a clade and then a birth-death process within the clade (and hence may miss some taxa). The original 100-taxon, 500-taxon, and 1000-taxon datasets had 6, 16, and 26 source trees respectively; the number of source trees was reduced to 6, 11, and 16 for the 500-taxon datasets, and 6, 11, 16, 21, and 26 for the 1000-taxon datasets. Sequences evolved down each scaffold and clade-based source tree under a GTR+Gamma model (selected from a set of empirically estimated parameters) with branch lengths that are deviated from the strict molecular clock. Maximum likelihood trees were estimated on each sequence alignment using RAxML under the GTRGAMMA model (with numeric parameters estimated by RAxML from the data), and used as source trees for the experiment. 10 replicates were analyzed for each scaffold factor of the 1000-taxon model condition.

Simulated Phylogenomic Datasets. The simulated phylogenomic datasets are from [16]; the gene trees were generated by SimPhy [14] and the sequences evolved down the gene trees using Indelible [5]. The species trees are randomly generated, and gene trees evolve within the species trees under the multi-species coalescent model; hence there is gene tree heterogeneity resulting from ILS in these datasets. Three levels of ILS were generated, characterized by the average normalized bipartition distance (AD) between true gene trees and true species trees: a moderate ILS condition (AD = 12%), a high ILS condition (AD = 31%), and a very high ILS condition (AD = 68%). These datasets have a speciation rate of 10^{-6}, resulting in speciation close to the tips of the model trees (recent divergence). Sequences evolved down each gene tree under a GTR+Gamma model with branch lengths that are deviated from the strict molecular clock.

These datasets were then modified for the purposes of this study. These datasets originally had 200 taxa each, but were randomly reduced to 50 taxa each to reduce the running time. The original datasets had variable length loci between 300 and 1500 bp, and were truncated for this experiment to 150 bp to produce datasets with properties that are consistent with empirical phylogenomic datasets (which frequently have very low phylogenetic signal). Each replicate was evaluated with 5, 10, and 25 loci. We evaluated model conditions where each gene contained all 50 taxa, as well as model conditions where each gene contained 10, 20, or 30 taxa chosen at random from the taxon set.

Gene trees were estimated on each sequence alignment using RAxML [23] under the GTRGAMMA model (with numeric parameters estimated by RAxML), and we analyzed 25 replicates for each model condition (defined by the ILS level, number of loci, and amount of missing data).

Overall, we examined 900 simulated phylogenomic datasets and 600 simulated supertree datasets.

Biological Phylogenomic Datasets. We analyzed two phylogenomic datasets on which ASTRAL had at least two optimal trees: a Sigmontidine rodent dataset

[13] with 285 species and 11 genes, and a Hymenoptera dataset [22] with 21 species and 24 genes. The Sigmontidine rodent dataset has 72 optimal ASTRAL trees and the Hymenoptera dataset has 4 optimal ASTRAL trees.

Performance Criteria. For the simulated datasets, we evaluate accuracy of the strict consensus trees in comparison to the accuracy of a single optimal tree returned by the default usage of either FastRFS or ASTRAL. We report both the false negative (FN) rate and the false positive (FP) rate, with respect to the model tree; the FN rate is the number of bipartitions in the model tree that are missing from the estimated tree and the FP rate is the number of bipartitions in the estimated tree that are not in the model tree, each divided by $n-3$ where n is the total number of leaves in the model tree. For each tree estimation method (i.e., ASTRAL and FastRFS), we report Delta-Error, which is the difference between the average of its FN and FP error rates, and the error rate of the strict consensus of the optimal trees. Hence, when Delta-Error is negative, the strict consensus has overall lower error than a single optimal tree. For the biological datasets, since topological accuracy cannot be assessed completely, we describe differences between the consensus trees we compute using SIESTA and trees computed using other techniques. We report the number of optimal trees for the optimization problems on all the datasets we examine. DendroPy [24] was used to measure tree error.

3 Results

3.1 Simulated Supertree Data

Topological Accuracy of Estimated Supertrees. Figure 1 shows a comparison on 1000-taxon simulated supertree datasets between the strict consensus tree computed by FastRFS-SIESTA and a single best FastRFS tree; note the FastRFS-SIESTA is more accurate than FastRFS for all scaffold factors, with the largest improvements when the scaffold factor is the smallest. The same trends hold on the 100- and 500-taxon datasets, as seen in Supplementary Fig. 7.

Figure 2 shows FN and FP rates separately for the 1000-taxon supertree datasets, and how they are impacted by the number of optimal trees. As expected, the FP rates decrease and the FN rates increase using the strict consensus tree as the number of optimal trees increases. However, as the number of optimal trees increases, the decrease in FP rate is substantially larger than the increase in FN rate. As a result, while there is generally a benefit in using FastRFS-SIESTA, the benefit increases with the number of optimal trees. The same trends hold on the 100- and 500-taxon datasets, as seen in Supplementary Figs. 7 and 8.

Number of Optimal FastRFS Trees. Both variants of FastRFS (the enhanced and basic forms) have a large number of optimal trees on these supertree datasets, as seen in Table 4. Datasets with 100 taxa typically have tens or hundreds of optimal solutions, but the number of optimal trees increases with the number

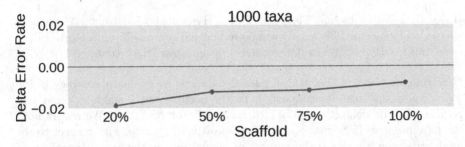

Fig. 1. FastRFS-SIESTA is more accurate than FastRFS. We show Delta-error (change in mean topological error between FastRFS and the strict consensus of FastRFS trees) on simulated supertree datasets with 1000 species; values below 0 indicate that the strict consensus FastRFS is more accurate (i.e., it has lower error) than FastRFS. The figure shows how the percentage of taxa in the scaffold source tree impact accuracy, averaged over 10 replicates. Error bars indicate the standard error; the topological error is the average of the FN and FP error rates. Results on other numbers of species show the same trends.

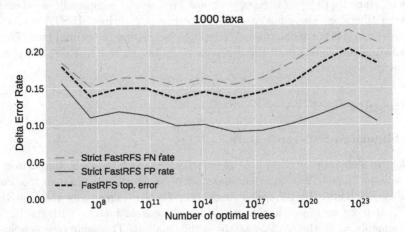

Fig. 2. Mean error rates for a single FastRFS tree and the strict consensus of all FastRFS trees on the supertree datasets with 1000 species, compared to the number of optimal trees. We show FP and FN rates (maximum error rate is 1.0) for each method; these are equal for default FastRFS (because it is always binary), but can be different for the strict consensus of the FastRFS trees. The decrease in the FP rate is larger than the increase in the FN rate for the strict consensus of the FastRFS trees, as the number of optimal trees increases, explaining why the average error for the strict consensus of the FastRFS trees is lower than for a single FastRFS tree (as shown in Fig. 1). Results for 193 replicates are shown.

of species, so that with 1000 species there are up to 10^{18} optimal trees. Most of the supertree datasets have a sparse scaffold and not too many source trees, and these factors generally (but not always) increase the number of optimal trees.

3.2   Simulated Phylogenomic Data

We examined two types of simulated datasets: the first type is where all the gene trees are complete (i.e., all species are present in all the genes), and the other type is where we deleted random species from the genes, so that all genes are missing the same number of species.

As shown in Supplementary Tables 1, 2 and 3, when all the gene trees are complete, nearly all the analyses produced only one optimal ASTRAL tree, and when more than one tree was produced it was typically a very small number (often just two). For these datasets, there was essentially no difference between the strict consensus and a single ASTRAL tree, as the strict consensus tree usually lost only one edge, and whether it was a false positive or a true positive the error rate was changed only slightly.

The situation changes for the datasets with incomplete gene trees: there are many optimal ASTRAL trees (see Supplementary Tables 1, 2 and 3). Furthermore, when there are many optimal trees, the average error rates for the strict consensus of the ASTRAL trees are lower than the error rate for a single ASTRAL tree: Fig. 3 shows results under the highest ILS condition as a function of the amount of missing data, and Fig. 4 shows this as a function of the number of optimal trees. The trends are the same under lower ILS conditions (Supplementary Figs. 9 and 10).

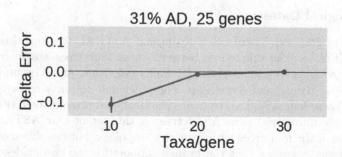

Fig. 3. The strict consensus of ASTRAL trees is more accurate than ASTRAL. We show Delta-error (change in mean topological error between FastRFS and the strict consensus of FastRFS trees) on simulated phylogenomic datasets with 25 incomplete gene trees with three different ILS levels; values below 0 indicate that the strict consensus ASTRAL is more accurate (i.e., it has lower error) than ASTRAL. Note how the percentage of taxa in each gene tree impact accuracy. We show results for 25 replicates. Error bars indicate the standard error; topological error is the average of the FN and FP error rates.

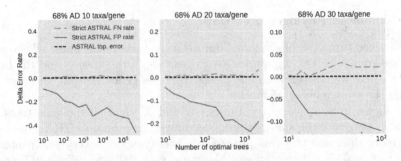

Fig. 4. Mean change in error between the strict consensus of the ASTRAL trees compared to a single ASTRAL tree on the 50-taxon phylogenomic datasets with high ILS and varying degrees of missing data, as a function of the number of optimal trees. Values below zero indicate that the strict consensus tree has better accuracy (lower error) than a single ASTRAL tree. We show the change in FP rates (blue, solid line) and in FN rates (red, dashed); the black line represents the baseline. This figure shows that the strict consensus has lower false positives than a single ASTRAL tree and higher false negative, but also that the reduction in false positives is larger than the increase in false negatives. The figure also shows that the reduction in false positives increases with the number of optimal trees. (Color figure online)

3.3   Biological Datasets

Hymenoptera Dataset. There are four optimal ASTRAL trees on this dataset (shown in Fig. 5). The differences between these four trees are restricted to two clades with three species each: (1) Solenopsi, Apis, and Vesputal_C, and (2) Acyrthosi, Myzus, and Acyrthosp. The strict and majority consensus trees (Fig. 6) on these four ASTRAL trees are identical, and present these two groups as completely unresolved. The MCC tree on this set of four ASTRAL trees matches one of the four trees with respect to topology, but has different branch support on the edges, so that the branch support for the two clades in question are halved in comparison to the four ASTRAL trees; thus, the MCC tree correctly identifies these clades as having very low support.

Sigmontidine Rodent Dataset. The species tree computed on this dataset in [13] was a concatenated Bayesian tree using MrBayes [20], with branch support based on posterior probabilities. We used the approach detailed in Sect. 6.3 to further analyze the Sigmontidine rodent dataset [13], which had 72 optimal ASTRAL trees. The ASTRAL MCC tree is highly unresolved after collapsing edges with less than 75% support. This dataset has 285 taxa, meaning that a fully resolved tree would have 282 internal edges; the collapsed ASTRAL MCC tree has only 74 internal nodes. By comparison, the MrBayes tree has 223 internal nodes after collapsing edges with less than 75% support.

Comparing the MrBayes tree with the ASTRAL MCC tree, we find that 64 bipartitions are present and highly supported in both trees. After collapsing

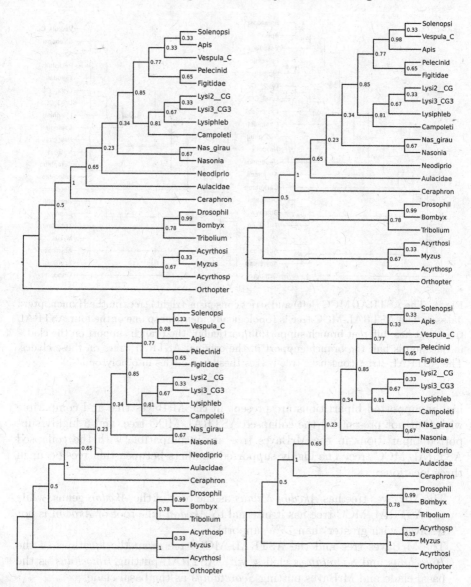

Fig. 5. The four optimal ASTRAL trees on the Hymenoptera dataset, each rooted at the outgroup, and given with local posterior probabilities for branch support. The four trees differ only in two groups: (1) Solenopsi, Apis, and Vesputal_C, and (2) Acyrthosi, Myzus, and Acyrthosp.

the edges with lower support, we are left with only the high support edges. Six highly supported bipartitions are present in the ASTRAL MCC tree and compatible with the collapsed MrBayes tree, and three bipartitions are present in the ASTRAL MCC tree and incompatible with the collapsed MrBayes tree. 153

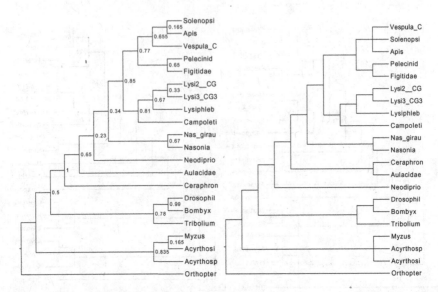

Fig. 6. The ASTRAL MCC (left) and strict consensus (right) trees on the Hymenoptera dataset. The ASTRAL MCC tree is topologically identical to one of the four ASTRAL trees, but has different branch support; in particular, the branch support on the clades in question is half the branch support in the original ASTRAL trees on these clades. The ASTRAL strict consensus tree makes these two clades into polytomies.

highly supported bipartitions are present in the MrBayes tree and compatible with (but not present in) the collapsed ASTRAL MCC tree, and 5 highly supported bipartitions in the MrBayes tree are incompatible with the collapsed ASTRAL MCC tree. The highly supported conflicts between the trees occur in three locations:

1. The MrBayes tree has *Akodon Mimus* as the root of the *Akodon* genus, while the ASTRAL MCC tree has it internal to *Akodon* (the root of *Akodon* is not resolved with greater than 75% support).
2. The MrBayes tree and the ASTRAL MCC tree swap the locations of the *Holochilus* and *Sooretamys* clades, with ASTRAL putting *Holochilus* as the basal clade and MrBayes putting *Sooretamys* as the basal clade.
3. The ASTRAL MCC tree and the MrBayes tree disagree about some resolutions within the *Oligoryzomys* clade.

These placements are in general not well established in the literature [1,6,11], and so it is not clear which of the two trees is more likely to be correct.

4   Discussion

We studied SIESTA in conjunction with ASTRAL and FastRFS on a collection of biological and simulated datasets, mainly focusing on using SIESTA to compute the strict consensus of the set of optimal trees. This study showed that using SIESTA to compute the strict consensus produced a benefit for some methods in some cases, but not in all. The trends we observed clearly indicate that when there are many optimal trees, the use of the strict consensus tree results in a substantial reduction in the false positive rate and a lesser increase in the false negative rate, for an overall reduction in topological error. Conversely, when there are only a small number of optimal trees, there is little change between the strict consensus tree and any single optimal tree. Thus, the impact of using the strict consensus depends on the number of optimal solutions. We also saw that the number of optimal trees depends on the amount of missing data, so that the benefit of using SIESTA to compute the strict consensus seems to be reliable only when there is missing data.

The study also showed that FastRFS typically benefited from using the strict consensus tree, while ASTRAL's benefit varied with the dataset. To some extent this is a natural consequence of using FastRFS on the supertree datasets, all of which have substantial amounts of missing data, and we used ASTRAL on the phylogenomic datasets, most of which had no missing data. However, a comparison of ASTRAL and FastRFS on the same datasets shows that ASTRAL tends to have fewer optimal trees than FastRFS (Table 4).

The reason that FastRFS tends to have more optimal solutions than ASTRAL, even on the same datasets, is probably that the number of possible FastRFS scores is substantially smaller than the number of possible ASTRAL scores. Specifically, if n is the number of species and k is the number of source trees, the FastRFS scores are all integers in the range $[0, (n-3)k]$, while the possible ASTRAL scores are integers in the range $[0, k\binom{n}{4}]$. Therefore, the frequency of multiple trees with the same optimal score is higher for FastRFS than for ASTRAL.

5   Conclusions

SIESTA is a simple technique for computing a data structure that implicitly represents a set of optimal trees found during the dynamic programming algorithms used by ASTRAL and FastRFS, but SIESTA is generalizable to any algorithm that uses the same basic dynamic programming structure. Once the data structure is computed, it can be used in multiple ways to explore the solution space. In particular, it can be used to count the number of optimal solutions and determine the support for a particular bipartition, thus enabling the estimation of the support on branches for a given optimal tree that takes into account the existence of other optimal trees. This study showed that SIESTA generally improves topological accuracy when the number of optimal trees is not too small, and that otherwise it allows the user to confirm that the solution that is returned is highly supported.

An interesting application of SIESTA is to produce better statistical support values on the edges of ASTRAL trees. In the current implementation of ASTRAL, the support values are obtained using posterior probabilities based on quartet trees around an edge in a single optimal tree. However, a simple example can explain why this can be misleading. Suppose T_1 and T_2 are the only trees that are optimal for ASTRAL, and that T_1 has a split π that T_2 does not have. Then under the assumption that T_1 and T_2 are both equally likely to be the true species tree, the *maximum* probability that π can be a true split is 0.5 – since it is in only one optimal tree. It is easy to see that any support value greater than 0.5 produced when T_1 is examined is inflated, and that a correction must be made that takes into consideration that T_2 is also an optimal tree. SIESTA's way of calculating support explicitly enables this correction, since it explicitly considers the support of each bipartition obtained from the entire set of optimal trees.

6   The SIESTA Algorithm

6.1   The Dynamic Programming Approach to Constrained Optimization

We begin with a review of the fundamentals of the dynamic programming algorithms for the constrained optimization problems. Recall that in the constrained optimization approach, the input is a set of source trees (estimated gene trees in the case of ASTRAL, generic source trees in the case of FastRFS) as well as a set X of allowed bipartitions of the set S of species. Given this set X of allowed bipartitions, we define a set $\mathcal{C}$ of "allowed clades" by taking the two halves of each bipartition, and we also include the set S; thus, $\mathcal{C} = \{A : [A|S \setminus A] \in X\} \cup \{S\}$.

We also form a set $TRIPS$ of "allowed tripartitions", as follows. $TRIPS$ contains all ordered 3-tuples (A, B, C) of allowed clades that are pairwise disjoint, that union to S, and where $A \cup B$ is also an allowed clade. We require that A and B be non-empty, but we allow C to be empty.

The purpose of creating this set is that it allows us to perform the dynamic programming algorithm to find optimal solutions for some optimization problems. To see this, consider an unrooted binary tree T that is a feasible solution to the constrained optimization problem under consideration. Now root the tree T arbitrarily and pick some internal node v defining clade c. Since T is a feasible solution to the optimization problem, all the clades in $T^{(r)}$ (the rooted version of T) are allowed clades, and every vertex v defining clade c that is not a leaf has two major subclades A and B defined by its two children. The 3-tuple (A, B, C) where $C = S \setminus (A \cup B)$ is the tripartition associated to node v (equivalently, associated to clade c). If v is the root of T, then C will be empty. The set of "allowed tripartitions" is defined to ensure that it includes all 3-tuples that could be formed in this way. Finally, by construction, we consider (A, B, C) and (B, A, C) to be equivalent tripartitions.

Similarly, given a rooted binary tree $T^{(r)}$ on leafset S, each non-leaf node v in $T^{(r)}$ defines a tripartition (A, B, C) where A and B are the clades (i.e., leafsets) below the two children of v, and $C = S \setminus (A \cup B)$. We refer to the set of tripartitions of a rooted binary tree $T^{(r)}$ by $trips(T^{(r)})$.

Hence, the objective of the constrained optimization problems is to find an unrooted tree T^* on leafset S that optimizes a function $F(\cdot)$ defined on unrooted trees, subject to T^* drawing its bipartitions from X. Hence, if we root T^*, we obtained a rooted tree $T^{*(r)}$ in which the non-leaf nodes define allowed tripartitions.

ASTRAL and FastRFS are each algorithms that find optimal binary trees for some optimization problem, subject to the constraint that the tree draw its bipartitions from a set X of allowed bipartitions. These algorithms reframe the problem by seeking a rooted tree that draws its clades (i.e., subsets of leaves defined by internal nodes) from the set $\mathcal{C}$ of allowed clades, and use the dynamic algorithm design that we will now describe.

For both ASTRAL and FastRFS, it is possible to define a function w on allowed tripartitions such that for any unrooted binary tree T on leafset S, letting T^r denote a rooted version of T (obtained by rooting T on any edge),

$$F(T) = \sum_{t \in trips(T^r)} w(t) \qquad (1)$$

where $F(T)$ is the optimization score for tree T.

The existence of a function w that is defined on tripartitions and that satisfies Eq. 1 is the key to these dynamic programming algorithms. Given function w that is defined on tripartitions, we define a recursive function f that is defined on clades that we can then use to find optimal solutions. We show how to define f for a maximization problem; defining it for a minimization problem is equivalently easy.

The calculation of $f(c)$ for a given allowed clade c given w and X uses the following recursion (phrased here in terms of maximization):

$$f(c) = \begin{cases} \max\{f(a) + f(b) + w(a,b,x) | (a,b,x) \in TRIPS, a \cup b = c\}, & |c| > 1 \\ 0, & |c| = 1 \end{cases}$$

By Eq. 1, $f(S) = F(T^*)$, where T^* is the optimal solution to the constrained optimization problem.

Hence, we can solve the optimization problem using dynamic programming. We compute all the $f(c)$ from the smallest clades to the largest clade S. To construct the optimal solution T^*, when we compute $f(c)$ for a clade c, we record how we obtained this best score (i.e., we record the unordered pair (a, b) of clades whose union is c achieving this optimal score), and we use backtracking to construct the rooted version of T^*. Then we unroot the rooted tree.

6.2 The SIESTA Data Structure

SIESTA modifies these algorithms so they output a set containing all the optimal trees that contain only clades in $\mathcal{C}$. When computing $f(c)$, instead of recording a single split of the clade c into two subclades that generates an optimal score, we record every such split of c. To achieve this, we show how we can represent the entire set of optimal trees computed during the algorithm with a novel data structure.

A rooted binary tree can be stored as a collection of nodes, where each node contains either two pointers (one to each of its two children, if it is an internal node) or a taxon label (if it is a leaf node). Since each node in a rooted binary tree with leaves labelled by S can be represented by a clade, this representation of a tree can be seen as having pointers from each clade c (with at least two species) to a pair of disjoint clades c_1 and c_2, whose union is c.

We modify this representation to compactly represent a set of rooted binary trees, using the correspondence between nodes in rooted trees and clades, as follows. Instead of having each clade have a pair of pointers to two sub-clades, we have each clade have a *set* of pairs of pointers to a potentially large number of sub-clades. We denote the set of pairs of pointers for clade c by $\mathcal{I}[c]$. Thus, the entire data structure is the array $\mathcal{I}$ indexed by the clades in $\mathcal{C}$.

Given such a representation, it is easy to generate any single tree by following a path from the entry $\mathcal{I}[S]$ down to the leaves, and at each clade corresponding to a non-leaf node, choosing one of the pairs of pointers in its set.

The asymptotic running time of this phase is equal to the asymptotic running time of the original DP algorithm, which is $O(|X|^2\alpha)$, where α is the time required to calculate w for a single tripartition [15]. Storing the entire data structure requires $O(|X|^2)$ space in the extreme case where every tree has the same score, but in many real-world cases will require less.

6.3   Using SIESTA

We show how we can use SIESTA in various ways, including counting the number of optimal trees, generating greedy, strict, and majority consensus trees, and computing the maximum clade credibility tree.

Counting the Number of Optimal Trees. We traverse the collection of allowed clades from smallest to largest, calculating for each allowed clade c the number of optimal rooted binary trees that contain exactly the taxa in c. Obviously, clades of size 1 have exactly one optimal rooted binary tree. For larger clades c, the following expression gives the number of optimal subtrees:

$$optimalsubtrees(c) = \sum_{(x,y)\in\mathcal{I}[c]} optimalsubtrees(x) \cdot optimalsubtrees(y) \quad (2)$$

The number of optimal rooted binary trees is $optimalsubtrees(S)$, where S is the entire set of species. For the algorithms we consider (ASTRAL and FastRFS), all rootings of a particular unrooted tree have the same criterion score, and so this quantity should be divided by $2n-3$, where $n = |S|$ is the number of species, to get the number of optimal unrooted trees.

Calculating Consensus Trees. A particular clade c is present in fraction A_c of the optimal trees, where

$$A_c = \frac{optimalsubtrees(c) * optimalsubtrees(S \setminus c)}{optimalsubtrees(S)} \tag{3}$$

For $\alpha \geq 0.5$, the α-consensus tree is the unique tree that contains exactly those bipartitions that occur in more than fraction α of the optimal trees. For smaller values of α, we can still construct a consensus tree, but the set of bipartitions that appear with frequency greater than α may not form a tree. To construct the α-consensus tree, we sort the clades in descending order by A_c, restricted only to those clades c with $A_c > \alpha$, and construct a greedy consensus tree using this ordering. The asymptotic running time of this phase is $O(|X| \log |X|)$.

The ASTRAL Maximum Clade Credibility Tree. ASTRAL-2 uses a quartet-based local posterior probability (PP) measure [21] to assign support values to edges. We can enhance this technique by outputting every tree in the space of optimal trees, assigning support local PP values to their edges using ASTRAL-2, then computing the average support of each clade (where a tree without a certain clade contributes a support of zero), and taking a greedy consensus of the resulting clades ranked by their average support over all optimal trees. In other words, we greedily compute a maximum clade credibility tree over all optimal trees, and we refer to this as the ASTRAL MCC tree.

Acknowledgments. We thank the anonymous reviewers for their helpful criticisms on an earlier draft, which greatly improved the manuscript. We also thank Erin Molloy, Sarah Christensen, and Siavash Mirarab, for feedback on the initial results.

Funding. This study made use of the Illinois Campus Cluster, a computing resource that is operated by the Illinois Campus Cluster Program in conjunction with the National Center for Supercomputing Applications and which is supported by funds from the University of Illinois at Urbana-Champaign. This work was partially supported by U.S. National Science Foundation Graduate Research Fellowship Program under Grant Number DGE-1144245 to PV and U.S. National Science Foundation grant CCF-1535977 to TW.

Supplementary Materials

Table 1. We show the mean number of optimal trees for ASTRAL, averaged over 25 replicates of 50-taxon simulated datasets with 5 genes that vary in the level of missing data. AD12 is moderate ILS, AD31 is high ILS, and AD68 is very high ILS.

# genes	5	5	5	5
# taxa per gene	10	20	30	50
50tx-AD12	286.7	707.4	24.1	2.1
50tx-AD31	171.5	210.2	15.5	1.6
50tx-AD68	176.1	154.9	12.2	1.2

Table 2. We show the mean number of optimal trees for ASTRAL, averaged over 10 replicates of 50-taxon simulated datasets with 10 genes that vary in the level of missing data. AD12 is moderate ILS, AD31 is high ILS, and AD68 is very high ILS.

# genes	10	10	10	10
# taxa per gene	10	20	30	50
50tx-AD12	132715.2	700.7	17.0	1.1
50tx-AD31	81694.2	612.2	15.8	1.0
50tx-AD68	16673.0	192.5	3.6	1.1

Table 3. We show the mean number of optimal trees for ASTRAL, averaged over 10 replicates of 50-taxon simulated datasets with 25 genes that vary in the level of missing data. AD12 is moderate ILS, AD31 is high ILS, and AD68 is very high ILS.

# genes	25	25	25	25
# taxa per gene	10	20	30	50
50tx-AD12	17958863.0	46.8	1.8	1.0
50tx-AD31	278584.5	10.3	1.4	1.0
50tx-AD68	107973.8	24.2	1.4	1.0

Table 4. Number of optimal trees (in scientific notation) for ASTRAL, FastRFS-basic, and FastRFS-enhanced on SMIDgen simulated supertree data sets with varying numbers of taxa and genes, and differing scaffold factors. ASTRAL has several orders of magnitude fewer optimal trees than FastRFS-basic and FastRFS-enhanced.

# taxa	# genes	Scaffold	ASTRAL	FastRFS-basic	FastRFS-enh
100	6	20%	9.36	3.52×10^2	2.23×10^3
100	6	50%	4.00	1.31×10^2	8.66×10^3
100	6	75%	1.72	7.27×10^1	1.70×10^2
100	6	100%	1.04	2.49×10^1	3.54×10^1
500	6	20%	2.72×10^2	3.17×10^7	1.53×10^9
500	6	50%	7.93×10^1	1.27×10^9	1.60×10^{10}
500	6	75%	1.09×10^1	5.16×10^9	8.84×10^{10}
500	6	100%	1.00	8.24×10^7	1.56×10^8
500	11	20%	5.18×10^2	8.18×10^7	1.07×10^{10}
500	11	50%	4.91×10^1	1.40×10^8	5.64×10^9
500	11	75%	2.92×10^1	1.89×10^8	1.32×10^{10}
500	11	100%	1.00	7.61×10^7	1.28×10^8
500	16	20%	1.62×10^3	6.09×10^7	4.91×10^{10}
500	16	50%	3.94×10^1	1.97×10^8	2.20×10^9
500	16	75%	4.23×10^1	1.37×10^8	1.37×10^9
500	16	100%	1.00	5.36×10^6	2.60×10^7
1000	6	20%	3.28×10^2	6.26×10^6	4.47×10^{11}
1000	6	50%	3.62×10^2	1.46×10^{11}	1.40×10^{12}
1000	6	75%	8.52×10^1	3.47×10^{11}	2.46×10^{12}
1000	6	100%	1.00	2.77×10^{11}	5.96×10^{11}
1000	11	20%	2.85×10^3	1.61×10^{10}	5.39×10^{16}
1000	11	50%	3.72×10^2	1.29×10^{14}	8.28×10^{16}
1000	11	75%	2.54×10^2	1.95×10^{13}	1.11×10^{15}
1000	11	100%	1.00	1.39×10^{14}	4.18×10^{14}
1000	16	20%	1.08×10^5	1.77×10^{17}	5.70×10^{25}
1000	16	50%	3.92×10^3	9.50×10^{17}	1.59×10^{20}
1000	16	75%	2.59×10^2	4.22×10^{15}	2.33×10^{18}
1000	16	100%	1.00	4.19×10^{14}	2.05×10^{15}
1000	21	20%	2.92×10^5	2.73×10^{16}	2.94×10^{22}
1000	21	50%	2.43×10^4	3.70×10^{14}	2.17×10^{20}
1000	21	75%	5.35×10^2	2.09×10^{14}	1.51×10^{20}
1000	21	100%	1.00	8.28×10^{13}	7.21×10^{14}
1000	26	20%	6.48×10^5	2.32×10^{15}	8.48×10^{20}
1000	26	50%	3.60×10^4	9.17×10^{14}	1.89×10^{23}
1000	26	75%	5.67×10^2	2.51×10^{14}	5.96×10^{19}
1000	26	100%	1.00	1.97×10^{13}	1.39×10^{14}

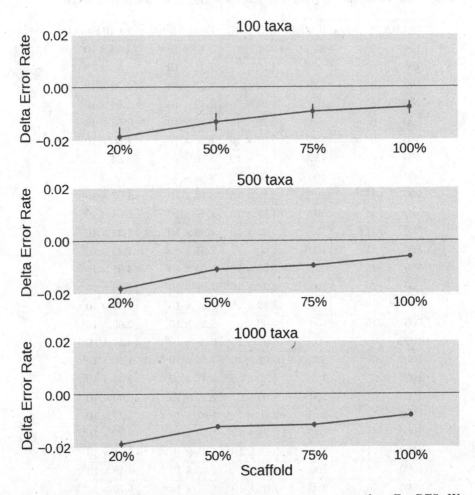

Fig. 7. The strict consensus of FastRFS trees is more accurate than FastRFS. We show Delta-error (change in mean topological error between FastRFS and the strict consensus of FastRFS trees) on simulated supertree datasets with 100, 500, and 1000 species; values below 0 indicate that the strict consensus FastRFS is more accurate (i.e., it has lower error) than FastRFS. The figure shows how the percentage of taxa in the scaffold source tree impact accuracy, averaged over 10 replicates for 1000-taxon data and 25 replicates for 100- and 500-taxon data. Error bars indicate the standard error; the topological error is the average of the FN and FP error rates.

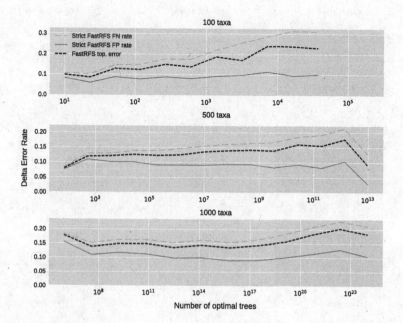

Fig. 8. Mean error rates for a single FastRFS tree and the strict consensus of all FastRFS trees on the supertree datasets with 100, 500, and 1000 species, compared to the number of optimal trees. We show FP rate and FN rates for each method; these are equal for default FastRFS (because it is always binary), but different for the strict consensus of the FastRFS trees. As the number of optimal trees increases, the decrease in the FP rate is larger than the increase in the FN rate for the strict consensus of the FastRFS trees, explaining why the average error for the strict consensus of the FastRFS trees is lower than for a single FastRFS tree (as shown in Fig. 7). Results for 193 replicates are shown on 1000-taxon data, results for 312 replicates are shown on 500-taxon data, and results for 104 replicates are shown on 100-taxon data.

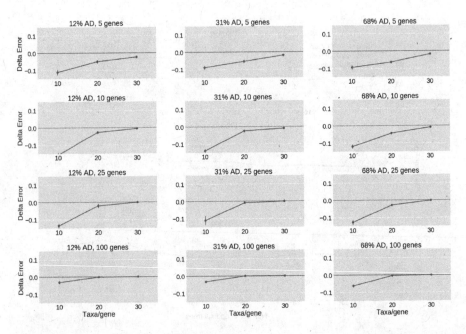

Fig. 9. The strict consensus of ASTRAL trees is more accurate than ASTRAL when gene trees are incomplete. We show Delta-error (change in mean topological error between FastRFS and the strict consensus of FastRFS trees) on simulated phylogenomic datasets with varying numbers of incomplete gene trees on 50-species datasets with three different ILS levels; values below 0 indicate that the strict consensus ASTRAL is more accurate (i.e., it has lower error) than ASTRAL. Note that there is a big advantage in computing the strict consensus tree of the optimal ASTRAL trees instead of a single ASTRAL tree under the highest amount of missing data, and that the advantage decreases as the amount of missing data decreases. We show results for 25 replicates. Error bars indicate the standard error; topological error is the average of the FN and FP error rates.

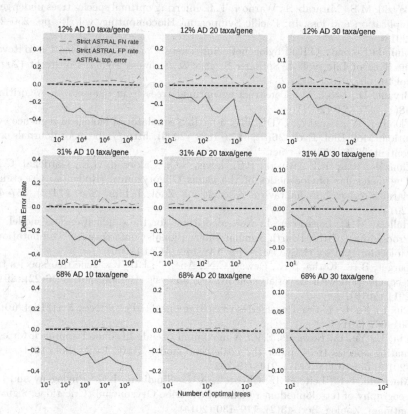

Fig. 10. Mean change in error between the strict consensus of the ASTRAL trees compared to a single ASTRAL tree on the 50-taxon phylogenomic datasets with varying degrees of missing data and ILS, as a function of the number of optimal trees. Values below zero indicate that the strict consensus tree has better accuracy (lower error) than a single ASTRAL tree. We show the change in FP rates (blue, solid line) and in FN rates (red, dashed); the black line represents the baseline. This figure shows that the strict consensus has lower false positives than a single ASTRAL tree and higher false negatives, but also that the reduction in false positives is larger than the increase in false negatives. The figure also shows that the reduction in false positives increases with the number of optimal trees. (Color figure online)

References

1. Alvarado-Serrano, D.F., D'Elía, G.: A new genus for the Andean mice Akodon latebricola and A. bogotensis (Rodentia: Sigmodontinae). J. Mammal. **94**(5), 995–1015 (2013)
2. Bayzid, M.S., Mirarab, S., Warnow, T.J.: Inferring optimal species trees under gene duplication and loss. In: Pacific Symposium Biocomputing, vol. 18, pp. 250–261 (2013)
3. Bininda-Emonds, O.R.: Phylogenetic Supertrees: Combining Information to Reveal the Tree of Life, vol. 4. Springer Science & Business Media, Dordrecht (2004). doi:10.1007/978-1-4020-2330-9
4. Bryant, D., Steel, M.: Constructing optimal trees from quartets. J. Algorithms **38**(1), 237–259 (2001)
5. Fletcher, W., Yang, Z.: INDELible: a flexible simulator of biological sequence evolution. Mol. Biol. Evol. **26**(8), 1879–1888 (2009). http://mbe.oxfordjournals.org/content/26/8/1879.abstract
6. González-Ittig, R.E., Rivera, P.C., Levis, S.C., Calderón, G.E., Gardenal, C.N.: The molecular phylogenetics of the genus Oligoryzomys (Rodentia: Cricetidae) clarifies rodent host-hantavirus associations. Zool. J. Linn. Soc. **171**(2), 457–474 (2014)
7. Hallett, M.T., Lagergren, J.: New algorithms for the duplication-loss model. In: Proceedings of the Fourth Annual International Conference on Computational Molecular Biology (RECOMB), pp. 138–146. ACM (2000)
8. Larget, B.R., Kotha, S.K., Dewey, C.N., Ané, C.: BUCKy: gene tree/species tree reconciliation with Bayesian concordance analysis. Bioinformatics **26**(22), 2910–2911 (2010)
9. Liu, L., Yu, L.: Estimating species trees from unrooted gene trees. Syst. Biol. **60**(5), 661–667 (2011)
10. Liu, L., Yu, L., Edwards, S.V.: A maximum pseudo-likelihood approach for estimating species trees under the coalescent model. BMC Evol. Biol. **10**(1), 1–18 (2010). doi:10.1186/1471-2148-10-302
11. Machado, L.F., Leite, Y.L., Christoff, A.U., Giugliano, L.G.: Phylogeny and biogeography of tetralophodont rodents of the tribe Oryzomyini (Cricetidae: Sigmodontinae). Zoolog. Scr. **43**(2), 119–130 (2014)
12. Maddison, W.: Gene trees in species trees. Syst. Biol. **46**(3), 523–536 (1997). doi:10.1093/sysbio/46.3.523
13. Maestri, R., Monteiro, L.R., Fornel, R., Upham, N.S., Patterson, B.D., Freitas, T.R.O.: The ecology of a continental evolutionary radiation: is the radiation of sigmodontine rodents adaptive? Evolution **71**(3), 610–632 (2017)
14. Mallo, D., Martins, L.D.O., Posada, D.: SimPhy: phylogenomic simulation of gene, locus, and species trees. Syst. Biol. **65**(2), 334–344 (2016). doi:10.1093/sysbio/syv082
15. Mirarab, S., Reaz, R., Bayzid, M.S., Zimmermann, T., Swenson, M.S., Warnow, T.: ASTRAL: genome-scale coalescent-based species tree estimation. Bioinformatics **30**(17), i541–i548 (2014)
16. Mirarab, S., Warnow, T.: ASTRAL-II: coalescent-based species tree estimation with many hundreds of taxa and thousands of genes. Bioinformatics **31**(12), i44–i52 (2015)
17. Mossel, E., Roch, S.: Incomplete lineage sorting: consistent phylogeny estimation from multiple loci. IEEE/ACM Trans. Comput. Biol. Bioinform. (TCBB) **7**(1), 166–171 (2010)

18. Nguyen, N., Mirarab, S., Warnow, T.: MRL and SuperFine+MRL: new supertree methods. Algorithms Mol. Biol. **7**(1), 3 (2012)
19. Roch, S.: A short proof that phylogenetic tree reconstruction by maximum likelihood is hard. IEEE/ACM Trans. Comput. Biol. Bioinform. (TCBB) **3**(1), 92 (2006)
20. Ronquist, F., Teslenko, M., Van Der Mark, P., Ayres, D.L., Darling, A., Höhna, S., Larget, B., Liu, L., Suchard, M.A., Huelsenbeck, J.P.: MrBayes 3.2: efficient Bayesian phylogenetic inference and model choice across a large model space. Syst. Biol. **61**(3), 539–542 (2012)
21. Sayyari, E., Mirarab, S.: Fast coalescent-based computation of local branch support from quartet frequencies. Mol. Biol. Evol. **33**(7), 1654–1668 (2016)
22. Sharanowski, B.J., Robbertse, B., Walker, J., Voss, S.R., Yoder, R., Spatafora, J., Sharkey, M.J.: Expressed sequence tags reveal Proctotrupomorpha (minus Chalcidoidea) as sister to Aculeata (Hymenoptera: Insecta). Mol. Phylogenet. Evol. **57**(1), 101–112 (2010)
23. Stamatakis, A.: RAxML version 8: a tool for phylogenetic analysis and post-analysis of large phylogenies. Bioinformatics **30**(9) (2014). doi:10.1093/bioinformatics/btu033
24. Sukumaran, J., Holder, M.T.: Dendropy: a python library for phylogenetic computing. Bioinformatics **26**(12), 1569–1571 (2010)
25. Swenson, M.S., Barbançon, F., Warnow, T., Linder, C.R.: A simulation study comparing supertree and combined analysis methods using SMIDGen. Algorithms Mol. Biol. **5**, 8 (2010)
26. Szöllősi, G.J., Rosikiewicz, W., Boussau, B., Tannier, E., Daubin, V.: Efficient exploration of the space of reconciled gene trees. Syst. Biol. **62**, 901–912 (2013)
27. Than, C., Nakhleh, L.: Species tree inference by minimizing deep coalescences. PLoS Comput. Biol. **5**(9), e1000501 (2009). doi:10.1371/journal.pcbi.1000501.g016
28. Vachaspati, P.: Simulated data for siesta paper (2017). doi:10.6084/m9.figshare.5234803.v1. Accessed 21 July 2017
29. Vachaspati, P., Warnow, T.: ASTRID: accurate species TRees from internode distances. BMC Genom. **16**(10), 1–13 (2015). doi:10.1186/1471-2164-16-S10-S3
30. Vachaspati, P., Warnow, T.: FastRFS: fast and accurate Robinson-Foulds Supertrees using constrained exact optimization. Bioinformatics **33**(5), 631–639 (2017)
31. Yu, Y., Warnow, T., Nakhleh, L.: Algorithms for MDC-based multi-locus phylogeny inference: beyond rooted binary gene trees on single alleles. J. Comput. Biol. **18**(11), 1543–1559 (2011)

On the Rank-Distance Median of 3 Permutations

Leonid Chindelevitch[1] and João Meidanis[2(✉)]

[1] Simon Fraser University, Burnaby, Canada
leonid@sfu.ca
[2] University of Ottawa, Ottawa, Canada
meidanis@ic.unicamp.br

Abstract. Recently, Pereira Zanetti et al. [9] have proposed a new definition of a distance for which the computational complexity of the median
problem is currently unknown. In this formulation, each genome is represented as a permutation on n elements that is the product of disjoint
cycles of length 1 (*telomeres*) and length 2 (*adjacencies*). The permutations are converted into their matrix representation, and the *rank distance* d is used to define the median.

In their paper, the authors provide an $O(n^3)$ algorithm for determining three candidate medians, prove the tight approximation ratio $\frac{4}{3}$, and
provide a sufficient condition for their candidates to be true medians.
They also conduct some experiments that suggest that their method is
accurate on simulated and real data.

In this paper, we extend their results and provide the following:
- 2 invariants characterizing the problem of finding the median of 3
  matrices
- a strengthening of one of the results in the work of Pereira Zanetti
  et al.
- a sufficient condition for optimality that can be checked in $O(n)$ time
- a faster, $O(n^2)$ algorithm for determining the median under this
  condition
- a new heuristic algorithm for this problem based on compressed
  sensing.

1 Introduction

The genome median problem asks, given as input three genomes G_1, G_2, G_3 and
a distance (metric) d, to find the *median* genome G minimizing $\sum_{i=1}^{3} d(G, G_i)$.
This problem has been extensively studied due to its many applications in phylogenetics and ancestral genome reconstruction [1–3], and is NP-hard for all
but a few known distances d [4–7]. Many approximation algorithms have been
developed for this problem [4,8], and some heuristic approaches have also been
successfully applied to this problem [3].

Recently, Pereira Zanetti et al. [9] have proposed a new definition of a distance for which the computational complexity of the median problem is currently

J. Meidanis is on Leave from University of Campinas, Campinas, Brazil.

© Springer International Publishing AG 2017
J. Meidanis and L. Nakhleh (Eds.): RECOMB CG 2017, LNBI 10562, pp. 256–276, 2017.
DOI: 10.1007/978-3-319-67979-2_14

unknown. In this formulation, each genome is represented as a permutation on n elements that is the product of disjoint cycles of length 1 (*telomeres*) and length 2 (*adjacencies*). The permutations are converted into their matrix representation, and the *rank distance* d is used to define the median.

In their paper, the authors provide an $O(n^3)$ algorithm for determining three candidate medians, prove the tight approximation ratio $\frac{4}{3}$, and provide a sufficient condition for their candidates to be true medians. They also conduct some experiments that suggest that their method is accurate on simulated data.

In this paper, we extend their results and provide the following:

- 2 invariants characterizing the problem of finding the median of 3 matrices
- a strengthening of one of the results in the work of Pereira Zanetti et al.
- a sufficient condition for optimality that can be checked in $O(n)$ time
- a faster, $O(n^2)$ algorithm for determining the median under this condition
- a new heuristic algorithm for this problem based on compressed sensing

Our results suggest that the problem is easy in a certain regime (determined by the values of the invariants). However, on the negative side, we show that the fraction of random instances in the favorable regime tends to 0 as the genome size grows. While making substantial progress towards a solution, our work therefore leaves open the main problem of determining the complexity of the median of 3 genomic permutation matrices with respect to the rank distance.

2   Definitions and Invariants

Let $\mathbb{F}$ be a field (here, we use $\mathbb{F} = \mathbb{R}$, the field of real numbers, although the problem can also be posed in finite fields). Let $n \in \mathbb{N}$ be an integer and let $\mathbb{F}^{n \times n}$ be the ring of $n \times n$ matrices with entries in $\mathbb{F}$.

2.1   Genomic Matrices

Following [9], we say that a matrix $M \in \mathbb{F}^{n \times n}$ is *genomic* if the following conditions hold:

$$M_{i,j} \in \{0,1\} \ \forall \ 1 \leq i,j \leq n; \ M^T = M^{-1} = M.$$

Note that, unlike [9], we do not require n to be even for M to be genomic.

The first of the conditions above means that M is a 0–1 matrix, and the condition $M^T = M^{-1}$ means that M is an *orthogonal* matrix; together, they imply that M is a *permutation* matrix, which has a single 1 in each row and each column. It therefore defines a permutation π via the relationship

$$\pi(i) = j \iff M_{i,j} = 1.$$

The condition $M^{-1} = M$ ensures that $M^2 = I$; in this case, the corresponding permutation π is also its own inverse, so it must have order 2 and hence is a product of disjoint cycles of length 1 and 2. The cycles of length 1 correspond to *telomeres* while the cycles of length 2 correspond to *adjacencies*. The correspondence between a genome G and a genomic matrix M is defined by

$$M_{i,j} = 1 \iff (i,j) \text{ is an adjacency in } G, \text{ or } i = j \text{ and } i \text{ is a telomere in } G.$$

There is also a graphical way of depicting one or more genomes. When we use the rank distance, we assume all genomes have the same genes, and the rows and columns of the matrices correspond in fact to the extremities of these common genes [9]. Consider a graph on n vertices, each vertex associated to one of these gene extremities in a one-to-one fashion. Then we can draw the adjacencies of genomes as edges in this graph, using a different color for each genome. If we draw just one genome, we have a *matching* on this graph, and vice-versa, that is, each matching defines a unique genome. Telomeres correspond to unsaturated vertices in this matching. If we draw two genomes, since each one is a matching, we end up with a collection of paths and cycles. We can draw any number of genomes in this way. We call such a graph a *multi-genome breakpoint graph*, in analogy to the breakpoint graph used to study pairs of genomes. Notice, however, that our multi-genome breakpoint graphs do not contain caps to close linear chromosomes. Our representation is cap-free.

2.2   Rank Distance

The rank distance $d(\cdot, \cdot)$ [10] is defined on $\mathbb{F}^{n \times n}$ via

$$d(A, B) = r(A - B),$$

where $r(X)$ is the rank of the matrix X. The fact that it is a metric follows from the following lemma, which also establishes necessary conditions for equality in the triangle inequality that we are going to use later on.

Lemma 2.1. *The rank distance d is a metric. If $d(A, B) + d(B, C) = d(A, C)$, then there exists a projection matrix P such that $B = A + P(C - A)$.*

Proof. Clearly, $r(X) \geq 0 \ \forall X$, and $r(A - B) = 0 \iff A - B = 0 \iff A = B$. The symmetry follows from $d(A, B) = r(A - B) = r(B - A) = d(B, A)$.

The triangle inequality follows from the fact that the rank of X is the vector space dimension of its *image* $\text{Im}(X) := \{Xv | v \in \mathbb{F}^n\}$, so it suffices to show that

$$\dim(\text{Im}(X)) + \dim(\text{Im}(Y)) \geq \dim(\text{Im}(X + Y))$$

and substitute $X = B - A$ and $Y = C - B$ to conclude that

$$d(A, B) + d(B, C) \geq d(A, C) \ \forall \ A, B, C \in \mathbb{F}^{n \times n}.$$

Now, it is easy to see that

$$\text{Im}(X + Y) = \{(X + Y)v | v \in \mathbb{F}^n\} = \{(Xv + Yv | v \in \mathbb{F}^n\}$$
$$\subseteq \{Xv | v \in \mathbb{F}^n\} + \{Yv | v \in \mathbb{F}^n\} = \text{Im}(X) + \text{Im}(Y),$$

where the addition in the second line is the addition of vector spaces. Therefore,

$$r(X+Y) = \dim(\mathrm{Im}(X+Y)) \le \dim(\mathrm{Im}(X) + \mathrm{Im}(Y)) = -\dim(\mathrm{Im}(X) \cap \mathrm{Im}(Y))$$
$$+ \dim(\mathrm{Im}(X)) + \dim(\mathrm{Im}(Y)) \le \dim(\mathrm{Im}(X)) + \dim(\mathrm{Im}(Y)) = r(X) + r(Y),$$

where the equality follows from properties of vector space addition [13]. In particular, equality happens if and only if both of the following conditions hold:

$$\dim(\mathrm{Im}(X+Y)) = \dim(\mathrm{Im}(X) + \mathrm{Im}(Y)) \iff \mathrm{Im}(X+Y) = \mathrm{Im}(X) + \mathrm{Im}(Y);$$
$$\dim(\mathrm{Im}(X) \cap \mathrm{Im}(Y)) = 0 \iff \mathrm{Im}(X) \cap \mathrm{Im}(Y) = \{0\}.$$

These two conditions are equivalent to the vector space $\mathrm{Im}(X+Y)$ being the *direct sum* of $\mathrm{Im}(X)$ and $\mathrm{Im}(Y)$, written as $\mathrm{Im}(X+Y) = \mathrm{Im}(X) \oplus \mathrm{Im}(Y)$.

To finish the proof, let us pick a basis $\mathbb{B}_X$ for $\mathrm{Im}(X)$ and a basis $\mathbb{B}_Y$ for $\mathrm{Im}(Y)$. From the directness of the sum, it follows that every vector $u \in \mathrm{Im}(X+Y)$ can be uniquely written as $u = v + w$, with $v \in \mathrm{Im}(X)$ and $w \in \mathrm{Im}(Y)$. Since this is true for any vector in $\mathrm{Im}(X+Y)$, this is in particular true for the n columns of $X+Y$, say $c_1, \ldots, c_n$.

But we also have

$$c_i = x_i + y_i \ \forall \, 1 \le i \le n, \tag{1}$$

where x_i (respectively y_i) is the i-th column of X (respectively Y). Therefore, since $x_i \in \mathrm{Im}(X)$ and $y_i \in \mathrm{Im}(Y)$, Eq. (1) gives the desired decomposition for each c_i. Let $\mathbb{B} := \mathbb{B}_X \cup \mathbb{B}_Y$ be the basis of $\mathrm{Im}(X+Y)$ obtained by joining the bases $\mathbb{B}_X$ and $\mathbb{B}_Y$, and let $\mathbb{B}^+ := \mathbb{B} \cup \mathbb{B}'$ be an extension of the basis $\mathbb{B}$ to a basis of the entire space $\mathbb{F}^n$. Let P be the matrix of the projection operator onto X with respect to the basis $\mathbb{B}^+$, with all the elements in $\mathbb{B}' = \mathbb{B}^+ - \mathbb{B}$ being projected onto the 0 vector. Then

$$P c_i = x_i \ \forall \, 1 \le i \le n,$$

so that $P(X+Y) = X$. By substituting $X = B - A$ and $Y = C - B$, we see that the desired equality is only possible if $B = A + P(C - A)$, as required.

We also state a helpful.

Corollary 2.1. *If $x \in \mathbb{F}^n$ is a vector such that $Ax = Cx$ and $d(A, B) + d(B, C) = d(A, C)$, then $Bx = Ax$ as well. In particular, if A and C have all row sums equal to 1, so does B.*

Proof. The distance condition implies that $B = A + P(C - A)$ for some P, so we can compute $Bx = Ax + P(C - A)x = Ax + P(Cx - Ax) = x$, so $Bx = Ax$ as claimed. The second statement follows from the first one by taking $x = e$, the vector of n 1's, since Ae is the vector containing the row sums of A.

The rank distance is equivalent to the Cayley distance between permutations. We will develop this connection further in Sect. 3.

2.3 Biological Motivation

Although the rank distance has been known and studied for a long time, its use on genome matrices brings an additional motivation for its study, based on biological grounds. Indeed, the most frequent genome rearrangements occurring in genome evolution, such as inversions, transpositions, translocations, fissions, and fusions, correspond to genomes with very low rank distances. This suggests that the rank distance may be a good indicator of the amount of evolution that separates two genome matrices. Table 1 shows the rank distance associated to some of the most common genome rearrangement operations. Notice that the term "transposition" has a different meaning in biology than it does in permutation group theory. In this section we are using the biological meaning, but the rest of the paper follows the mathematical meaning.

Table 1. Frequent rearrangement operations and rank distance between genomes that differ by the corresponding operation.

Operation	Rank distance
Inversion	2
Transposition	4
Translocation	2
Fission	1
Fusion	1

2.4 Relationship with DCJ and SCJ

In addition, the rank distance is closely related to other distances that have traditionally been used in genome studies, such as the double-cut-and-join (DCJ) distance [11] and the single-cut-or-join (SCJ) distance [7]. If we use the multi-genome breakpoint graph for two genomes A and B, we can derive formulas for the DCJ, SCJ, and rank distances based on graph elements, as follows.

The SCJ distance is defined as the number of adjacencies that belong to exactly one of the two genomes. Therefore, we can compute the SCJ distance by counting all adjacencies (edges) in the graph and subtracting the number of common adjacencies. A common adjacency will appear in the graph as a 2-cycle, that is, a cycle composed of two parallel edges. The formula for the SCJ distance is then:

$$d_{SCJ}(A, B) = \#\text{edges} - 2C_2,$$

where C_2 is the number of 2-cycles in the graph.

The multi-genome breakpoint graph for A and B is a collection of paths and cycles. It can be shown that each path contributes its number of edges to the rank distance, and each cycle contributes its number of edges minus 2 to the rank distance. Therefore, the rank distance is:

$$d(A, B) = \#\text{edges} - 2C,$$

where C is the number of cycles of any length in the graph. From these equations it is easy to derive the following relationship between the SCJ and rank distances:

$$d(A, B) \leq d_{SCJ}(A, B) \leq 2d(A, B).$$

These inequalities are tight, as witnessed by cases in which the graph has no cycles (for the leftmost inequality), and graphs composed solely of 4-cycles (for the rightmost inequality).

With respect to the DCJ distance, it can be shown that

$$d_{DCJ}(A, B) = \frac{1}{2}\#\text{edges} - C + \frac{1}{2}P_{odd},$$

where P_{odd} is the number of paths of odd length (number of edges) in the graph. From these equations it is easy to derive the following relationship between the DCJ and rank distances:

$$d(A, B) \leq 2d_{DCJ}(A, B) \leq 2d(A, B).$$

These inequalities are tight as well, as witnessed by graphs with no paths (leftmost inequality), and by graphs composed solely of paths of length 1 (rightmost inequality).

2.5 Median of 3 Matrices

Given three matrices A, B, C, the median M is defined as a global minimizer of the *score* function

$$d(M; A, B, C) := d(A, M) + d(B, M) + d(C, M). \tag{2}$$

Since d is a metric, we can use symmetry and the triangle inequality to see that the score has a simple lower bound:

$$d(M; A, B, C) = \frac{d(A, M) + d(M, B)}{2} + \frac{d(B, M) + d(M, C)}{2} + \frac{d(C, M) + d(M, A)}{2}$$
$$\geq \frac{d(A, B)}{2} + \frac{d(B, C)}{2} + \frac{d(C, A)}{2} = \frac{1}{2}[d(A, B) + d(B, C) + d(C, A)],$$

with equality if and only if

$$d(X, M) + d(M, Y) = d(X, Y) \text{ for any distinct } X, Y \in \{A, B, C\}. \tag{3}$$

2.6 The First Invariant

We now define the first invariant of the median-of-three problem via

$$\beta(A, B, C) := \frac{1}{2}[d(A, B) + d(B, C) + d(C, A)].$$

It is easy to see that this is indeed an invariant in the sense that it does not change under permuting of the three matrices, or permuting the rows or the columns of all the matrices in the same way. Namely,

$$\beta(A, B, C) = \beta(A, C, B) = \beta(B, C, A) = \beta(B, A, C) = \beta(C, B, A) = \beta(C, A, B),$$

and, for any $n \times n$ permutation matrices P and Q,

$$\beta(A, B, C) = \beta(PAQ^T, PBQ^T, PCQ^T).$$

The fact that d is a metric allows us to establish a first (trivial) approximation algorithm with an approximation ratio of $\frac{4}{3}$ [1] - namely, pick the matrix among A, B, C with the smallest score. The approximation ratio follows from

$$\min_{X \in \{A,B,C\}} d(X; A, B, C) \leq \frac{1}{3} \sum_{X \in \{A,B,C\}} d(X; A, B, C)$$

$$= \frac{1}{3}[d(B, A) + d(C, A) + d(A, B) + d(C, B) + d(A, C) + d(B, C)] = \frac{4}{3}\beta(A, B, C).$$

We note that any matrix with score β is necessarily a median, and that for any matrix M that attains this score, we necessarily have

$$d(A, M) = \beta - d(B, C); \quad d(B, M) = \beta - d(A, C); \quad d(C, M) = \beta - d(A, B).$$

However, in the general case, it is not possible to attain the lower bound β. For instance, if

$$A = \begin{pmatrix} -1 & 0 \\ 0 & -1 \end{pmatrix}, B = \begin{pmatrix} 0 & 0 \\ 0 & 0 \end{pmatrix}, C = \begin{pmatrix} 1 & 0 \\ 0 & 1 \end{pmatrix},$$

then $d(A, B) = d(B, C) = d(C, A) = 2$, so $\beta = 3$, but it is easy to check that no matrix is simultaneously at unit rank distance from all three of these matrices, and the minimum score is 4 (attained by, for instance, any diagonal matrix with diagonal entries in $\{-1, 0, 1\}$).

2.7   The Second Invariant

The second invariant, which was already identified in [9] as playing an important role in the median problem, is

$$\alpha(A, B, C) := \dim(V_1), \text{ where } V_1 := \{x \in \mathbb{F}^n | Ax = Bx = Cx\}. \tag{4}$$

Once again, it is easy to check that it is invariant under permutations of the three matrices, or permutations of the rows or the columns of all the matrices.

3   Permutation Matrices

Let us now consider the special case of A, B, C being permutation matrices. While, as our example showed, the lower bound $\beta(A, B, C)$ for the score cannot always be attained, there is a possibility that the lower bound can always be attained for permutation matrices, meaning that the equality conditions in Eq. (3) can always be satisfied. The present paper is motivated by an attempt to establish whether this is indeed the case.

3.1   Integrality of the First Invariant

Let us denote by S_n the group of permutations on n elements. Pereira Zanetti et al. [9] have already shown that, for any two permutations σ and τ in S_n, the *transposition distance* $d_T(\sigma, \tau)$, also known as Cayley distance [14] and counting the minimum number of transpositions (switches) needed to transform σ into τ, equals $d(S, T)$, where S and T are the permutation matrices corresponding to σ and τ, respectively, and d is the rank distance.

Let us begin by showing that, if A, B, C are permutation matrices, $\beta(A, B, C)$ is always an integer; this is also not the case in general, as can be seen in the one-dimensional example of three different scalars, for which $\beta(A, B, C) = 3/2$.

To this end, we recall that any permutation $\tau \in S_n$ can be written as a product of disjoint cycles, uniquely up to order of the cycles and order of the elements within the cycle, provided that fixed points are represented by cycles of length 1. We define the *cycle counter* $c(\tau)$ as the number of disjoint cycles in a disjoint cycle representation of τ. For instance, if $\tau = (12)(34)(5)$, then $c(\tau) = 3$.

Lemma 3.1. *If A, B are permutation matrices corresponding to permutations ρ, σ, respectively, then $d(A, B) = n - c(\rho^{-1}\sigma)$.*

Proof. It was already shown in [9] that $d(A, B) = d_T(\rho, \sigma)$, where d_T is the transposition distance. Since d_T is left invariant [14], we have

$$d_T(\rho, \sigma) = d_T(\rho^{-1}\rho, \rho^{-1}\sigma) = d_T(e, \rho^{-1}\sigma).$$

It remains to show that the minimum number of transpositions needed to transform a permutation τ into the identity is $n - c(\tau)$. This follows from the facts that a k-cycle needs exactly $k - 1$ transpositions to transform into the identity, the total length of all the cycles (including the fixed points) is n, and the optimal set of transpositions affects one cycle at a time.

Corollary 3.1. *If A, B are permutation matrices corresponding to permutations ρ, σ, respectively, then the nullspace of $A - B$ is spanned by the indicator vectors of the cycles of $\rho^{-1}\sigma$ (each taking value 1 on the cycle and 0 outside it).*

This corollary, which follows from Lemma 3.1 and the rank-nullity theorem, could also have been deduced directly from the following:

Remark 3.1. If the permutation matrix A corresponds to ρ, $Ax = [x_{\rho(1)}, \ldots, x_{\rho(n)}]^T$.

Lemma 3.2. *If A, B, C are permutation matrices, $\beta(A, B, C)$ is an integer.*

Proof. Let ρ, σ, τ be the permutations corresponding to A, B, C, respectively. From the foregoing discussion, $d_T(\rho, \sigma)$ is the smallest number of transpositions needed to transform ρ into σ. Note that transforming ρ into σ and then σ into τ also transforms σ into τ. In addition, it is known that the number of transpositions needed to transform one permutation into another has a fixed parity that only depends on the *signs* of the permutations [14]. Therefore

$$d(A, B) + d(B, C) + d(C, A) = d_T(\rho, \sigma) + d_T(\sigma, \tau) + d_T(\rho, \tau) \equiv$$
$$\equiv d_T(\rho, \sigma) + d_T(\sigma, \tau) + [d_T(\rho, \sigma) + d_T(\sigma, \tau)] \mod 2 \equiv 0 \mod 2,$$

so that $\beta(A, B, C)$ is indeed an integer.

An alternative proof can be obtained by noting that $(-1)^{d_T(\rho,\sigma)} \equiv \det(A^{-1}B)$ mod 2, where A, B are the permutation matrices for ρ, σ respectively; therefore

$$(-1)^{d_T(\rho,\sigma)+d_T(\sigma,\tau)+d_T(\tau,\rho)} \equiv \det(A^{-1}B)\det(B^{-1}C)\det(C^{-1}A) =$$
$$= \det((A^{-1}B)(B^{-1}C)(C^{-1}A)) = \det(A^{-1}(BB^{-1})(CC^{-1})A) =$$
$$= \det(I) = 1 \mod 2 \implies d_T(\rho, \sigma) + d_T(\sigma, \tau) + d_T(\tau, \rho) \equiv 0 \mod 2.$$

3.2   Fast Computation of the Invariants

We now show how to compute the invariants α and β in $O(n)$ time given the three permutations ρ, σ, τ represented by A, B, C, respectively. For β, it suffices to use the identity

$$\beta(A, B, C) = \frac{1}{2}\big(3n - [c(\rho^{-1}\sigma) + c(\sigma^{-1}\tau) + c(\tau^{-1}\rho)]\big),$$

obtained using Lemma 3.1 and the definition of β. Permutations can be multiplied and inverted in $O(n)$ time using standard algorithms, and their cycles can be counted in $O(n)$ time by using their graph representation (with a directed edge from i to $\pi(i)$ for every $1 \leq i \leq n$) and identifying the weakly connected components. Therefore, the computation of $\beta(A, B, C)$ takes $O(n)$ time overall.

For α, we note that, by Corollary 3.1,

$$x \in V_1 \iff Ax = Bx = Cx \iff Ax = Bx \text{ and } Bx = Cx \iff$$
$$\iff x \text{ is constant on the cycles of } \rho^{-1}\sigma \text{ and on the cycles of } \sigma^{-1}\tau.$$

The computation of $\rho^{-1}\sigma$ and $\sigma^{-1}\tau$ is the same as the one performed for computing β, and the dimension of V_1 is then equal to the number of weakly connected components of the union of their graph representations described above.

Indeed, if C_1, C_2 are 2 disjoint weakly connected components of the graph representation of $\rho^{-1}\sigma$ and there is an edge between C_1 and C_2 in the graph representation of $\sigma^{-1}\tau$, then any vector $x \in V_1$ must be constant on $C_1 \cup C_2$. By iterating this reasoning, we conclude that α is precisely the number of weakly connected components of the union of the graph representations of $\rho^{-1}\sigma$ and $\sigma^{-1}\tau$. Each graph can be computed in $O(n)$ time, and so can their union and its connected components, so computing α also requires $O(n)$ time overall.

3.3 Subspace Dimensions in Terms of Invariants

Pereira Zanetti et al. [9] showed how to decompose the space $\mathbb{F}^n$ into a direct sum of five subspaces, and expressed their median candidates using the projection operators of these subspaces. We now show how to express the dimensions of these subspaces using the invariants α and β, and deduce a sufficient condition for their median candidate to be a true median. Readers familiar with this paper will recognize our use of the dot notation for partitions introduced there (e.g., $.AB.C$). However, all subspaces needed here are defined here as well, so the exact meaning of this notation is not relevant in our context.

The subspace $V_1 = V(.A.B.C.)$ is defined in Eq. (4), and is the subspace of all vectors on which all three matrices agree. Its dimension is α, by definition.

The subspace $V_2 := V(.AB.C.) \cap V_1^\perp$ is defined via V_1 and the subspace

$$V(.AB.C) := \{x \in \mathbb{F}^n | Ax = Bx\}.$$

The dimension of $V(.AB.C)$ is precisely $c(\rho^{-1}\sigma)$, where ρ and σ are the permutations corresponding to A and B, respectively. This follows from Corollary 3.1 which tells us that

$$Ax = Bx \iff A^{-1}Bx = x \iff x \text{ is constant on every cycle of } \rho^{-1}\sigma.$$

Since $V_1 \subseteq V(.AB.C)$, it follows that $V(.AB.C.)^\perp \subseteq V_1^\perp$ and

$$\dim(V_2) = \dim(V(.AB.C.) \cap V_1^\perp) = \dim((V(.AB.C.)^\perp + V_1)^\perp) =$$
$$= n - \dim((V(.AB.C.)^\perp + V_1) = n - \dim(V(.AB.C.)^\perp) - \dim(V_1) =$$
$$= n - (n - c(\rho^{-1}\sigma)) - \alpha = c(\rho^{-1}\sigma) - \alpha,$$

where the dimension of the sum of vector subspaces in the second line splits into the sum of the individual dimensions due to their trivial intersection.

We can apply a similar reasoning to the subspaces $V_3 := V(.A.BC.) \cap V_1^\perp$ and $V_4 := V(.AC.B) \cap V_1^\perp$ to get

$$\dim(V_2) = c(\rho^{-1}\sigma) - \alpha; \ \dim(V_3) = c(\sigma^{-1}\tau) - \alpha; \ \dim(V_4) = c(\tau^{-1}\rho) - \alpha.$$

Since V_5 is the last term in the direct sum decomposition of $\mathbb{F}^n$, we get that

$$\dim(V_5) = n - \sum_{i=1}^{4} \dim(V_i) = n + 2\alpha - (c(\rho^{-1}\sigma) + c(\sigma^{-1}\tau) + c(\tau^{-1}\rho)) =$$
$$= n + 2\alpha(A, B, C) - (3n - 2\beta(A, B, C)) = 2(\alpha(A, B, C) + \beta(A, B, C) - n).$$

From this, we immediately deduce the following.

Corollary 3.2. $\alpha(A, B, C) + \beta(A, B, C) \geq n$ for permutation matrices A, B, C, with equality if and only if $V_5 = \{0\}$.

In addition, we can now combine this with the expression in [9] for the score of their median candidates M_A, M_B, M_C to deduce that

$$d(M_A; A, B, C) = 2\dim(V_5) + \dim(V_2) + \dim(V_3) + \dim(V_4) = n - \dim(V_1) +$$
$$+ \dim(V_5) = n - \alpha + 2(\alpha + \beta - n) = \beta + (\alpha + \beta - n) = \beta + \frac{1}{2}\dim(V_5).$$

As expected, the median M_A achieves the lower bound if and only if $V_5 = \{0\}$.

3.4   A New Algorithm

Next we introduce a new algorithm that can be used to identify another candidate median, that in general differs from the candidates M_A, M_B, M_C identified in previous work [9]. Although we are not able to prove any approximation ratio on its performance, it does allow us to establish the uniqueness of the median in the special case of equality in Corollary 3.2, and gain insight into why this case is special, in addition to obtaining a faster $O(n^2)$ running time for it.

To this end, let us consider the necessary conditions from Lemma 2.1 that must be satisfied in order for the matrix M to attain the lower bound β:

$$M = A + S(B - A) = B + T(C - B) = C + U(A - C), \tag{5}$$

where S, T, U are some projection matrices. This triple equality follows from the facts that Eq. (3) must be satisfied for each pair among A, B and C.

Let us count the independent equations and non-redundant unknowns in this system. We note that it suffices to consider the equivalent system

$$\begin{aligned} A + S(B - A) = B + T(C - B) &\iff A - B = T(C - B) - S(B - A); \\ B + T(C - B) = C + U(A - C) &\iff B - C = U(A - C) - T(C - B), \end{aligned} \tag{6}$$

since the third equality automatically follows from the first two. Furthermore, we will not enforce the condition of S, T, U being projection matrices, since a projection matrix is defined by $P^2 = P$ and this results in a set of conditions quadratic in the entries of P.

Consider the effect of multiplying a matrix S by a permutation matrix A corresponding to the permutation ρ. It is easy to see that this results in permuting the columns of S according to ρ, so that, denoting by s_i the i-th column of S, we get $SA = [s_{\rho(1)} s_{\rho(2)} \ldots s_{\rho(n)}]$. Therefore, the i-th column of $S(B - A)$ will simply be the difference between $s_{\rho(i)}$ and $s_{\sigma(i)}$. For each cycle C of $\rho^{-1}\sigma$, the corresponding columns of $S(B - A)$ will add up to 0.

Thus, changing variables to the "difference variables" of the form

$$s_i' := s_{\rho(C[i])} - s_{\sigma(C[i])}, \tag{7}$$

where $C[i]$ denotes the i-th element in the cycle C, we can see that $S(B - A)$ will have precisely $n - c(\rho^{-1}\sigma)$ linearly independent columns, and one column per cycle will be linearly dependent on the others in the same cycle. By applying the same argument to $T(C - B)$ and $U(A - C)$, we get a grand total of

$$n - c(\rho^{-1}\sigma) + n - c(\sigma^{-1}\tau) + n - c(\tau^{-1}\rho) = 2\beta(A, B, C)$$

linearly independent (non-redundant) columns, each containing n free variables.

We note that the system of Eq. (6), rewritten with respect to the non-redundant difference variables, splits into n independent linear systems, one per row, with identical left-hand sides and only differing by their right-hand sides:

$$\overset{d(A,B)}{\underset{i=1}{\sum}} p_i s_i + \overset{d(B,C)}{\underset{i=1}{\sum}} q_i t_i = a_{ji} - b_{ji} \; ; \; \overset{d(C,A)}{\underset{i=1}{\sum}} r_i u_i - \overset{d(B,C)}{\underset{i=1}{\sum}} q_i t_i = b_{ji} - c_{ji} \; , \quad (8)$$

where the p_i, q_i, r_i are coefficients that only depend on the column (variable) index i, but not on the row j.

Next we count the linearly independent equations within each system. In the first half of the system (with right-hand sides coming from the j-th row of $A - B$), linear dependence between the left-hand sides of the equations can only arise from vectors y such that

$$T(C - B)y = 0 = S(B - A)y \ \forall \ S, T,$$

since S and T represent the variables in the system and those variables are distinct. Similarly, in the second half of the system, they can only arise from vectors y such that

$$U(A - C)y = 0 = T(C - B)y \ \forall \ T, U,$$

since T and U represent the variables in the system and those variables are distinct. But then, for any such vector y it must be the case that

$$Ay = By = Cy \iff y \in V_1,$$

meaning that there are exactly $\alpha(A, B, C)$ dependence relationships between the left-hand sides in each of the half-systems since that is the dimension of the susbspace V_1.

Furthermore, all such dependence relationships lead to the tautology $0 = 0$ rather than the contradiction $0 = 1$, because every row of $A - B$ and $B - C$ is orthogonal to any $y \in V_1$, by definition of V_1. Lastly, the two half-systems are linearly independent from one another since the condition on their linear dependence is subsumed by the condition of the linear dependence within the half-systems; more precisely, since the t-variables in the first half-system appear with coefficients that are exactly the negative of the coefficients in the second half-system, a linear dependence relationship between the half-systems would have to arise from a vector y such that

$$U(A - C)y = 0 = S(B - A)y \ \forall \ S, U \iff Ay = By = Cy \iff y \in V_1.$$

It follows that in after eliminating the redundancies in each half-system, exactly β variables and $n - \alpha$ equations remain. Since $\alpha + \beta \geq n$ the system is always under-constrained except in the special case of $\alpha + \beta = n$, in which it has a unique solution (since the remaining equations are linearly independent).

In the special case when $\alpha + \beta = n$, we can furthermore choose to eliminate precisely those redundant equations that contain the "composite" difference variables of the form $-s_1 - s_2 - \cdots - s_k$, corresponding to a cycle of length $k + 1$ in the appropriate permutation. In this way, the remaining equations will only have two active variables (with non-zero coefficients) each, so the algorithm by Aspvall and Shiloach [15] can be applied to solve the resulting system in $O(n)$ time. It follows that the algorithm will only require $O(n^2)$ time to find the median in the special case.

Although the system is under-constrained outside of the special case, we can use ideas from the field of *compressed sensing* [16] to find a solution that is likely to be sparse, and hence hopefully low rank. Namely, we seek the solution containing as many zeros as possible. While this is a hard problem in general, many instances are solvable by using the L_1 norm minimization, which can be achieved by using linear programming. The appropriate linear program becomes

$$\min \sum_i y_i \text{ subject to } -y_i \leq x_i \leq y_i \; \forall \, i \text{ and } Ux = b,$$

where $Ux = b$ is the original system of equations, and the y_i serve as the absolute values of the x_i whose sum is to be minimized. Such linear programs can be readily solved using existing off-the-shelf solvers such as lpsolve [17] or CPLEX [18].

3.5 An Example of the Algorithm

To clarify the procedure, we now illustrate the running of our algorithm on $\rho = (12)(34), \sigma = (13)(24), \tau = (14)(23)$, with $n = 4$. First, we note that the product of any two of these equals the third, and they are each their own inverse. Hence we can compute

$$\beta(A, B, C) = \frac{1}{2}\left(3n - [c(\rho^{-1}\sigma) + c(\sigma^{-1}\tau) + c(\tau^{-1}\rho)]\right) = \frac{1}{2}[12 - 3 \cdot 2] = 3$$

and

$$\alpha(A, B, C) = 1$$

since the union of the graphs of $\rho^{-1}\sigma = \tau$ and $\sigma^{-1}\tau = \rho$ has a unique weakly connected component. Therefore, $\alpha(A, B, C) + \beta(A, B, C) = n$ and, as we will prove below, the algorithm will in fact produce the unique median of A, B, C.

We start by forming the equations in system (6):

$$A - B = [t_4 - t_3, t_3 - t_4, t_2 - t_1, t_1 - t_2] \quad -[s_3 - s_2, s_4 - s_1, s_1 - s_4, s_2 - s_3]$$
$$B - C = [u_2 - u_4, u_1 - u_3, u_4 - u_2, u_3 - u_1] \quad -[t_4 - t_1, t_3 - t_4, t_2 - t_1, t_1 - t_2].$$

We now define the "difference variables"

$$s_1' := s_3 - s_2; s_2' := s_4 - s_1; t_1' := t_4 - t_3; t_2' := t_2 - t_1; u_1' := u_2 - u_4; u_2' := u_1 - u_3,$$

and express our system in terms of those:

$$A - B = [t'_1, -t'_1, t'_2, -t'_2] - [s'_1, s'_2, -s'_2, -s'_1]$$
$$B - C = [u'_1, u'_2, -u'_1, -u'_2] - [t'_1, -t'_1, t'_2, -t'_2].$$

For convenience we will use the same name (without row superscripts) for the variables in each row, to emphasize that the system of equations for each row has the same left-hand side. For row j it has $2n = 8$ equations that read:

$$t'_1 - s'_1 = A_{j1} - B_{j1} \tag{9.1}$$
$$-t'_1 - s'_2 = A_{j2} - B_{j2} \tag{9.2}$$
$$t'_2 + s'_2 = A_{j3} - B_{j3} \tag{9.3}$$
$$-t'_2 + s'_1 = A_{j4} - B_{j4} \tag{9.4}$$
$$u'_1 - t'_1 = B_{j1} - C_{j1} \tag{9.5}$$
$$u'_2 + t'_1 = B_{j2} - C_{j2} \tag{9.6}$$
$$-u'_1 - t'_2 = B_{j3} - C_{j3} \tag{9.7}$$
$$-u'_2 + t'_2 = B_{j4} - C_{j4} \tag{9.8}$$

As expected from our counting argument above, the only linear dependencies here are that the first 4 equations add up to 0 and the second 4 equations add up to 0 (both their left-hand sides and their right-hand sides), so after eliminating, say, Eqs. (9.4) and (9.8), we end up with a consistent and non-redundant system of $2(n-\alpha) = 6$ equations in $2\beta = 6$ unknowns, which therefore has a unique solution. We illustrate this system for the first row, $j = 1$, since it looks identical for all the other rows except for changes in its right-hand side.

$$t'_1 - s'_1 = 0 \tag{10.1}$$
$$-t'_1 - s'_2 = 1 \tag{10.2}$$
$$t'_2 + s'_2 = -1 \tag{10.3}$$
$$u'_1 - t'_1 = 0 \tag{10.4}$$
$$u'_2 + t'_1 = 0 \tag{10.5}$$
$$-u'_1 - t'_2 = 1 \tag{10.6}$$

Since each equation has exactly two variables, we can use the method of [15] to solve them in $O(n)$ time, which means that the total time to solve the system for all the n rows is $O(n^2)$. In the case of this system, we see that the solution for the first row is

$$s'_1 = -\frac{1}{2}, s'_2 = -\frac{1}{2}, t'_1 = -\frac{1}{2}, t'_2 = -\frac{1}{2}, u'_1 = -\frac{1}{2}, u'_2 = \frac{1}{2}.$$

After solving the system for all the other rows in the same way, we conclude that the unique median of A, B, C in this case is $\frac{1}{2}J - I$, where $J = ee^T$ is the matrix of all 1's and I is the identity matrix. In particular, this example shows that the unique median of three genomic matrices may not itself be genomic.

3.6   Example of the Compressed Sensing Approach

We now consider a different example, one which does not fall into the special case $\alpha + \beta = n$. We take $n = 3$ and $\rho = (12), \sigma = (13), \tau = (23)$. In this case, we have $d(A, B) = d(B, C) = d(C, A) = 2$ so $\beta(A, B, C) = 3$ and $\alpha(A, B, C) = 1$. We know that the system of Eq. (6) will be under-constrained in this case. We start by forming this system of equations:

$$A - B = [t_1 - t_3, t_3 - t_2, t_2 - t_1] - [s_3 - s_2, s_2 - s_1, s_1 - s_3]$$
$$B - C = [u_2 - u_1 u_1 - u_3 u_3 - u_2] - [t_1 - t_3, t_3 - t_2, t_2 - t_1]$$

and introduce the difference variables

$$s_1' := s_3 - s_2, s_2' := s_2 - s_1, t_1' := t_1 - t_3, t_2' := t_3 - t_2, u_1' := u_2 - u_1, u_2' := u_1 - u_3$$

to rewrite it as 3 systems (one for each row) of the form

$$t_1' - s_1' = A_{1j} - B_{1j} \tag{11.1}$$
$$t_2' - s_2' = A_{2j} - B_{2j} \tag{11.2}$$
$$-(t_1' + t_2') + (s_1' + s_2') = A_{3j} - B_{3j} \tag{11.3}$$
$$u_1' - t_1' = B_{1j} - C_{1j} \tag{11.4}$$
$$u_2' - t_2' = B_{2j} - C_{2j} \tag{11.5}$$
$$-(u_1' + u_2') + (t_1' + t_2') = B_{3j} - C_{3j} \tag{11.6}$$

As in the previous example, we can eliminate the redundant Eqs. (11.3) and (11.6) from each system, leaving us with a total of 4 equations in 6 variables.

Let us now show the compressed sensing formulation for this system for row $j = 1$. We get the linear program

minimize $y_1 + y_2 + y_3 + y_4 + y_5 + y_6$

subject to

$$- [y_1, y_2, y_3, y_4, y_5, y_6] \leq [s_1', s_2', t_1', t_2', u_1', u_2'] \leq [y_1, y_2, y_3, y_4, y_5, y_6]$$
$$t_1' - s_1' = 0; \quad t_2' - s_2' = 1; \quad u_1' - t_1' = -1; \quad u_2' - t_2' = 0.$$

The optimal solution to this linear program is

$$s_1' = 0, s_2' = -1, t_1' = 0, t_2' = 0, u_1' = -1, u_2' = 0.$$

By repeating this for the other two rows we obtain the solution $M = [0 \ 0 \ e]$, which unfortunately yields a suboptimal score of 6, whereas the optimal solutions, given by the identity matrix, either of the 3-cycles (123) or (132), or a subset of the affine combinations of those matrices, yield a score of $\beta = 3$. This shows that the compressed sensing approach is not guaranteed to be optimal, or even better than the algorithm that picks the best "corner" option among A, B, C.

3.7 Proof of Uniqueness for the Special Case

We now prove that if $\alpha(A, B, C) + \beta(A, B, C) = n$, then there is a unique median, and both the $O(n^3)$ algorithm by Pereira Zanetti et al. [9] as well as our $O(n^2)$ algorithm proposed here correctly identify it.

Theorem 3.1. *Suppose $\alpha(A, B, C) + \beta(A, B, C) = n$. Then there is a unique median minimizing $d(M; A, B, C)$, found by our algorithm and the one in [9].*

Proof. By the calculations in [9] recapitulated in Corollary 3.2, the median M_A achieves the lower bound β. Furthermore, by the calculations in the analysis of the system (6), we see that there exists a unique matrix M that simultaneously satisfies the necessary conditions for attaining the lower bound β. Since M_A attains this lower bound, M_A also satisfies these necessary conditions; by uniqueness, $M_A = M$, so our algorithm also finds a median, and it is unique. ∎

3.8 Rarity of the Special Case

We now use some asymptotic results from analytic combinatorics to show that the probability of three random genomic matrices satisfying the optimality conditions in Corollary 3.2 tends to 0 as n increases. Recall that an *involution* is a permutation that is its own inverse; this is precisely the class of permutations defined by genomic matrices. We begin by restating, without proof, the following result from [19]:

Theorem 3.2. *If σ and τ are random involutions, then the mean number of cycles of $\sigma\tau$ is $\sqrt{n} + \frac{1}{2}\log n + O(1)$. If σ and τ are constrained to be fixed-point free, then the distribution of the number of cycles of $\sigma\tau$ is asymptotically normal with mean $\log n$ and variance $2\log n$.*

Now we can immediately conclude the following:

Corollary 3.3. *If A, B, C are the genomic matrices corresponding to random involutions (respectively random involutions with no fixed points, i.e. telomeres), then $\beta \sim \frac{3}{2}(n - \sqrt{n})$ or $\beta \sim \frac{3}{2}(n - \log n)$, respectively. In particular, the probability of these matrices satisfying the optimality conditions in Corollary 3.2 tends to 0 as n increases.*

4 Experimental Results

We tested our algorithm on two datasets - first, a simulated one obtained by applying rearrangement operations to a starting genome, and second, a real one obtained by taking three genomes at a time from a family of plants. In this section, we describe the performance of our algorithm as well as our observations.

4.1 Implementation

For the implementation, we use the R statistical computing language [20] as well as the CPLEX linear programming solver [18], with which we interface via the command line. Specifically, our program first computes the invariants α and β, and then branches into either an exact solution if $\alpha + \beta = n$, or the compressed sensing heuristic if not. In the latter case, R writes the linear program for finding the solution of system (6) of minimum L_1 norm into a file, the CPLEX solver processes this file, and R parses the solution to obtain the median candidate.

We use the igraph package [21] to quickly compute the invariants as well as the cycle decompositions of the permutations involved in the system 6. The cycle decompositions allow us to decide which variables to include in the system 6, and which equations to exclude to make it non-redundant. The resulting system always has $2(n - \alpha)$ equations in 2β variables. In order to try to make the system as sparse as possible even when $\alpha + \beta > n$, we make the variable corresponding to the last (highest) element of each cycle non-basic by expressing it in terms of the others, as per Eq. (7). Furthermore, we eliminate the equation corresponding to the last (highest) element of each connected component in the union graph defining α; since each such connected component is a disjoint union of cycles (of each of the three permutations $\rho^{-1}\sigma, \sigma^{-1}\tau, \tau^{-1}\rho$), this guarantees that fewer composite variables remain present, leading to a sparser system (6).

In the special case $\alpha + \beta = n$, we use the *solve* function from the *Matrix* package [22], and do not implement the linear-time algorithm by Aspvall and Shiloach [15]. Therefore, strictly speaking our implementation is currently not guaranteed to run in $O(n^2)$ time despite having a sparse coefficient matrix with all non-zero coefficients being ± 1. However, since *solve* is able to take advantage of the sparsity of the system, the special case runs extremely efficiently even for the largest input size, $n = 500$.

4.2 Numerical Stability

Computing the score of the median candidates, as defined in Eq. (2), requires a rank computation, which is known to be numerically challenging [23]. In fact, Zanetti Pereira et al. report that despite all their median candidates being expected to have the same score, this is not true in practice for random permutation inputs in about 10% of the cases when using GNU Octave or MATLAB [9].

In order to circumvent this challenge we adopt several measures. First, we use the combinatorial expression from Corollary 3.2 for the score of the median candidates proposed in [9] in all our comparisons, which does not require any rank computation, only a graph-based analysis of the underlying permutations. Second, to score the candidate median produced by the algorithm presented here, we use the *rankMatrix* function with the QR decomposition method from the *Matrix* package [22], with a tolerance of $\epsilon = 10^{-12}$. Lastly, we round any entry of the median that happens to be within ϵ of an integer (0 or ± 1) to this integer. While we cannot be completely sure that this bypasses all numerical stability issues, we observe that in all instances with $\alpha + \beta = n$, on which our algorithm should produce a median achieving the lower bound β, this is indeed the case.

4.3 Simulated Dataset

The simulated dataset consists of a collection of 540 genomic inputs, ranging in size from 6 to 250 genes (i.e. $n = 12$ to $n = 500$), a subset of the simulated dataset used by Pereira Zanetti et al. [9]. We generate the simulated instances as follows. We start with a unichromosomal linear genome with n genes and apply a random number between $\frac{rn}{2}$ and $\frac{3rn}{2}$ DCJ operations [11] to obtain each of the three input genomes, where r is a fraction between 0 and 1 (we refer to r as the *rearrangement rate*). After that we cut any circular chromosomes so that the resulting instance has three multi-chromosomal linear genomes. We use values of r ranging between 0.05 and 0.3, as higher values of r may require a distance correction and lead parsimony-based methods to produce incorrect results [24]. For each setting of n and r we generate 10 instances, and report the averages.

First, we observe that the exact $O(n^2)$ algorithm for the case $\alpha + \beta = n$ is extremely fast, requiring less than 45 s in total for all the 473 inputs (or 87.5%) that belong to it, i.e. less than 0.1 s per instance on average. The compressed sensing algorithm is somewhat slower, requiring a total of 105 s for the 67 inputs that it ran on, for an average of just over 1.5 s per instance. However, the time for all but the largest instances is in fact dominated by writing and reading the linear program file, not the actual solution. For instance, reading the file and solving the linear program each take CPLEX around 1.5 s when $n = 500$. In short, producing the median candidate using our method is extremely efficient relative to both the $O(n^3)$ computation proposed by Pereira Zanetti et al. [9] as well as the exact and heuristic methods they compared it to.

Second, we observe that the vast majority of the inputs produce median candidates that are genomic matrices. More specifically, only 12 out of 540 outputs contain fractional values (and all of these are actually optimal as they fall into the case $\alpha + \beta = n$); these fractional values are $\pm\frac{1}{2}, \pm\frac{1}{3}, \frac{2}{3}, \pm\frac{1}{4}$ and $\frac{3}{4}$. The remaining 528 out of 540 outputs contain only integer values, among which 5 contain a -1, and it is a single -1 in all cases (none of these are optimal in the sense of attaining the lower bound β). The other 523 are binary (have all entries in $\{0, 1\}$), and of those, 34 are not permutation matrices; as expected from Corollary 2.1 they all contain a single 1 per row, but they each contain multiple 1's in 1 or 2 columns (and none of these are optimal). Of the final 489, 3 are permutation matrices that are not involutions (i.e. genomic matrices), and interestingly, all of these are optimal and are only found by our algorithm, not the one in [9]. The final 486 are genomic matrices, and are optimal in all except 7 cases; in those 7 cases, both our algorithm and the one in [9] are off by 1 from the optimal bound β. The final 479 outputs are both genomic and optimal.

Third, we observe that our algorithm produces strictly more optimal solutions than the one in [9], namely, 493 instead of 473 - this is reassuring as it shows that our algorithm can also be optimal in cases where the original one fails (the other

Table 2. Average percentage excess over the lower bound β; the first number denotes our algorithm, while the second one (in brackets) represents the algorithm by Pereira Zanetti et al. [9]

n,r	0.05	0.1	0.15	0.2	0.25	0.3
12	0 (0)	3.3 (1.7)	0 (0)	2 (2.9)	3.64 (1.8)	9.3 (5.1)
16	0 (0)	0 (0)	0 (0)	0 (0)	3 (2.3)	0.7 (1.4)
20	0 (0)	0 (0)	0 (0)	2.3 (1.5)	2.1 (2.8)	6.4 (2.9)
30	3.3 (1.1)	2.2 (1.1)	2 (0.7)	1.9 (1)	1.1 (0.9)	3.7 (2.3)
50	0 (0)	0 (0)	1.1 (0.4)	0.9 (0.3)	1.6 (1)	1.8 (1.5)
100	0 (0)	0.8 (0.3)	0 (0)	0.8 (0.4)	0 (0.2)	0.8 (0.3)
200	0 (0)	0 (0)	0 (0)	0.3 (0.2)	0.1 (0.2)	0.2 (0.3)
300	0 (0)	0 (0)	0.1 (0.1)	0 (0.1)	0.4 (0.2)	0.1 (0.1)
500	0 (0)	0 (0)	0.1 (0.1)	0 (0)	0 (0.1)	0.1 (0.1)

way around is not possible due to Theorem 3.1). However, in the non-optimal cases, the approximation ratio of the original algorithm tends to be lower; this occurs in 39 cases, and 8 other cases result in ties. This is described in Table 2.

As can be seen from this table, both algorithms tend to be very close to the optimal, but there is no consistent winner between them; therefore, it might make sense to pick the better-scoring candidate among their respective outputs when the best median candidate is desired.

4.4   Real Dataset

The real dataset consists of a set of 12 Campanulaceæ chloroplast genomes as well as the Tobacco choloplast genome. We create all possible triples of inputs from this dataset, for a total of 286 input samples; each input had 105 genes, or $n = 210$ extremities.

The total time required for processing all the samples was 75 s, or less than 0.3 s per sample on average, which is consistent with the running times we obtained on simulated data.

Among the 286 test cases, 103 had some fractional output values. A total of 2448 entries among them, or 0.05% of the total, were fractions, and they included $\pm\frac{1}{2}, \pm\frac{1}{4}, \pm\frac{1}{5}, \frac{2}{5}, \frac{3}{5}$ and $\frac{3}{4}$. Just over half of them, 52 out of 103, had a score that attained the lower bound β.

Of the remaining 183 median candidates, 3 had a single -1 value in the output and were not optimal. Another 15 were binary but not permutation matrices (most with multiple 1s in 1 or 2 columns, and one occurrence in which there were multiple 1s in 3, 4 and 5 columns, respectively), and those were also not optimal. The remaining 165 were genomic matrices (there were no non-genomic permutation matrices), and all of these were optimal.

On real data, our algorithm again outperformed the one in [9] in terms of the number of optimal medians (those with score β) found - 217 vs. 189 out of 286;

however, it did not perform as well in terms of the average ratio between the obtained score and the lower bound β - the average was 3% above β for us vs. 2% above β for the original algorithm. Our algorithm had a higher score more often, 57 vs. 32 out of 286 times, with the remaining 197 being ties. Once again, the choice of algorithm depends on the user's preference for a higher chance of getting an exact median vs. a better approximation ratio, and the optimal method seems to be to pick the best-scoring output among the two algorithms.

5 Conclusion and Future Work

In this paper we introduced a new algorithm for the median-of-three problem based on a necessary condition for attaining the lower bound, and used it to prove the uniqueness of the median in a favorable regime. In addition, we tested this algorithm on both simulated and real genomes, and saw that while it was more likely to find an exact median than the one by Pereira Zanetti et al. [9], it generally had a worse approximation ratio than theirs did.

There are several remaining open questions, which we list here.

- What is the approximation ratio of the compressed sensing algorithm here?
- Are there optimality conditions less restrictive than the one we found?
- Are there triples of permutations for which the bound β is unattainable?
- What is the complexity of finding the median of 3 genomic matrices?

Acknowledgments. The authors would like to thank Cedric Chauve and Pedro Feijão for helpful discussions. LC would like to acknowledge financial support from NSERC, CIHR, Genome Canada and the Sloan Foundation. JM would like to acknowledge financial support from NSERC.

References

1. Sankoff, D., Blanchette, M.: Multiple genome rearrangement and breakpoint phylogeny. J. Comput. Biol. **5**(3), 555–570 (1998)
2. Moret, B.M., Wang, L.S., Warnow, T., Wyman, S.K.: New approaches for reconstructing phylogenies from gene order data. Bioinformatics **17**, 165–173 (2001)
3. Bourque, G., Pevzner, P.A.: Genome-scale evolution: reconstructing gene orders in the ancestral species. Genome Res. **12**(1), 26–36 (2002)
4. Caprara, A.: The reversal median problem. INFORMS J. Comput. **15**(1), 93–113 (2003)
5. Fertin, G., Labarre, A., Rusu, I., Tannier, E., Vialette, S.: Combinatorics of Genome Rearrangements. MIT Press, Cambridge (2009)
6. Tannier, E., Zheng, C., Sankoff, D.: Multichromosomal median and halving problems under different genomic distances. BMC Bioinform. **10**, 120 (2009)
7. Feijao, P., Meidanis, J.: SCJ: a breakpoint-like distance that simplifies several rearrangement problems. Trans. Comput. Biol. Bioinform. **8**, 1318–1329 (2011)
8. Pe'er, I., Shamir, R.: Approximation algorithms for the median problem in the breakpoint model. In: Sankoff, D., Nadeau, J.H. (eds.) Comparative Genomics, pp. 225–241. Springer, Berlin (2000)

9. Pereira Zanetti, J.P., Biller, P., Meidanis, J.: Median approximations for genomes modeled as matrices. Bull. Math. Biol. **78**(4), 786–814 (2016)
10. Delsarte, P.: Bilinear forms over a finite field, with applications to coding theory. J. Comb. Theory A **25**(3), 226–241 (1978)
11. Yancopoulos, S., Attie, O., Friedberg, R.: Efficient sorting of genomic permutations by translocation, inversion and block interchange. Bioinformatics **21**, 3340–3346 (2005)
12. Feijao, P., Meidanis, J.: Extending the algebraic formalism for genome rearrangements to include linear chromosomes. Trans. Comput. Biol. Bioinform. **10**, 819–831 (2012)
13. Roman, S.: Advanced Linear Algebra. Graduate Texts in Mathematics. Springer, New York (2008)
14. Arvind, V., Joglekar, P.S.: Algorithmic problems for metrics on permutation groups. In: Geffert, V., Karhumäki, J., Bertoni, A., Preneel, B., Návrat, P., Bieliková, M. (eds.) SOFSEM 2008. LNCS, vol. 4910, pp. 136–147. Springer, Heidelberg (2008). doi:10.1007/978-3-540-77566-9_12
15. Aspvall, B., Shiloach, Y.: A fast algorithm for solving systems of linear equations with two variables per equation. Linear Algebra Appl. **34**, 117–124 (1980)
16. Donoho, D.L.: Compressed sensing. IEEE Trans. Inf. Theory **52**(4), 1289–1306 (2006)
17. LPsolve Team: lp_solve 5.5. http://lpsolve.sourceforge.net/. Accessed 22 July 2017
18. IBM: CPLEX Optimizer. http://www-01.ibm.com/software/commerce/optimization/cplex-optimizer/. Accessed 22 July 2017
19. Lugo, M.: The cycle structure of compositions of random involutions (2009). https://arxiv.org/abs/0911.3604
20. R Core Team: R: a language and environment for statistical computing. R Foundation for Statistical Computing, Vienna, Austria. http://www.R-project.org/
21. Csardi, G., Nepusz, T.: The igraph software package for complex network research. InterJournal Complex Syst., 1695 (2006). http://igraph.org
22. Bates, D., Maechler, M.: Matrix: Sparse and Dense Matrix Classes and Methods. R package version 1.2-10. http://CRAN.R-project.org/package=Matrix
23. Trefethen, L.N., Bau, D.: Numerical Linear Algebra, 1st edn. SIAM: Society for Industrial and Applied Mathematics, Philadelphia (1997)
24. Biller, P., Guéguen, L., Tannier, E.: Moments of genome evolution by double cut-and-join. BMC Bioinform. **16**(Suppl 14), S7 (2015)

Statistical Consistency of Coalescent-Based Species Tree Methods Under Models of Missing Data

Michael Nute[1]([⊠]) and Jed Chou[2]

[1] Department of Statistics, University of Illinois at Urbana-Champaign,
725 S. Wright St., Champaign, IL 61820, USA
nute2@illinois.edu

[2] Department of Mathematics, University of Illinois at Urbana-Champaign,
1409 W. Green St., Urbana, IL 61801, USA

Abstract. The estimation of species trees from multiple genes is complicated by processes such as incomplete lineage sorting, duplication and loss, and horizontal gene transfer, that result in gene trees that differ from the species tree. Methods to estimate species trees in the presence of gene tree discord resulting from incomplete lineage sorting (ILS) have been developed and proved to be statistically consistent when gene tree discord is due only to ILS and every gene tree has the full set of species. Here we address statistical consistency of coalescent-based species tree estimation methods when gene trees are missing species, i.e., in the presence of missing data.

1 Introduction

The estimation of a species phylogeny from multiple loci is confounded by biological processes such as horizontal gene transfer and incomplete lineage sorting that cause individual gene tree topologies to differ from that of the overall species tree [19]. Incomplete lineage sorting (ILS), which is modeled by the well-studied multi-species coalescent (MSC) model, is considered to be perhaps the major cause for this discordance [5]. Many methods have been developed to estimate the species tree in the presence of ILS in a statistically consistent manner, which means that as the amount of data increases, the species tree topology estimated by the method converges in probability to the true species tree topology. Examples of methods for species tree estimation that are statistically consistent under the MSC include ASTRAL [20,21], ASTRID [31], *BEAST [7], BEST [15], the population tree in BUCKy [12], GLASS [22], MP-EST [17], METAL [3], NJst [16], SMRT [4], SNAPP [1], STEAC [18], STAR [18], STEM [11], and SVDquartets [2]. Some of these methods (e.g., ASTRAL, ASTRID, BUCKy-pop, and NJst) estimate just the species tree topology but not the branch lengths in coalescent units, while others (e.g., BEST, *BEAST, and MP-EST) also estimate the branch lengths. In this paper, we will refer to all methods that have been proven to be statistically consistent under the MSC as "coalescent-based species tree estimation methods".

© Springer International Publishing AG 2017
J. Meidanis and L. Nakhleh (Eds.): RECOMB CG 2017, LNBI 10562, pp. 277–297, 2017.
DOI: 10.1007/978-3-319-67979-2_15

One of the key assumptions in the proofs of statistical consistency for standard methods is that every gene is present in every species. This assumption is unrealistic for many empirical datasets (e.g., the plant transcriptome dataset studied in [32]), which can have substantial missing data. The impact of missing data on species tree estimation has mostly been investigated from an empirical rather than theoretical standpoint. Early studies focused on the impact of missing data on the estimation of individual gene trees [6,24,38], while later studies examined the impact on multi-locus species tree estimation but without any gene tree discord [14,33–35]. Four recent studies have examined the impact of missing data on species tree estimation using multiple loci, when gene trees can differ from the species tree due to ILS [8,29,31,36]. These studies have largely focused on whether it is better to include taxa and/or genes that have substantial amounts of missing data (e.g., taxa that are absent for 50% or more of the genes), and the relative performance of different coalescent-based species tree estimation methods in the presence of missing data. In general these studies have shown that although deleting whole genes from the overall data reduces accuracy compared to having no missing data, including gene data (even if they are highly incomplete) may be on the whole beneficial to species tree estimation efforts, at least for many coalescent-based species tree estimation methods.

Yet, the question of whether coalescent-based methods are statistically consistent under the MSC in the presence of missing data has not been addressed. This paper examines whether standard coalescent-based species tree estimation methods remain statistically consistent in the presence of missing data. We explore this question under a simple $i.i.d.$ model of missing data (where every species is missing from every gene with the same probability $p > 0$, and denoted M_{iid}), and also under a more general model of missing data where, for some constant k, each subset of k species has non-zero probability of being present in a randomly selected gene. We refer to this as the "full subset coverage" model (denoted M_{fsc}). The M_{fsc} model includes the simpler $i.i.d.$ model as a special case, but also includes the models of taxon deletion considered in [8,36].

In this study, we address the question of whether coalescent-based species tree estimation methods are statistically consistent under the M_{iid} or M_{fsc} models of taxon deletion. We focus on coalescent-based species tree methods that operate by computing summary statistics for subsets of the taxon set and using those summary statistics to estimate the species tree. We show that whenever these calculated summary statistics are not impacted by deleting species outside the subset of interest, then the coalescent-based species tree method will be statistically consistent under the M_{fsc} model of taxon deletion. We also discuss taxon-deletion models under which species tree estimation methods cannot be statistically consistent, and we finish by discussing the impact of missing data on species tree estimation in practice.

2   Background

2.1   Problem Statement and Notation

The multi-species coalescent is a population genetics model that describes the evolution of individual genes within a population-level species phylogeny [10]. Specifically, a species phylogeny $\mathcal{T} = (T, \Theta)$ with topology T and branch lengths Θ is given (but unknown) on a set of n-taxa, $\mathcal{X} = \{x_i\}_{i=1}^{n}$, where the branch lengths are denominated in "coalescent units." This species tree then parameterizes a probability density function for a random variable $G(\mathcal{T})$ defined over all possible phylogenies of $\mathcal{X}$. For a gene tree $g \sim G(\mathcal{T})$, an additional assumption can be made regarding a sequence evolution model which may generate a set of sequences $s_g = (s_{g1}, \ldots, s_{gn})$ for each taxon in $\mathcal{X}$. Let the leaf set of gene tree g be denoted as $\mathcal{L}(g)$. Given a collection of genes $g_1, \ldots, g_m \sim G(\mathcal{T})$, the coalescent-based species tree estimation problem is the challenge of estimating the topology T from the input data I_m, which may include the gene trees, the accompanying sequences or both.

Thus coalescent-based species tree estimation methods can work with a variety of different types of inputs. Usually such methods assume that the estimation of gene trees given sequence data can be done with statistical consistency, which is true in the case of the most common models [27]. In this paper, we will consider the input data I to include, broadly, the gene trees themselves (one per gene), with or without branch lengths, or the multiple sequence alignments (one per gene), or both depending on the method. In either case, it is natural to consider the input data I as being potentially restricted to a subset $\mathcal{X}'$ of the taxa by considering, respectively, the sub-trees of each gene tree restricted to the leaves with taxa in $\mathcal{X}'$, or the multiple sequence alignment of only the sequences corresponding to taxa in $\mathcal{X}'$. We will refer occasionally to this restricted data as $I|_{\mathcal{X}'}$. In contexts where the number of genes may vary and is indexed by m, the input data I on m genes is correspondingly indexed as I_m.

Tuple-Based Methods. We will establish properties about statistical consistency in the presence of missing data for a class of coalescent-based species tree estimation methods that we collectively refer to as "tuple-based methods". As we will show, nearly all coalescent-based species tree estimation methods that have been proven to be statistically consistent under the MSC are tuple-based, so this restriction covers most of the methods in use.

A coalescent-based method is a "tuple-based" method if there is some $\ell \in \mathbb{Z}_{\geq 2}$ such that the method operates by computing a set of summary statistics from the input I for every subset of ℓ species, and then uses these summary statistics to compute the species tree. Furthermore a tuple-based method is called an ℓ-tuple-based method (or more simply an ℓ-tuple method) to reflect the specific value of ℓ on which it bases its summary statistics. We write each tuple-based method as a pair (F, α), with F the function that computes the set of summary statistics from I, and α the function that computes a species tree given $F(I)$. Also, the set of summary statistics computed by an ℓ-tuple method includes one statistic for every tree topology (possibly rooted) on every subset of ℓ species.

Since a "tree" on two species is just a path, the 2-tuple methods compute pairwise distances for every pair of species. Examples of 2-tuple methods include NJst and ASTRID, which operate by computing the "average internode distance" between every pair of species. Other 2-tuple based methods include GLASS [22] and its variants (e.g., [9]), METAL [3], STAR [18], and STEAC [18], which also compute a pairwise distance between every pair of species, but use a different technique to do the calculation. 2-tuple methods then compute a tree on the matrix of pairwise distances, using methods such as Neighbor Joining (NJ) [26] or FastME [13]; thus, NJ and FastME serve as the function α in the 2-tuple method.

MP-EST and SMRT are 3-tuple methods. MP-EST requires rooted gene trees (and so depends on the strict molecular clock), and uses the frequency of each rooted 3-leaf tree t induced in the input set of gene trees as the summary statistic for t. It then seeks the model species tree (topology and branch lengths) that is most likely to produce the observed distribution of rooted 3-leaf gene tree frequencies. SMRT is a site-based method that estimates rooted three-leaf subtrees from the concatenated gene sequence alignments, and so depends on the strict molecular clock. SMRT then combines the rooted three-leaf subtrees into a tree on the full set of taxa using the Modified Min Cut algorithm [23].

In contrast to 3-tuple methods (e.g., MP-EST and SMRT), 4-tuple methods operate on unrooted gene trees, and so do not depend on the strict molecular clock. 4-tuple methods begin by computing either the most likely tree on every four leaves, or by computing some real-valued statistic for each unrooted tree on every four leaves. An example of a 4-tuple method is ASTRAL [20,21], which uses the frequency of quartet tree t induced in the input gene trees as the real-valued support for t. Other 4-tuple methods include the population tree in BUCKy [12] (called BUCKy-pop in [37]) and the implementation of SVDquartets [2] within PAUP* [30]; in these two cases, the real-valued support for a quartet tree is either 1 or 0 (i.e., the best quartet tree on every four leaves is determined). All these 4-tuple methods then compute a species tree by applying some quartet amalgamation method to the set of quartet trees, weighted by their support values. For these 4-tuple methods, α is the quartet amalgamation technique used to construct the tree T from the set of weighted quartet trees.

The number of summary statistics that each type of method computes depends on the value for ℓ and the number n of species: 2-tuple methods compute $\binom{n}{2}$ summary statistics (one for each pair of species), 3-tuple methods compute $3\binom{n}{3}$ summary statistics (one for each rooted three-leaf tree), and 4-tuple methods compute $3\binom{n}{4}$ summary statistics (one for each unrooted four-leaf tree).

The proofs of statistical consistency for ℓ-tuple-based methods have the following basic steps: first, they show that as the number m of genes increases, the vector of summary statistics computed by F on input data I_m converges in probability to a constant vector (which we will refer to as F_0). Second, they show that $\alpha(F_0) = T$, where T is the topology of the true species tree. Third, they show that there is some $\delta > 0$ so that whenever $L_\infty(F_1, F_0) < \delta$ then $\alpha(F_1) = \alpha(F_0) = T$ (here L_∞ is the infinity-norm, i.e. the maximum absolute difference of individual vector components). It follows that the algorithm $A = (F, \alpha)$ is statistically consistent under the MSC. Therefore, when we refer to a statistically consistent ℓ-tuple method, we will assume that these properties hold for the method, and then study the impact of missing data on the method.

Proofs of statistical consistency for many coalescent-based methods typically require several extra conditions. For example, the proofs of statistical consistency of SVDquartets, MP-EST, STEM, STAR, and SMRT require that sequences evolve under the strict molecular clock. Similarly, the proofs of statistical consistency for nearly all methods that operate by combining gene trees require completely correct gene trees (for an exception to this rule see [25]), and it is unknown whether any standard coalescent-based methods that estimate species trees by combining gene trees are statistically consistent in the presence of gene tree estimation error. Another complication in the proofs of statistical consistency is the typical requirement that α provide an exact solution to an optimization problem (e.g., finding the species tree that maximizes some optimization criterion with respect to the input gene data). This is generally not an issue for 2-tuple methods, which use methods like neighbor joining [26] to compute trees from distance matrices, but can be a problem for 3-tuple and 4-tuple methods. For example, 4-tuple methods tend to have two steps, where the first step computes a set of quartet trees (using F) and the second step computes a tree from the set of quartet trees using α. Since quartet tree compatibility is NP-hard [28], quartet amalgamation methods are typically heuristics that have no guarantees (the dynamic programming algorithm in ASTRAL is one of the few exceptions to this), and may not even be guaranteed to return a tree T when given its set of quartet trees. Thus, statistical consistency of coalescent-based methods is complicated, even when there are no missing data.

Extension of Tuple-Based Methods to Missing Data. These tuple-based methods are defined and described on the assumption of gene trees or sequence alignments without missing data, and the statistics or the algorithm may not be fully defined if not all taxa are present. For example, for a given gene tree, the topology for quartet $ijkl$ does not exist if one or more of the species is missing from the gene. Intuitively, if the method would have called for the calculation of a statistic on a particular set of taxa for a particular gene, it is not possible to calculate this if any taxon in the set is not present, so that gene should be excluded for purposes of that statistic. Thus, the natural extension of a tuple-based method (F, α) to inputs with missing data (species missing from genes) is as follows:

Definition 1. *Let $A = (F, \alpha)$ be an ℓ-tuple species tree estimation method. The* **natural extension** *of A computes the summary statistics for a given set B of ℓ species based only on those genes that contain all the species in B.*

Type 1 and Type 2 ℓ-Tuple Methods. Since the set of summary statistics includes a real number for every tree t on ℓ species, we will let $F_t(I)$ denote the summary statistic computed by the function F for tree t given input I. For a set B of

ℓ species drawn from the full set $\mathcal{X}$ of species, let $I|_B$ denote the input set I restricted to B; thus, all species in $\mathcal{X} \setminus B$ are deleted entirely from the input. Then tuple-based methods can be characterized further depending on how they behave on such inputs. Specifically, we will partition ℓ-tuple methods (F, α) into two categories:

- Type 1: For all inputs I, all sets B of ℓ species from $\mathcal{X}$, and all trees t on B, $F_t(I) = F_t(I|_B)$.
- Type 2: There is at least one input I, one set B of ℓ species, and one tree t on B such that $F_t(I) \neq F_t(I|_B)$.

Thus, a Type 1 ℓ-tuple method has the property that deleting taxa from outside a set B does not impact the summary statistics it computes for any tree on B. Note that taxon deletion impacts both Type 1 and Type 2 methods, in that if enough taxa are deleted from enough genes then accuracy must decrease. As we will see, Type 1 methods are easier to analyze than Type 2 methods, and in particular it is easy to prove that a Type 1 method remains statistically consistent in the presence of missing data for some models of random taxon deletion. Most statistically consistent coalescent-based methods are Type 1 tuple-based methods; for example, ASTRAL, GLASS, METAL, MP-EST, STEAC, and SVDquartets are all Type 1 tuple-based methods. ASTRID, NJst, and STAR are Type 2 methods.

2.2 Taxon Deletion Models

Let $\mathcal{T}$ be a species tree on a set $\mathcal{X}$ of n species, with $\mathcal{X} = \{x_i\}_{i=1}^n$, and let m gene trees evolve within $\mathcal{T}$ under the multi-species coalescent model. We denote the set of gene trees by $\mathbf{T} = \{T_i\}_{i=1}^m$, and the set of genes by $\mathcal{G} = \{g_i\}_{i=1}^m$. To model taxon deletion, we let g_i denote an arbitrary gene, and $Y_i = [Y_{i1}, \ldots, Y_{in}]^T$ be a random n-dimensional vector where

$$Y_{ij} = \mathbb{I}_{\{x_j \text{ is present in } g_i\}} \tag{1}$$

Here each individual Y_{ij} is a binary random variable that represents whether a species x_j is present for a random gene g_i.

Exchangeability. For the following lemma, we will assume that if T_i is generated before Y_i, then the post-deletion tree $T_i^* = T_i|Y_i = y_i$ is obtained by taking the sub-tree of T_i restricted to the set of leaves $\{x_j|y_{ij} = 1\}$, that is, the set of leaves whose taxa have not been deleted. If Y_i is generated before T_i, then T_i^* is obtained by taking the same sub-tree of the species tree $\mathcal{T}$, $\mathcal{T}|Y_i$ and simulating a gene tree within this species sub-tree under the multi-species coalescent.

Lemma 2. *If T_i and Y_i are independent, the two variables are exchangable and the distribution of T_i^* does not depend on the sequence.*

Proof. If Y_i is generated first, then the conditional distribution of T_i^* is equal to the distribution of gene trees under the multi-species coalescent on $\mathcal{T}|Y_i$, by definition.

If T_i is generated first, then the pruning operations described above mean that T_i^* will lie entirely within the subtree $\mathcal{T}|Y_i$. It remains to show that the probability of any given pattern of coalescence on the remaining branches is identical to the MSC under $\mathcal{T}|Y_i$. This follows from the memoryless property of coalescence under the MSC: the probability of any two lineages originating within $\mathcal{T}|Y_i$ coalescing at any given point is not dependent on either lineage's coalescent history.

It should be noted by this model description, taxa are absent or present independently of the generation of the gene data, including tree topology and sequence evolution, and the two processes are exchangeable. Also, as is the case with the general multi-species coalescent model, gene trees evolve under a process that is *i.i.d.* with respect to one another.

We will now define the two models for taxon deletion described briefly earlier.

The i.i.d. Model (M_{iid}). M_{iid} is a family of models parameterized by p, with $0 < p < 1$, where p is the probability that a random gene is present in a random species. For the M_{iid} model for parameter p, we assume that $Y_{ij} \sim Bernoulli(p)$ for all genes i and all taxa j, and that Y_{ij} and Y_{kj} are independent for $k \neq i$. (By extension of the statement earlier that genes evolve independently of one another, this also implies that Y_{ij} and Y_{ik} are independent for genes j, k where $j \neq k$.)

The Full Subset Coverage Model (M_{fsc}). M_{fsc} is a family of models parameterized by $k \geq 2$. We assume that the taxon deletion process is *i.i.d.* across the genes but we do not assume that it is *i.i.d.* across species. An M_{fsc} model for parameter k satisfies the property that for any subset B of at most k species there is a strictly positive probability p_B (that can depend on B) so that given a random gene, every member of B is present in the data for that gene with probability p_B. Since the number of taxon sets of size at most k is finite, $p^* = \min\{p_B : B \subseteq \mathcal{X}, |B| \leq k\} > 0$; hence, every taxon subset of size at most k appears in a random gene with probability at least p^*. Note that every M_{iid} model satisfies the property of being an M_{fsc} model for every k.

Comparison to Previous Models. Most prior studies of the impact of missing data on phylogenomic analysis have been performed under the M_{iid} model; this model is referred to as **R** in [36] and as the "random allocation" model in [8]. [36] also considered the **G** model, where missing data are concentrated in a subset of randomly chosen genes, and then ingroup taxa are deleted under an *i.i.d.* process from these genes. [36] also studied the **S** model, where missing data are allowed only in a subset of randomly chosen ingroup species, and that the genes are deleted from the selected species under an *i.i.d.* process. Note that the **S** and **G** models studied in [36] are M_{fsc} models.

3 Results

3.1 Results under an adversary model of taxon deletion

Theorem 3. *Let taxon deletion be dependent on gene tree topology. There exists a dependency structure under which no method is statistically consistent.*

Proof. Let T and T' be possible species trees; note that T' has a topology that appears with strictly positive probability under the MSC for species tree T. For each gene g_i (whose true gene tree topology we'll denote as t_i for clarity), consider the dependency structure where all taxa are present in the data for g_i with probability 1 if the topology of t_i is identical to T', and all taxa are absent with probability 1 otherwise. Effectively, gene g_i is observed if and only if it has topology identical to T'. Then the distribution of observed gene data is not unique to the species tree and identifiability of the species tree is lost, so no method can be statistically consistent.

The theorem above demonstrates that a dependence between gene tree topology and taxon presence can quickly unravel statistical consistency guarantees in the absence of additional assumptions. But such a dependence may exist for some realistic models of gene presence/absence, including a birth/death type model where a gene may be present only for a clade of the tree. Such models are interesting and unsolved, but are beyond the scope of this paper.

3.2 Results for Type 1 Methods Under M_{fsc}

We now discuss the statistical consistency guarantees of Type 1 tuple-based methods. As we will see, most of the tuple-based methods remain statistically consistent even in the presence of missing data, as long as the process that generates the missing taxa is well-behaved (e.g., not generated by an adversary that biases the method towards the wrong tree).

Let $A = (F, \alpha)$ be a Type 1 ℓ-tuple method that satisfies the following properties:

- (i) For all model species trees $T = (T, \Theta)$, as the number m of genes increases, $F(I_m) \xrightarrow{p} F_0$, where F_0 is a constant vector parameterized by T.
- (ii) There exists $\delta > 0$ such that for all vectors of summary statistics F_1 satisfying $L_\infty(F_1, F_0) < \delta$, $\alpha(F_1) = \alpha(F_0) = T$.

Theorem 4. *Let $A = (F, \alpha)$ be a Type 1 ℓ-tuple species tree estimation method satisfying the two properties (i) and (ii) above, and assume that the number of species is at least ℓ. The natural extension of A is statistically consistent under M_{fsc} with parameter $k \geq \ell$, and thus also under M_{iid} for any parameter p.*

Proof. Let $T = (T, \Theta)$ be the model species tree, I_m be the input dataset containing m genes, C be the number of summary statistics computed by algorithm $A = (F, \alpha)$ on input I_m. Since $A = (F, \alpha)$ satisfies condition (i) when there are no missing data, then as the number of genes m increases, $F(I_m) \xrightarrow{p} F_0$, where F_0 is a vector of constants. We will denote the ith summary statistic computed on input I_m by $F_i(I_m)$ and the i-th component of F_0 as F_{0_i}. We write $F(I_m) = (F_1(I_m|_{\mathbf{x}_1}), \ldots, F_C(I_m|_{\mathbf{x}_C}))$ where $\mathbf{x}_i$ denotes a particular set of ℓ taxa. In other words, since A satisfies condition (i) when there are no missing data, for all $i = 1, \ldots, C$ there exist a constant F_{0_i} such that $F_i(I_m|_{\mathbf{x}_i}) \xrightarrow{p} F_{0_i}$ as $m \to \infty$. Since the data for each gene are independent of all others, to prove statistical consistency under the M_{fsc} model we merely require that $I_m|_{\mathbf{x}_i}$ include an infinite number of genes as $m \to \infty$. Under the M_{fsc} model, $Pr[\mathbf{x}_i \subseteq L(g)] > 0$ for every gene g (where $\mathcal{L}(g)$ denotes the set of species for gene g). Hence, by the Borel-Cantelli lemma, the number of genes that include all ℓ taxa in $\mathbf{x}_i$ will also approach infinity. Thus $I_m|_{\mathbf{x}_i}$ will include an infinite number of genes, and $F(I_m|_{\mathbf{x}_i}) \xrightarrow{p} F_{0_i}$. By the definition of the natural extension of A, α does not change under deleted taxa. Since A satisfies condition (ii), $\exists \delta > 0$ such that $\forall F_1$ with $L_\infty(F_1, F_0) < \delta$, $\alpha(F_1) = T$, and so the natural extension of A is statistically consistent under M_{fsc}. Since M_{iid} is a subset of M_{fsc}, it is also statistically consistent under M_{iid}.

Corollary 5. *ASTRAL and METAL are statistically consistent under the MSC even when taxa are deleted under an M_{fsc} model, provided that each is run in exact mode and so finds globally optimal species trees. MP-EST and STEM are statistically consistent under the MSC even when taxa are deleted under an M_{fsc} model, if sequence evolution is under a strict molecular clock and they find globally optimal species trees. SVDquartets is statistically consistent under the MSC even when taxa are deleted under an M_{fsc} model, if sequence evolution is under a strict molecular clock and the quartet amalgamation heuristic used is modified to ensure that it returns a compatibility tree when the input set of quartets is compatible. SMRT is statistically consistent under the MSC even when taxa are deleted under an M_{fsc} model, if sequence evolution is under the symmetric two-state model with a strict molecular clock.*

3.3   Statistical Consistency of Versions of ASTRAL Under M_{fsc}

ASTRAL-1 [20] (and its improved version, ASTRAL-2 [21]) are coalescent-based methods for estimating species trees that take unrooted gene trees as input, and return a tree that minimizes the quartet tree distance to the input gene trees. Each can be run in exact mode, which guarantees that the tree that is returned has the minimum distance to the input gene trees. However, the exact versions are computationally intensive (running in time that grows exponentially with the number of species), and so heuristic versions are also available. These heuristic versions operate by constraining the search space using the input set of gene trees, and then guarantee that an optimal tree is returned within the search space. The important difference between the two methods is how the search space is constrained, and ASTRAL-2 explicitly enlarges the space compared to ASTRAL-1 when the input gene trees can be incomplete (i.e., when some species are missing from some gene trees). Because the search space is constrained using the input gene trees, the two ASTRAL algorithms depends on the input in a way that makes the analysis of their statistical guarantees non-trivial.

This section shows that both ASTRAL-1 and ASTRAL-2 are statistically consistent under the M_{iid} model, but not under any M_{fsc} model. We then present a modification to ASTRAL-1, denoted by ASTRAL*, and which differs from ASTRAL-1 only in how the search space is constrained. We then show that ASTRAL* is statistically consistent under many M_{fsc} models.

ASTRAL-1. We begin with a formal description of the ASTRAL-1 algorithm.

Notation. We let $\mathcal{X}$ denote the full set of species, and $\mathcal{X}'$ denote an arbitrary subset of $\mathcal{X}$. Every tree t we consider is assumed to be a binary unrooted tree with leaves taken from a subset of $\mathcal{X}$, and as earlier we denote the leafset of t by $\mathcal{L}(t)$. Each edge of t defines a **bipartition** of the set $\mathcal{L}(t)$ (denoted by $B|B'$, for some set $B \subseteq \mathcal{X}$ and $B' = \mathcal{L}(t) \setminus B$) obtained by deleting the edge but not its endpoints from t. We will refer to the set of all these bipartitions as $Bip(t)$, and the set of halves of the bipartitions of t as the **clades** of t. (Note that the term "clades" is normally used only in the context of rooted trees, but we extend the term here to allow us to refer to halves of bipartitions using the same term.) We let $T_{\mathcal{X}}(X)$ denote the set of unrooted binary trees on leafset $\mathcal{X}$ that satisfy $Bip(t) \subseteq X$. If X is not provided, then we assume the set of unrooted binary trees is not constrained, and let $T_{\mathcal{X}}$ denote the set of all unrooted binary trees on leafset $\mathcal{X}$.

We let $Q(t)$ denote all 4-leaf homeomorphic subtrees of t induced by a set of four leaves in t, and we note that when t is binary (i.e., fully resolved), then $Q(t)$ contains only binary quartet trees. Let $\mathcal{Q}$ be the set of all $\binom{n}{4}$ 4-taxon subsets of the taxon set $\mathcal{X}$. Let $q \in \mathcal{Q}$, let t be an arbitrary tree topology on $\mathcal{X}$ and let $Top(q, t)$ denote the induced quartet subtree topology for quartet q in t.

Definition 6. *ASTRAL Optimization Problem*
   Input: *Taxon set* $\mathcal{X} = \{x_i\}_{i=1}^n$, *gene trees* $t_1, \ldots, t_m$, *and set X of allowed bipartitions of $\mathcal{X}$.*
   Output: *Binary tree T where*

$$T = \arg \max_{t \in T_{\mathcal{X}}(X)} \sum_{q \in \mathcal{Q}} \sum_{i=1}^m \mathbb{I}_{\{Top(q,t)=Top(q,g_i)\}}$$

ASTRAL-1 will specifically return the tree that maximizes the optimization criteria in the innermost summation subject to the constraint that every bipartition in the output tree be included in the set $\mathcal{X}$. Therefore, to show consistency under a model of missing data it is necessary to show not only that the optimization criteria still works, but also that the true topology will be still be included in this constrained search space.

ASTRAL-1 and ASTRAL-2 differ in how they define the default set X of allowed bipartitions, and also use slightly different dynamic programming techniques to assemble the optimal tree from the bottom up. Note that to run

ASTRAL-1 or ASTRAL-2 in exact mode, the set X is defined to be all bipartitions on $\mathcal{X}$. In the default version of ASTRAL-1 (referred to as the "heuristic version"), X is the set of all bipartitions that appear in any gene tree. Hence, when there are no missing data, then as the number of genes increases, the set X will include all possible bipartitions on the taxon set with probability converging to 1 (and hence in particular the bipartitions in the true species tree). However, when there are missing data, then proving that the set X contains all the bipartitions in the species tree takes some care. In particular, if every gene tree is incomplete, then no bipartition in any gene tree is a bipartition of the full set of taxa, and so this default setting will not enable a statistically consistent estimation method.

ASTRAL*: We will modify ASTRAL-1 by changing how it defines the set X of allowed bipartitions, and refer to this modification as ASTRAL*. Specifically, we will add bipartitions to the default setting computed by ASTRAL-1. Hence, ASTRAL*'s extra bipartitions could also be added to ASTRAL-2.

Note that ASTRAL-1, ASTRAL-2, and hence also ASTRAL, when run in heuristic mode, are different from the species tree estimation methods described previously, in that α depends not only on the summary statistics $F(I)$ but also on the input data I. Therefore, we denote the output of the function by $\alpha(F, I)$.

For every clade $C \subset \mathcal{X}$ occurring in a gene tree, we require that ASTRAL* adds the bipartition $C|C'$ where $C' = \mathcal{X} \setminus C$, to its set X. (Note that since the trees in this problem are unrooted, a clade and one half of a bipartition are equivalent concepts.) This is a trivial extension of the algorithm for a model of incomplete genes and one that strengthens the conditions under which the method is consistent, as we will see below.

Theorem 7. *(1) ASTRAL*, as well as ASTRAL-1 and ASTRAL-2 run in default heuristic mode, are statistically consistent under the MSC for any M_{iid} model of taxon deletion. (2) ASTRAL-1 is not statistically consistent under an M_{fsc} model with parameter k if the number of species is greater than k and ASTRAL-1 is run in default heuristic mode. (3) ASTRAL* is statistically consistent under any M_{fsc} model of taxon deletion with parameter k if the number n of species is at most $2k$.*

Proof

(1) Let $\mathcal{T}$ be a model species tree, and consider taxon deletion under some M_{iid} model. We will show that there is non-zero probability that every bipartition in the species tree appears in the search space computed by ASTRAL-1 in its default setting. Since the search space computed by ASTRAL-1 is a subset of the search space computed by ASTRAL-2 and ASTRAL*, the result will follow. Recall that ASTRAL-1 includes all bipartitions $C|C'$ that appear in any input gene tree. Under the MSC model, every bipartition appears in some gene tree with probability increasing to 1 as the number of genes increases. Under any M_{iid} model, for every subset of taxa, the probability that none of the taxa in the subset are deleted is strictly greater than 0.

Hence, under M_{iid}, the set X of bipartitions allowed in the ASTRAL-1 search space will converge to the set of all possible bipartitions. Therefore, ASTRAL-1 is statistically consistent under M_{iid}. Since the sets of bipartitions computed by ASTRAL* and ASTRAL-2 contain the set of bipartitions computed by ASTRAL-1, ASTRAL-2 and ASTRAL* are also statistically consistent under M_{iid}.

(2) Now consider the M_{fsc} model with parameter k: let $n > k$, and further let the taxon deletion process be such that every gene has exactly k taxa, (e.g. k taxa sampled uniformly, a valid model under M_{fsc}). Then if ASTRAL-1 is run in heuristic mode, it will compute a set X that contains no bipartitions on the set $\mathcal{X}$ of species, and so cannot be statistically consistent under all M_{fsc} models.

(3) We now show that ASTRAL* run in heuristic mode is statistically consistent under any M_{fsc} model with parameter k when $n \leq 2k$. Let $C|C'$ be an arbitrary bipartition on $\mathcal{X}$, and assume without loss of generality that $|C| \leq k$. Hence, under M_{iid}, the probability that C appears in a random gene tree is strictly positive. Under the MSC, any bipartition on $\mathcal{X}$ appears in a given true gene tree with strictly positive probability. Since this process is independent from the removal of taxa, and since there is non-zero probability that all members of a clade appear in the gene tree, the probability is non-zero that the set C appears as a clade in a random gene tree.

Hence, as the number m of gene trees increases, the probability approaches 1 that C appears as a clade in at least one gene tree. Thus the probability approaches 1 that the set X computed by ASTRAL* will contain $C|C'$, where $C' = \mathcal{X} \setminus C$. Therefore, ASTRAL*, run in heuristic mode, will be statistically consistent under the M_{fsc} model with parameter k, provided that the number of species $n \leq 2k$.

Theorem 8. *ASTRAL-1 and ASTRAL*, when run in heuristic mode, are not statistically consistent under the MSC_{fsc} class of models with parameter k, if the number of species $n > 2k$.*

Proof. Consider a model of taxon deletion where every gene tree has exactly k taxa, selected at random from the full set of taxa. This model satisfies the conditions of the M_{fsc} models with parameter k. Now assume $k < \lfloor n/2 \rfloor$.

Let $\mathcal{T}$ be a caterpillar tree on a set $\mathcal{X}$ of n taxa. Then $\mathcal{T}$ contains a clade B of size $\lfloor n/2 \rfloor$ whose complement is at least as large; hence both B and $\mathcal{X} \setminus B$ have more than k species. Hence, under this model of taxon deletion, neither B nor $\mathcal{X} \setminus B$ will be in any gene tree. Hence, the bipartition $B|B'$ (where $B' = \mathcal{X} \setminus B$) will not be in X (the constraint on the search space) as computed by ASTRAL and ASTRAL*. Hence, neither ASTRAL nor ASTRAL* can recover the true species tree under this random taxon deletion model.

3.4 Statistical Consistency of ASTRID and NJst Under M_{iid}

As noted earlier, ASTRID, NJst, and STAR are Type 2 methods, and proofs we provided of statistical consistency for Type 1 tuple-based methods do not apply to these methods (or other Type 2 methods). In this section we will show ASTRID and NJst remain statistically consistent under the M_{iid} models of taxon deletion. However, the statistical consistency of these methods under the more general M_{fsc} models is unknown.

NJst is a distance-based method that uses the average topological "internode" distance between taxa in the gene trees. The internode distance between two taxa x_i and x_j in a tree is the count of individual nodes along the path from x_i to x_j, denoted $\rho(x_i, x_j)$. ASTRID is an extension of NJst with the averaging redefined to better accommodate missing taxa and is precisely the natural extension of NJst as defined earlier, where the statistic for each pair is calculated as usual but restricted to the genes in which both members of the pair appear. However, NJst and ASTRID are not Type 1 methods, under the definition provided above, because the internode distance for two taxa x_i and x_j can be affected by the presence or absence of a third taxon.

Distance methods are formally 2-tuple methods, and use an algorithm such as neighbor joining [26] to return a tree topology and branch lengths given $\binom{n}{2}$ pairwise distances (collectively, the distance "matrix"). We now state some well-known properties of such methods for reference in the proof below. For a tree with topology $T = (V, E)$ on n taxa and edge weights l_e, $e \in E$, if the distance for any two taxa j and k is equal to the sum of the edge weights over edges in the shortest path between leaves j and k, then neighbor-joining will return a tree with topology T, and the distance matrix is said to be **additive** on the topology T. An equivalent definition of an additive matrix is as follows:

Definition 9. The Four Point Condition. *Let $T = (V, E)$ be a tree on n leaves labeled as $\mathcal{S} = \{s_i\}_{i=1}^n$ with positive edge weights l_e, $e \in E$. Let $D = d_{ij}$ be the matrix of pairwise distances between all pairs of taxa (i, j), $i, j \in \mathcal{S}$. The matrix D is additive on the topology T if and only if for all sets of four leaves $\{i, j, k, l\} \subset \mathcal{S}$ with quartet-subtree topology $ij|kl$ in T, without loss of generality, the following holds:*

$$d_{ij} + d_{kl} < d_{ik} + d_{jl} = d_{il} + d_{jk}.$$

Furthermore, if instead of D as above we are given $\hat{D} = d_{ij} + \varepsilon_{ij}$, where ε_{ij} is an unknown noise term such that for all $i, j \in \mathcal{S}$, $|\varepsilon_{ij}| < \frac{1}{2} \min_{a, b \in \mathcal{S}} d_{ab}$, then neighbor-joining applied to the matrix $\hat{D}$ will also return the topology T with probability 1. Therefore, since NJst is a distance-method, to prove statistical consistency it suffices to show that the metric given by the average internode distance collectively converges to an additive matrix on the true topology.

Theorem 10. *Assume that taxa are absent from the data for each gene according to the M_{iid} model. Then NJst and ASTRID are statistically consistent under the MSC.*

First we give a helpful lemma. Note that since $\rho(x_i, x_j)$ is undefined when either of x_i or x_j are removed, the expectation of $\rho(x_i, x_j)$ is formally undefined as long as the probability of either is nonzero. We nonetheless use the notation

$\mathbf{E}\left[\rho(x_i, x_j)\right]$ in the lemma and proofs below, which will refer implicitly to the conditional expectation on the event that neither x_i nor x_j are removed.

Lemma 11. *Under the MSC and M_{iid}, let a, b, c and d be four taxa, and consider the event in the coalescent probability space in which the lineages of these taxa have entered a common population and no pair have coalesced with one another, denoted as $\mathcal{E}_{abcd}$. Denote the points on each respective lineage in which they enter the common population as A, B, C, and D. Let Y be the random variable representing the taxon deletion process. Then for any two taxa $\{i, j\} \subset \{a, b, c, d\}$ and respectively $\{I, J\} \subset \{A, B, C, D\}$:*

$$\mathbf{E}\left[\rho(i, j)|\mathcal{E}\right] = \mathbf{E}_Y\left[\rho(i, I)|\mathcal{E}\right] + \mathbf{E}_Y\left[\rho(j, J)|\mathcal{E}\right] + K$$

where K is a constant that does not depend on the identities of i and j.

Proof (Lemma 11). Consider any gene tree g meeting the condition of event $\mathcal{E}$, and let G be the subtree of g sitting between the points A, B, C and D and the root. The topology of G in this model, given that $g \in \mathcal{E}$, is determined by a set of *i.i.d.* Exponentially distributed random variables corresponding to the pairwise times-to-coalescence of all remaining lineages concurrent with and including $\{A, B, C, D\}$. By De Finetti's theorem, the probability density function of G is unique up to a permutation of the indices of the random variables. Thus for any $\{I, J\} \subset \{A, B, C, D\}$, since $\rho(I, J)$ is dependent only on the topology of G, the probability density of $\rho(I, J)$ does not depend on the identity of I and J, and thus $\mathbf{E}\left[\rho(I, J)|\mathcal{E}\right] = K$.

Proof (Theorem 10). In order to show that the method is statistically consistent, we must show the following:

1. For a gene tree generated under the coalescent process in the MSC followed by the removal of taxa subject to M_{iid}, the expected value of the summary statistics $\rho(x_i, x_j)$ for each pair of taxa x_i, x_j form an additive matrix that defines the topology of the species tree.
2. The statistics themselves converge to their expectations as the number m of genes approaches infinity.

The second property follows from the *i.i.d.*-generation of gene trees and the weak law of large numbers; hence, we need only establish the first property.

Let G be a random gene tree on n taxa generated by the MSC on species tree $\mathcal{T}$, and let $Y_n \sim M_{iid}$ with probability of deletion p. We will show that for an arbitrary set of four taxa $\{x_1, x_2, x_a, x_b\}$, the expectations over G and Y_n of the internode distances are additive on the species tree, which we check by confirming that they obey the four point condition for the species tree $\mathcal{T}$. In what follows, we will refer to this by saying that the expectations "obey the four point condition", with the understanding that this is with reference to the species tree $\mathcal{T}$. Importantly, the event $\mathcal{E}$ defined in Lemma 11 includes all cases in which the quartet-subtree topology in the gene tree does not match that of the species tree, and implies that the four point condition holds in these cases.

As a result, it suffices to show that the four point condition holds when the quartet-subtree in G is identical to the topology in the species tree, which we assume without loss of generality has topology $x_1x_2|x_ax_b$. We will show this by induction on the number n of taxa, and begin with the smallest non-trivial case, $n = 4$.

Base Step: n = 4. For $n = 4$ we can write the expected values of the internode distances in closed form and check the four point condition directly:

$$\mathbf{E}\left[\rho(x_1, x_2)\right] = 2 - p^2$$
$$\mathbf{E}\left[\rho(x_a, x_b)\right] = 2 - p^2$$
$$\mathbf{E}\left[\rho(x_1, x_a)\right] = 3(1 - p)(1 - p) + 2(p + p - p^2) + 1(p^2)$$
$$= 3 - 3p - 3p + 3p^2 + 2p + 2p - 2p^2 + p62$$
$$= 3 - 2p + 2p^2$$
$$\mathbf{E}\left[\rho(x_2, x_b)\right] = 3 - 2p + 2p^2$$

Thus to test that the four point condition holds, we have:

$$\mathbf{E}\left[\rho(x_1, x_2)\right] + \mathbf{E}\left[\rho(x_a, x_b)\right] = 4 - p^2 - p^2$$
$$= 4 - 2p^2$$
$$\mathbf{E}\left[\rho(x_1, x_a)\right] + \mathbf{E}\left[\rho(x_2, x_b)\right] = 6 - 4p + 4(p^2)$$
$$\mathbf{E}\left[\rho(x_2, x_a)\right] + \mathbf{E}\left[\rho(x_1, x_b)\right] = 6 - 4p + 4(p^2)$$

Thus the four point condition holds for $n = 4$.

Induction Step: Assume that for a set S_n of n taxa, the expected value of the matrix $D = [\rho(x_i, x_j)]$ for a random gene tree G_n and taxon-removal variable Y_n is additive on the true species tree. That is, for any set of n-taxa under the MSC and M_{iid} and any quartet (a, b, c, d), the four point condition holds for $\mathbf{E}_n[\rho(i, j)]$, $i, j \in \{a, b, c, d\}$, a fact that we will use below. Here $\mathbf{E}_n$ denotes the expectation operator under n taxa for notational clarity. We will now show that the same holds for a set of size $n + 1$.

Let x_{n+1} be a new taxon and let G be generated under the MSC on $S_n \cup \{x_{n+1}\}$ with taxon-removal variable $Y_{n+1} = (Y_n, y_{n+1})$ where $y_{n+1} \sim Bernoulli(p)$ in accordance with M_{iid}. Our approach will be to define three cases, each of which have non-zero probability, and show that regardless of the placement of x_{n+1} in the species tree, the four point condition holds on $\{x_1, x_2, x_a, x_b\}$.

Case 1: x_{n+1} is deleted from G (i.e. $y_{n+1} = 1$). In this case the conditional values of $\mathbf{E}_{n+1}[\rho(x, x')]$ for $x \neq x'$, $x, x' \in \{x_1, x_2, x_a, x_b\}$ are identical to the n-taxa case, and thus by our assumption they obey the four point condition.

Case 2: x_{n+1} is not deleted and it coalesces on a branch that is *not* on the induced quartet-subtree of G for the quartet $\{x_1, x_2, x_a, x_b\}$. In this case, the coalescence event for x_{n+1} does not add to the internode distances along any of the branches connecting any two members of the quartet. As a result, again the conditional values of $\mathbf{E}[\rho(x, x')]$ for $x \neq x'$, $x, x' \in \{x_i, x_j, x_k, x_l\}$ are identical to the n-taxa case, where again they obey the four point condition, by the induction assumption.

Case 3: x_{n+1} is not deleted and coalesces directly with a branch on the induced quartet subtree of G for the quartet $\{x_1, x_2, x_a, x_b\}$. This case is non-trivial and requires some analysis. For shorthand, we will refer to the quartet-subtree of G restricted to $\{x_1, x_2, x_a, x_b\}$ as simply q. For a pair of taxa $i, j \in S_n$ the value of the expected internode distance in the presence of x_{n+1} is:

$$\mathbf{E}_{n+1}[\rho(i, j)] = \mathbf{E}_n[\rho(i, j)] + P(i, j) \tag{2}$$

where $P(i, j)$ denotes the probability that x_{n+1} coalesces on a branch in the path from i to j, which would cause $\rho(i, j)$ to increase by 1. This quantity is non-trivial and depends jointly on both the topology of G *and* the value of Y_n. However, since the first term obeys the four point condition by the induction hypothesis, the proof depends on showing that $P(i, j)$ does as well.

To do that, we will filrst partition the joint probability space of the MSC and M_{iid}, denoted as $\mathcal{G}$ and assign each part of this partition to one of the five branches in the subtree q, then show that for $\{i, j\} \subset \{x_1, x_2, x_a, x_b\}$, $P(i, j)$ can be expressed as the sum of probabilities assigned to the branches between i and j. We will label the branches as b_1, b_2, b_a, and b_b for the four outer branches and b_m for the middle branch.

Figure 1 describes the partition based on the branch with which x_{n+1} coalesces (in each column) and the outcome of the taxon-removal process (in each row). For the illustrations in the header of each row, a branch represented with a dotted line represents the case that **all** lineages coalescing along that branch other than x_{n+1} are fully removed by the taxon-removal process. The positions of the relative taxa are given in the illustration in the first row header, and all taxon-removal outcomes that allow at least one $\rho(i, j)$, $i, j \in \{x_1, x_2, x_a, x_b\}$ to be measured are represented in the rows. The assignment is not unique, as implied by the entries with offer two possibilities, but for the proof either will work so we consider the default to be the first branch listed.

For an event $(G, Y_{n+1}) \in \mathcal{G}$, denote the assignment of (G, Y_{n+1}) to branch b_e as $(G, Y_{n+1}) \to b_e$, and for that branch let

$$p_e = P(\{(G, Y_{n+1}) \in \mathcal{G} | (G, Y_{n+1}) \to b_e\}).$$

In this way, for any pair of taxa $i, j \in \{x_1, x_2, x_a, x_b\}$, the probability that $\rho(i, j)$ is incremented by 1 in the presence of x_{n+1} is given precisely by the sum of p_e for set of edges e between i and j in the quartet subtree q. This should be apparent by visual inspection of the table. Thus, $P(i, j)$ in (2) is the sum of positive edge weights on the topology $x_1 x_2 | x_a x_b$, which is also the species tree topology on these four taxa, and so is additive on the species tree. Hence, it meets the four point condition, completing the proof.

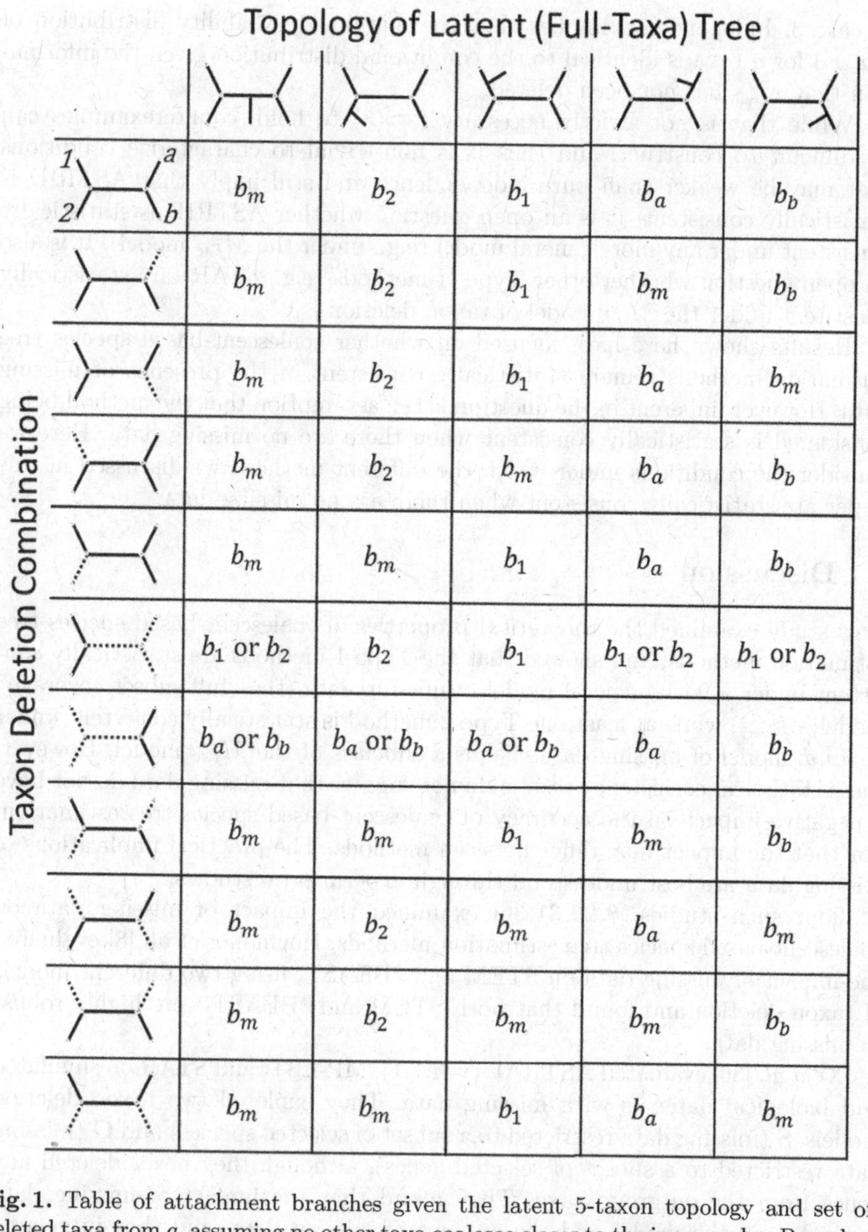

Topology of Latent (Full-Taxa) Tree

Taxon Deletion Combination

	Topology 1	Topology 2	Topology 3	Topology 4	Topology 5
	b_m	b_2	b_1	b_a	b_b
	b_m	b_2	b_1	b_m	b_b
	b_m	b_2	b_1	b_a	b_m
	b_m	b_2	b_m	b_a	b_b
	b_m	b_m	b_1	b_a	b_b
	b_1 or b_2	b_2	b_1	b_1 or b_2	b_1 or b_2
	b_a or b_b	b_a or b_b	b_a or b_b	b_a	b_b
	b_m	b_m	b_1	b_m	b_b
	b_m	b_2	b_m	b_a	b_m
	b_m	b_2	b_m	b_m	b_b
	b_m	b_m	b_1	b_a	b_m

Fig. 1. Table of attachment branches given the latent 5-taxon topology and set of deleted taxa from q, assuming no other taxa coalesce closer to the inner nodes. Branches (in rows) given by dotted lines are pruned as a result of taxon deletion.

This proof has been limited to ASTRID, and only provided under the M_{iid} model of taxon deletion because the independence of taxon deletion (between taxa as well as from gene tree generation) was used when we noted that

$$\mathbf{E}_{n+1}[\rho(x, x')] = \mathbf{E}_n[\rho(x, x')] + p_{xx'} \qquad (3)$$

in case 3. Independence implies that the marginal probability distribution of $\rho(x, x')$ for n taxa is identical to the conditional distribution given the information that x_{n+1} has not been deleted.

While that is not strictly necessary for (3) to hold, counterexamples can be difficult to construct, and thus it is non-trivial to characterize conditions that may be weaker than pure independence and still imply that ASTRID is statistically consistent. It is an open question whether ASTRID is statistically consistent under any more general model (e.g., under the M_{fsc} model). It is also an open question whether other Type II methods (e.g., STAR) are statistically consistent under the M_{iid} model of taxon deletion.

Results shown here have focused on whether coalescent-based species tree estimation methods remain statistically consistent in the presence of missing data. However, inherent in the question is the assumption that the method being considered is statistically consistent when there are no missing data. Here we consider the conditions under which the different methods we discussed in this paper are statistically consistent when there are no missing data.

4   Discussion

This study examined the theoretical properties of coalescent-based species tree estimation methods, and showed that the Type 1 methods are statistically consistent under a fairly general model of missing data (the "full subset coverage" model, M_{fsc}), while at least one Type 2 method is statistically consistent under an $i.i.d.$ model of missing data that is a subclass of the M_{fsc} model. However, these statistical consistency results do not suggest that missing data do not have a negative impact on the accuracy of coalescent-based species tree estimation, nor that the impact may differ between methods. The practical implications of missing data are best understood through experimental studies.

Four such studies [8,29,31,36] examined the impact of missing data on coalescent-based species tree estimation methods. Hovmöller et al. [8] evaluated the impact of missing data on STEM and *BEAST, using two different models of taxon deletion and found that both STEM and *BEAST were highly robust to missing data.

Xi et al. [36] evaluated ASTRAL (v. 4.7.1), MP-EST, and STAR on simulated and biological datasets with missing data. They explored two taxon deletion models, S (missing data restricted to a subset of selected species) and G (missing data restricted to a subset of selected genes), although they never deleted any genes from the outgroup taxa. They found that the impact of missing data depended on the model and the amount of missing data, and that in general ASTRAL and MP-EST were highly robust to missing data, while STAR was much more negatively impacted.

Streicher et al. [29] examined the impact of missing data on the branch support of species trees estimated using NJst and ASTRAL (v. 4.7.6) on a biological dataset of Iguanian lizards. They explored different amounts of missing data, and concluded that in general it is best to include all the data, except (perhaps) when the amount of missing data in the gene or species exceeds 50%.

Finally, Vachaspati and Warnow [31] studied the impact of missing data on ASTRID and ASTRAL-2 (v. 4.7.8) on simulated datasets with 50 species under a model in which all genes had missing data, but the genes that were deleted from each species were selected under an *i.i.d.* model. They observed that both ASTRID and ASTRAL were substantially impacted by missing data, so that error increased with taxon deletion. However, even with very high missing data rates (i.e., all genes missing 80% of the species), both ASTRID and ASTRAL-2 achieved good accuracy with a large enough number of genes.

Overall these studies suggest that missing data is not especially detrimental to accuracy using MP-EST, ASTRID, ASTRAL, STEM, and *BEAST.

However, one particularly challenge of missing data is the possibility that outgroup taxa may be the missing taxa. When the phylogenomic estimation method requires rooted gene trees, this creates a challenge of using a different rooting method (e.g., midpoint rooting, or maximum likelihood under a strict molecular clock assumption). These rooting techniques are not statistically consistent when the data does not follow a strict clock, however. Only two of the studies above [8,36] considered methods that require rooted gene trees, but both evolved sequences under a strict molecular clock. Xi et al. [36] used outgroup rooting but did not simulate under conditions where data was missing for the outgroup. Thus our understanding of the empirical impact of missing data on methods that require rooted gene trees would benefit from additional study.

5 Summary

We have shown that if taxon absence/presence is independent of gene tree topology, then it is exchangeable with the MSC. We have further shown that full subset coverage is sufficient for a tuple-based method to be statistically consistent, and that *i.i.d.* taxon deletion is a necessary condition for NJst and ASTRID to be statistically consistent.

Acknowledgments. MN was supported by NSF grants DBI-1461364, CCF-1535977 and AF:1513629 and by a fellowship from the CompGen initiative in the Coordinated Science Laboratory at UIUC. JC was supported by the Mathematics Department at UIUC.

A great deal of thanks is owed to our advisor, Dr. Tandy Warnow, who guided this manuscript from start to finish and pushed us to leave no stone unturned.

References

1. Bryant, D., Bouckaert, R., Felsenstein, J., Rosenberg, N.A., RoyChoudhury, A.: Inferring species trees directly from biallelic genetic markers: bypassing gene trees in a full coalescent analysis. Mol. Biol. Evol. **29**(8), 1917–1932 (2012)
2. Chifman, J., Kubatko, L.: Quartet inference from SNP data under the coalescent. Bioinformatics **30**(23), 3317–3324 (2014)

3. Dasarathy, G., Nowak, R., Roch, S.: Data requirement for phylogenetic inference from multiple loci: a new distance method. IEEE/ACM Trans. Comput. Biol. Bioinf. **12**(2), 422–432 (2015)
4. DeGiorgio, M., Degnan, J.H.: Fast and consistent estimation of species trees using supermatrix rooted triples. Mol. Biol. Evol. **27**(3), 552–569 (2010)
5. Edwards, S.V.: Is a new and general theory of molecular systematics emerging? Evolution **63**, 1–19 (2009)
6. Graybeal, A.: Is it better to add taxa or characters to a difficult phylogenetic problem? Syst. Biol. **47**(1), 9–17 (1998)
7. Heled, J., Drummond, A.J.: Bayesian inference of species trees from multilocus data. Mol. Biol. Evol. **27**(3), 570–580 (2010)
8. Hovmöller, R., Knowles, L.L., Kubatko, L.S.: Effects of missing data on species tree estimation under the coalescent. Mol. Phylogenet. Evol. **69**, 1057–1062 (2013)
9. Jewett, E., Rosenberg, N.: iGLASS: an improvement to the GLASS method for estimating species trees from gene trees. J. Comput. Biol. **19**(3), 293–315 (2012)
10. Kingman, J.F.C.: On the genealogy of large populations. J. Appl. Probab. **19**, 27 (1982)
11. Kubatko, L.S., Carstens, B.C., Knowles, L.L.: STEM: species tree estimation using maximum likelihood for gene trees under coalescence. Bioinformatics **25**(7), 971–973 (2009)
12. Larget, B.R., Kotha, S.K., Dewey, C.N., Ané, C.: BUCKy: gene tree/species tree reconciliation with Bayesian concordance analysis. Bioinformatics **26**(22), 2910–2911 (2010)
13. Lefort, V., Desper, R., Gascuel, O.: FastME 2.0: a comprehensive, accurate, and fast distance-based phylogeny inference program: table 1. Mol. Biol. Evol. **32**(10), 2798–2800 (2015)
14. Lemmon, A.R., Brown, J.M., Stanger-Hall, K., Lemmon, E.M.: The effect of ambiguous data on phylogenetic estimates obtained by maximum likelihood and Bayesian inference. Syst. Biol. **58**(1), 130–145 (2009)
15. Liu, L.: BEST: Bayesian estimation of species trees under the coalescent model. Bioinformatics **24**(21), 2542–2543 (2008)
16. Liu, L., Yu, L.: Estimating species trees from unrooted gene trees. Syst. Biol. **60**(5), 661–667 (2011)
17. Liu, L., Yu, L., Edwards, S.V.: A maximum pseudo-likelihood approach for estimating species trees under the coalescent model. BMC Evol. Biol. **10**(1), 302 (2010)
18. Liu, L., Yu, L., Pearl, D.K., Edwards, S.V.: Estimating species phylogenies using coalescence times among sequences. Syst. Biol. **58**(5), 468–77 (2009)
19. Maddison, W.P.: Gene trees in species trees. Syst. Biol. **46**(3), 523–536 (1997)
20. Mirarab, S., Reaz, R., Bayzid, M., Zimmermann, T., Swenson, M., Warnow, T.: ASTRAL: genome-scale coalescent-based species tree estimation. Bioinformatics **30**(17), i541–i548 (2014)
21. Mirarab, S., Warnow, T.: ASTRAL-II: coalescent-based species tree estimation with many hundreds of taxa and thousands of genes. Bioinformatics **31**(12), i44–i52 (2015)
22. Mossel, E., Roch, S.: Incomplete lineage sorting: consistent phylogeny estimation from multiple loci. IEEE/ACM Trans. Comput. Biol. Bioinf. **7**(1), 166–171 (2010)
23. Page, R.D.M.: Modified mincut supertrees. In: Guigó, R., Gusfield, D. (eds.) WABI 2002. LNCS, vol. 2452, pp. 537–551. Springer, Heidelberg (2002). doi:10.1007/3-540-45784-4_41
24. Pollock, D.D., Zwickl, D.J., McGuire, J.A., Hillis, D.M.: Increased taxon sampling is advantageous for phylogenetic inference. Syst. Biol. **51**, 664–671 (2002)

25. Roch, S., Warnow, T.: On the robustness to gene tree estimation error (or lack thereof) of coalescent-based species tree methods. Syst. Biol. **64**(4), 663–676 (2015)
26. Saitou, N., Nei, M.: The neighbor-joining method: a new method for reconstructing phylogenetic trees. Mol. Biol. Evol. **4**, 406–425 (1987)
27. Semple, C., Steel, M.: Phylogenetics. Oxford Lecture Series in Mathematics and its Applications. Oxford University Press, Oxford (2003)
28. Steel, M.: The complexity of reconstructing trees from qualitative characters and subtrees. J. Classif. **9**, 91–116 (1992)
29. Streicher, J.W., Schulte, J.A., Wiens, J.J.: How should genes and taxa be sampled for phylogenomic analyses with missing data? An empirical study in iguanian lizards. Syst. Biol. **65**(1), 128–145 (2016)
30. Swofford, D.: PAUP*: Phylogenetic analysis using parsimony (* and other methods) Ver. 4. Sinauer Associates, Sunderland, Massachusetts (2002)
31. Vachaspati, P., Warnow, T.: ASTRID: Accurate species trees from internode distances. BMC Genom. **16**(Suppl. 10), S3 (2015)
32. Wickett, N.J., Mirarab, S., Nguyen, N., Warnow, T., Carpenter, E., Matasci, N., Ayyampalayam, S., Barker, M.S., Burleigh, J.G., Gitzendanner, M.A., Ruhfel, B.R., Wafulal, E., Derl, J.P., Graham, S.W., Mathews, S., Melkonian, M., Soltis, D.E., Soltis, P.S., Miles, N.W., Rothfels, C.J., Pokorny, L., Shaw, A.J., De Gironimo, L., Stevenson, D.W., Sureko, B., Villarreal, J.C., Roure, B., Philippe, H., de Pamphilis, C.W., Chen, T., Deyholos, M.K., Baucom, R.S., Kutchan, T.M., Augustin, M.M., Wang, J., Zhang, Y., Tian, Z., Yan, Z., Wu, X., Sun, X., Wong, G.K.S., Leebens-Mack, J.: Phylotranscriptomic analysis of the origin and diversification of land plants. Proc. Nat. Acad. Sci. **111**(45), E4859–E4868 (2014)
33. Wiens, J.: Missing data, incomplete taxa, and phylogenetic accuracy. Syst. Biol. **52**, 528–538 (2003)
34. Wiens, J.: Missing data and the design of phylogenetic analyses. J. Biomed. Inform. **39**, 34–42 (2006)
35. Wiens, J.J., Morrill, M.C.: Missing data in phylogenetic analysis: reconciling results from simulations and empirical data. Syst. Biol. **60**, 719–731 (2011)
36. Xi, Z., Liu, L., Davis, C.C.: The impact of missing data on species tree estimation. Mol. Biol. Evol. **33**(3), 838–860 (2016)
37. Yang, J., Warnow, T.: Fast and accurate methods for phylogenomic analyses. BMC Bioinform. **12**(Suppl. 9), S4 (2011)
38. Zwickl, D.J., Hillis, D.M.: Increased taxon sampling greatly reduces phylogenetic error. Syst. Biol. **51**, 588–598 (2002)

Fast Heuristics for Resolving Weakly Supported Branches Using Duplication, Transfers, and Losses

Han Lai[1], Maureen Stolzer[1], and Dannie Durand[1,2(✉)]

[1] Department of Biological Sciences, Carnegie Mellon University,
Pittsburgh, PA 15213, USA
durand@cmu.edu

[2] Department of Computer Science, Carnegie Mellon University,
Pittsburgh, PA 15213, USA
http://www.cs.cmu.edu/~./durand/

Abstract. Weak branch supports in a gene tree suggest that the signal in sequence data is insufficient to resolve a particular branching order. One approach to reduce uncertainty takes the topology of the species tree into account. Under a maximum parsimony model, the best resolution of the weak branches is the binary tree that minimizes the cost of duplications, transfers, and losses. However, this problem is NP-hard, and the exact algorithm is limited to small, weakly supported areas.

We present an exact algorithm and several heuristic methods to resolve weak or non-binary gene trees given an undated species tree. These methods generate a set of optimal, binary resolutions that are temporally feasible, as well as event histories corresponding to each binary resolution. We compared the accuracy and runtime of these methods on simulated and biological datasets. The best of these heuristics provide close approximation to the event cost of the exact method and are much faster in practice. Surprisingly, a heuristic based on duplications and losses provides a good initialization for tree searching methods, even when transfers are present. Comparing event costs with RF distance, we observed that the two measures of distance captured very different information and are poorly correlated.

All methods are implemented in a new release of NOTUNG, a Java-based, cross-platform software for reconciling and resolving gene trees. NOTUNG is available at: http://www.cs.cmu.edu/~durand/Notung.

Keywords: Transfers · Resolve · Rearrange · Non-binary gene tree · Weak branches · Reconciliation · Gene tree corrections

1 Introduction

Gene trees disagree with species trees due to gene events, including duplication ($\mathcal{D}$), loss ($\mathcal{L}$), and transfer ($\mathcal{T}$). Given rooted gene and species trees, a mapping from extant genes to extant species, and an event model, the goal of reconciliation

© Springer International Publishing AG 2017
J. Meidanis and L. Nakhleh (Eds.): RECOMB CG 2017, LNBI 10562, pp. 298–320, 2017.
DOI: 10.1007/978-3-319-67979-2_16

is to infer the association between ancestral genes and species and the history of events that optimizes a combinatorial or probabilistic optimization criterion [13, 33–35]. Gene trees can also disagree with species trees due to phylogenetic error. Weak branch support in a gene tree suggests that the sequence data does not contain a signal strong enough to determine the branching order [1]. Uncertainty can also be represented by soft polytomies, i.e., non-binary nodes indicating that the true branching order cannot be resolved.

"Species-tree aware" approaches seek to improve the accuracy of gene tree inference by exploiting species tree information. Species-tree aware methods embody different trade-offs between speed and accuracy, ranging from detailed probabilistic models [5,16,40,42,43,48] to fast approaches based on parsimony or other criteria [6,7,23,25,28,30,37,41,45,49,51,52,54]. Some methods integrate species tree information into the gene tree reconstruction process, while others are designed to correct or resolve an uncertain gene tree.

Here, we focus on parsimony approaches designed to correct an unresolved gene tree. Corrective strategies are applied in two steps: First, a gene tree is constructed from sequence alignments using molecular phylogenetic methods. In a second, corrective step, species tree information is used to improve the quality of the gene tree. By separating consideration of sequence evolution and gene events, and focusing only on those areas where the sequence data cannot resolve the topology, the search space is substantially reduced. Corrective strategies support modularity. They can be applied to a corpus of pre-existing gene trees and be combined with other phylogenetic tools in customized pipelines. By treating correction as an independent step, the user has greater control over when and how species tree information is used in tree analysis.

The use of reconciliation to reduce gene tree uncertainty was first proposed in 2000 by Chen et al. [9]. This approach seeks the binary gene tree, or resolution, that minimizes an event-based parsimony criterion, while preserving the strongly supported branching order in the input tree. Rearrangement of weak branches and resolution of a non-binary gene tree are variants of the same problem. A binary gene tree can be converted into a non-binary gene tree by collapsing branches with support values below a minimum acceptable level of uncertainty. This approach can also be used to reconcile a non-binary gene tree without resolving it: Each polytomy is replaced by an optimal binary resolution. Following the reconciliation of this binary tree, the branches introduced during the binary reconciliation are collapsed to recover the original polytomies. All events associated with collapsed branches are mapped back to the original polytomy.

1.1   Background

Reconciliation takes as input a rooted gene tree $T_G = (V_G, E_G)$, a rooted species tree $T_S = (V_S, E_S)$, and a mapping $\mathcal{M}_L : L(T_G) \rightarrow L(T_S)$ between $L(T_G)$, the leaves of T_G, and $L(T_S)$, the leaves of T_S, that specifies the species from which each gene was sampled. The goal is to find the association between ancestral genes and species, expressed as a mapping $\mathcal{M} : V_G \rightarrow V_S$, and the events ($\mathcal{D}$, $\mathcal{T}$, and $\mathcal{L}$) that best explain this association with respect to an optimization criterion. The result is a gene tree annotated with the associated species and events. Under a maximum parsimony criterion, the event cost of reconciliation γ with the duplication-transfer-loss (DTL) model is:

$$\kappa(\gamma) = C_{\mathcal{D}} N_{\mathcal{D}} + C_{\mathcal{T}} N_{\mathcal{T}} + C_{\mathcal{L}} N_{\mathcal{L}} \tag{1}$$

where N_{ϵ} and $C_{\epsilon} > 0$ are the number and cost, respectively, of events of type $\epsilon \in \{\mathcal{D}, \mathcal{T}, \mathcal{L}\}$. Speciation events $(\mathcal{S})$ have zero cost. A most parsimonious reconciliation with a duplication-loss (DL) event model is an event history that minimizes Eq. 1 when $C_{\mathcal{T}} = \infty$. Here, we briefly review the reconciliation and resolution problems under both the DL and DTL event models.

DL-Reconciliation: In a gene family that evolved through a series of speciations, duplications, and losses, genes are inherited by vertical descent. Therefore, under the DL event model, the ancestors of $g \in V_G$ must be associated with species that are ancestors of $\mathcal{M}[g]$. As a result, the most parsimonious reconciliation of a binary gene tree is unique and can be calculated in a single a post-order traversal of T_G. At each node, $g \in V_G$, the mapping at g is the lowest common ancestor of its children (i.e., $\mathcal{M}[g] = LCA(\mathcal{M}[r(g)], \mathcal{M}[l(g)])$).

DL-Resolution: Let g be a polytomy with k_g children $g_1, \ldots, g_{k_g}$ mapped to species $\mathcal{M}[g_1], \ldots, \mathcal{M}[g_{k_g}]$. An optimal resolution of g is a binary tree with k_g leaves mapped to species $\mathcal{M}[g_1], \ldots, \mathcal{M}[g_{k_g}]$, that, when reconciled, minimizes Eq. 1. Under the DL model, there may be more than one most parsimonious set of events that results in a such tree.

Given a non-binary gene tree with one or more polytomies, a polynomial-time delay algorithm generates a most parsimonious binary resolution of T_G in two passes [14]. In the first pass, the algorithm visits each node, g, in post-order. If g is binary, the mapping at g is calculated from the mapping at its children, as in reconciliation. If g is a polytomy, additional calculations are performed that provide the information required to generate an optimal binary resolution of g. In the second pass, all optimal binary resolutions of T_G are generated in a series of pre-order traversals. The first exact DL-resolution algorithm required $O(|V_S||V_G|k_m{}^2)$ time per resolution reported. An algorithm to resolve a single polytomy in linear time followed [30] with further improvements in time complexity and generalizations of the event cost function [28, 37, 54] and the event model [7]. An expanded framework treats the gene tree correction and supertree problems as special cases of a more general problem formulation [15].

DTL-Reconciliation: In a gene family with a history that involves transfers, the ancestors of $g \in V_G$ are not necessarily associated with species that are ancestors of $\mathcal{M}[g]$; species $\mathcal{M}[g]$ may have acquired g via a transfer from distant species. This has several consequences for reconciliation: First, unlike the DL-model, it is not possible to determine the species associated with g directly from the species associated with its children. A value of $\mathcal{M}[g]$ that results in a suboptimal event cost at g may be required in order to obtain an optimal reconciliation at the root of T_G. Second, there may be more than one optimal DTL-reconciliation. Third, transfers introduce temporal constraints because the donor and recipient of a transfer must have co-existed. We say that an event history is *temporally infeasible* if it contains conflicting temporal constraints that could not be realized without time travel. Because the temporal constraints are a global property, i.e., they depend on all the inferred transfer events, conflicts cannot be determined locally. Given a dated species tree, an optimal DTL-reconciliation without temporal conflicts can be inferred in polynomial time [13,20,21].

With an undated species tree, DTL-reconciliation is NP-complete [18,38,50]; there is no known constructive algorithm for generating optimal, temporally feasible DTL-reconciliations. Instead, it is necessary to generate candidate minimum cost reconciliations and test each candidate for feasibility [12,38,44,50]. This procedure involves three steps: (1) a post-order traversal of T_G constructs a table at each node g, by pre-computing the cost associated with the possible species and event assignments at g; (2) a pre-order traversal constructs candidate optimal reconciliations from those cost tables; and (3) a test for temporal feasibility for each candidate reconciliation constructed. Only those candidates that are feasible are biologically valid event histories.

A DTL-reconciliation algorithm that returns all feasible, optimal reconciliations requires $O(|V_G||V_S|^2)$ complexity [44,50]; this can be reduced to $O(|V_G||V_S|)$ [2], at the expense of only returning only a single solution. Only a few programs generate all solutions [12,22,44]; most others generate a single solution, selected arbitrarily [2,11,38].

DTL-Resolution: The DTL-resolution algorithm is an extension of the algorithm to reconcile a binary tree, but with additional calculations at non-binary nodes. This algorithm, again, involves three steps. First, a post-order traversal of T_G constructs a table at each node g. If g is binary, the cost tables are calculated as with reconciliation. If g is a polytomy, candidate binary resolutions of g are enumerated and the costs associated with various candidate species and event assignments at g are pre-computed for each candidate binary resolution. The DTL-resolution problem is NP-hard [26]; in order to find optimal binary resolution(s) of g, all possible binary trees with k_g leaves must be considered in this step. In the second step, T_G is traversed in pre-order to construct candidate optimal binary resolutions from the cost tables. Each candidate resolution is reconciled and may also have more than one event history. Finally, each candidate resolution and event history must be tested for temporal feasibility.

Under the DL-model, the optimal binary resolution of each polytomy in an unresolved gene tree with multiple polytomies is independent of the optimal resolution of any other polytomy. Under the DTL-model, the cost tables at each node, whether binary and non-binary, depend only on the cost tables at the node's children. Once the cost tables have been instantiated, all information required to resolve a polytomy is acquired from its children (pass one) and its parent (pass two). With this information, each polytomy can be resolved by local calculations; no exchange of information from distant polytomies is required.

Therefore, under both models, it is not necessary to consider the combinatorial amalgam of all resolutions of all polytomies to resolve a non-binary tree.

The MowgliNNI [36] program was the first to tackle the DTL-resolution problem for a dated species tree. MowgliNNI uses a modified Nearest Neighbor Interchange (NNI) method to explore the space of binary resolutions of each polytomy. Subsequently, exact, fixed-parameter tractable algorithms for DTL-resolution were introduced for both dated and undated species trees: [23,25,41] and this work.

There are potentially an exponential number of DTL-resolutions. Moreover, for each resolution, there may be multiple optimal DTL-reconciliations (i.e., multiple minimum cost event histories). The simplest approach is to output a single resolution, selected arbitrarily. However, generating all solutions enables the user to choose the best resolution based on problem-specific criteria. Further, failure to consider all solutions can distort the results of a comparative performance evaluation, since two methods with the same behavior may appear to produce different results if only a subset of all possible solutions is returned by each program.

The problem of degeneracy has been approached from various perspectives. For example, the NOTUNG GUI provides a point-and-click interface that allows the users to investigate alternate resolutions interactively. This is a satisfying solution for exploratory analysis of a few families, but unmanageable for large phylogenomic analyses. Methods for summarizing a set of degenerate solutions have recently been proposed [32,41]. Another approach is to triage alternate binary resolutions by incorporating additional information, such as phylogenetic likelihood [3], conditional clade probabilities [23], bootstrap replicates [11] or genome neighborhoods [8,29].

1.2 Our Contributions

We present new methods for resolving a non-binary gene tree by minimizing inferred events in the DTL event model. An exact algorithm, based on exhaustive enumeration of the binary resolutions of each polytomy, generates all temporally feasible, optimal binary resolutions and event histories. Since the number of binary resolutions that must be evaluated grows exponentially with the size of the polytomy, heuristic methods to resolve trees with large polytomies are essential. We propose several heuristics for resolving a non-binary gene tree that are not guaranteed to find all (or any) optimal binary resolutions, but require substantially reduced running times.

These heuristics, as well as the exact algorithm, have been implemented in NOTUNG and are publicly available. All methods return only temporally feasible solutions. The exact algorithm reports all temporally feasible, optimal reconciliations for all minimum cost resolutions, up to a user-defined limit.

The accuracy and running time of the methods were investigated empirically on two publicly available, phylogenomic datasets [4,47]. The results not only demonstrate the relative effectiveness of the various heuristics, but also provide insight into the structure of the problem and the challenges of evaluating methods that are highly degenerate.

Notation: Given a phylogenetic tree, $T_i = (V_i, E_i)$, $L(T_i) \subset V_i$ designates the leaves of T_i and $\rho(T_i)$ designates its root node. $T_i(v)$ is the subtree of T_i rooted at node $v \in V_i$. The parent and children of node v are denoted $p(v)$ and $C(v)$, respectively. For each edge $(u, v) \in E_i$, u is the parent of v (i.e., $u = p(v)$). The out-degree of node v is designated k_v. If v is a binary node ($k_v = 2$), $l(v)$ and $r(v)$ refer to the left and right child of v, respectively. We denote a given gene node as $g \in V_G$ and a given species node as $s \in V_S$. Given nodes $u, v \in V_i$, if u is on the path from v to $\rho(T_i)$, then u is an ancestor of v, designated $u \geq_i v$, and v is a descendant of u, designated $v \leq_i u$. If u is neither an ancestor or descendant of v, i.e., $u \not\geq_i v$ and $v \not\leq_i u$, then u and v are incomparable, denoted $u \nleq_i v$.

We define a reconciliation of binary rooted T_G with T_S to be $\gamma(T_G, T_S) = (T_G; \mathcal{M}; \mathcal{E}; \Lambda)$, where $\mathcal{M} : V_G \to V_S$ is a mapping from each gene to the species that contained that gene; $\mathcal{E} : V_G \backslash L(T_G) \to \{\mathcal{S}, \mathcal{D}, \mathcal{T}\}$ is a function indicating the event that occurred at each gene node; and $\Lambda \subset E_G$ is the set of transfer edges. If edge $(g_1, g_2) \in \Lambda$, then the divergence at g_1 is due to a transfer (i.e., $\mathcal{E}[g_1] = \mathcal{T}$) from donor species $s_1 = \mathcal{M}[g_1]$ to recipient species $s_2 = \mathcal{M}[g_2]$, where $s_1 \nleq s_2$. For each $g \in V_G \backslash \{\rho_G\}$, the edge from $p(g)$ to g is annotated with $\mathcal{L}(g)$, the gene losses that occurred on that edge. Each loss is labeled with the species in which the loss occurred. $\mathcal{R}(T_G, T_S) = \{\gamma\}$ is the set of all reconciliations of T_G with T_S. A reconciliation γ^* is considered optimal if $\kappa(\gamma^*) = \min_{\gamma \in \mathcal{R}(T_G, T_S)}(\kappa(\gamma))$. The set of all optimal reconciliations is denoted $\mathcal{R}^*(T_G, T_S)$.

2   A Hybrid Approach to Solve DTL-Resolution

Here, we formally define the DTL-Resolution problem and present an exact algorithm and several new heuristic algorithms to correct incongruence in gene trees that may be caused by phylogenetic error.

The goal of resolving a non-binary gene tree is to find the binary resolution of the gene tree that minimizes the cost of event inference, while still retaining well-supported branches [6,14]. Formally, let $T_G = (V_G, E_G)$ be a non-binary, rooted gene tree. A binary, rooted gene tree, $T_G^b = (V_G^b, E_G^b)$, is a *binary resolution* of T_G, if for every binary node $g \in V_G$, there exists a node $g' \in V_G^b$, such that $L(T_G^b(l(g'))) = L(T_G(l(g)))$ and $L(T_G^b(r(g'))) = L(T_G(r(g)))$. The set of all possible binary resolutions is denoted $\mathfrak{T}(T_G)$. For a polytomy, $g \in V_G$, $\mathfrak{T}(g)$ denotes the set of all binary trees with k_g leaves, such that the leaf set is $C(g)$, the children set of g.

Problem statement: DTL-Resolution
Input: An unresolved gene tree $T_G = (V_G, E_G)$, a binary species tree T_S, a mapping $\mathcal{M}_L : L(T_G) \to L(T_S)$; and event costs C_ϵ for each event $\epsilon \in \{\mathcal{D}, \mathcal{T}, \mathcal{L}\}$.
Output: $\mathfrak{T}^*(T_G) = \{T_G^*\}$, the set of all binary resolutions of T_G, such that

$$T_G^* = argmin_b \kappa(\gamma(T_G^b, T_S)),$$

and $\gamma(T_G^*, T_S)$ is temporally feasible.

2.1   Exact Algorithm for DTL-Resolution

The algorithm to solve DTL-Resolution (Algorithm 1) consists of the same three steps as the DTL-Reconcile algorithm (described in detail in [44]). The key modification for DTL-Resolution is a procedure to calculate the cost tables when g is a polytomy. In the first step, all nodes in T_G are visited in post-order, and at each node g, all possible assignments of $\mathcal{M}[g]$ and $\mathcal{E}[g]$ are enumerated. The associated information is stored in two tables: $\mathcal{K}_g$, which stores the best cost for mapping g to each $s \in T_S$, and $\mathcal{H}_g$, which stores the corresponding book-keeping information necessary for calculating an event history in step two. If g is a binary gene node, the core subroutine that generates $\mathcal{K}_g$ and $\mathcal{H}_g$ is a function $costCalc(g, s_l, s_r)$ (line 9 in Algorithm 1, see also [44]). The cost of mapping gene node g to species node s only depends on the event at g (i.e. $\epsilon = \mathcal{E}[g]$), the mapping on children of g (i.e. $s_r = \mathcal{M}[l(g)]$ and $s_l = \mathcal{M}[r(g)]$), and the cost associated with these mappings $(\mathcal{K}_{l(g)}[s_l], \mathcal{K}_{r(g)}[s_r])$. This information is stored in $\mathcal{H}_g[s]$ in the form of a tuple: (ϵ, s_r, s_l).

If g is a polytomy, all possible binary resolutions in $\mathfrak{T}(g)$ are exhaustively enumerated, and $\mathcal{K}_g$ and $\mathcal{H}_g$ are calculated for each $T_g^b \in \mathfrak{T}(g)$. In each binary resolution, g and its children are replaced with a binary embedded subtree with k_g leaves, inserting $k_g - 1$ binary internal nodes (including the root that replaces g). Cost tables for these internal nodes are constructed, bottom up,

---

Algorithm 1. DTL Resolve

```
1 main(T_G, T_S):
2     K = {}, H = {}
3     DTLResolve(V_G, V_S)
4     {γ_1, γ_2...γ_m} = traceback(K, H, T_G)
5     {γ_1, γ_2...γ_n} = checkFeasibility({γ_1, γ_2...γ_m})
6 DTLResolve(V_G, V_S):
7     for each g ∈ V_G post order:
8         if g is binary:
9             for each s_1, s_2 ∈ V_S × V_S: costCalc(g, s_1, s_2)
10        else:
11            for each T_g^b = (V_g^b, E_g^b) ∈ 𝔗(g):
12                DTLResolve(V_g^b, V_S)
13                Merge(H_g^b, H_g, K_g^b, K_g, T_g^b)
14 Merge(H_g^b, H_g, K_g^b, K_g, T_g^b):
15     for each s ∈ V_S :
16         if K_g^b[s] < K_g[s] :
17             clear H_g[s]
18             for each (ε, s_l, s_r) in H_g^b[s]: enqueue (ε, s_l, s_r, T_g^b) in H_g[s]
19             K_g[s] = K_g^b[s]
20         else if K_g^b[s] == K_g[s]:
21             for each (ε, s_l, s_r) in H_g^b[s]: enqueue (ε, s_l, s_r, T_g^b) in H_g[s]
```

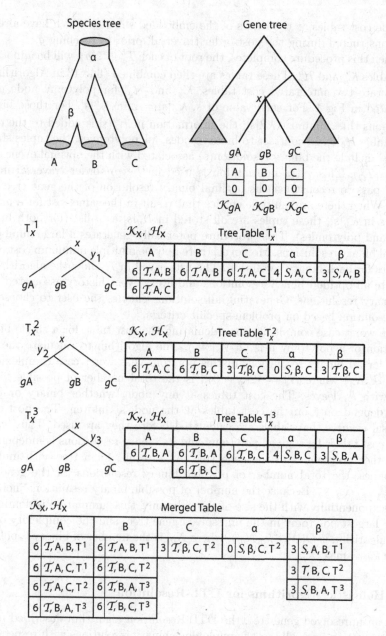

Fig. 1. Dynamic program to calculate a cost table for a non-binary gene node x. This process tries different binary resolutions for node x; for each binary resolution, it uses the same procedure to generate a cost table for x. Then, all cost tables for all binary resolutions are merged.

from the cost tables at the leaves of the embedded subtree, which have already been constructed during the post-order traversal, prior to reaching g.

When this procedure completes, the root of each $T_g^b \in \mathfrak{T}(g)$, will be annotated with tables $\mathcal{K}_g^b$ and $\mathcal{H}_g^b$. These tables are then combined (line 13 in Algorithm 1) to generate two integrated cost tables, $\mathcal{K}_g$ and $\mathcal{H}_g$, for polytomy node g, as illustrated in Fig. 1. For each value of s, $\mathcal{K}_g[s] = \min_b \mathcal{K}_g^b[s]$. For those binary resolutions that minimize $\mathcal{K}_g^b[s]$, the information in $\mathcal{H}_g^b[s]$ is added to the combined table $\mathcal{H}_g[s]$. In contrast to binary nodes, for polytomies, the tuples stored in $\mathcal{H}_g[s]$ include the binary resolution(s) associated with the optimal event cost: $(\epsilon, s_l, s_r, T_G^b)$. This extra information is used in the pre-order traversal in the second pass to reconstruct the optimal binary resolution of the gene tree as a whole. When there are multiple entries that result in the same cost for a pair of g and s in $\mathcal{K}_g[s]$, these tuples are all stored in $\mathcal{H}_g[s]$ as a list (for both binary nodes and polytomies). This procedure potentially generates a large number of optimal binary resolutions. Moreover, there may be multiple minimum cost event histories for each binary resolution. The lists saved in $\mathcal{H}$ allow the algorithm to generate all optimal binary resolutions and every event history associated with the binary resolutions. Generating all solutions enables the user to choose the best resolution based on problem-specific criteria.

The worst case complexity for calculating the cost table for a single binary resolution of a polytomy g is $O(|V_S|^2 k_g)$. The algorithm to calculate the cost tables for all binary resolutions of this polytomy has worst case complexity of $O(|V_S|^2 k_g N_g)$ where $N_g = (2k_g - 3)!!$ is the total number of possible binary trees with k_g leaves. The cost tables at any node, whether binary or polytomy, depend only on the cost tables at the node's children. The post-order traversal ensures that tables are calculated before they are used. Thus, calculating cost tables for a given polytomy and its binary resolutions is independent from other polytomies in the gene tree. As a result, for a tree with multiple polytomies, the total number of possible binary resolutions for the gene tree is $\sum_{g \in V_G \ni k_g > 2} N_g$. Because the number of possible binary resolutions increases super-exponentially with the size of the polytomy, this summation is dominated by the largest polytomy in the unresolved gene tree, and the complexity of the exact algorithm is $O(|V_S|^2 k_m N_m)$, where N_m is the number of binary topologies of size $k_m = \max_g k_g$.

2.2 Heuristic Algorithms for DTL-Resolution

Given an unresolved gene tree, the DTL-Resolution algorithm described in the previous section finds all most parsimonious binary resolutions with respect to a given set of event costs. However, finding the optimal solution requires enumeration of all possible binary resolutions of each polytomy in T_G, which is intractable for larger polytomies. For example, evaluating the 10,395 possible binary topologies for a polytomy with 7 leaves requires $\sim$20 s on a 2.4 GHz processor. For a polytomy of size 8, there are over 135k possible resolutions and the running time on the same machine is approximately 5 min. For phylogenomic analyses with thousands of gene trees, where many gene trees may contain large polytomies, heuristics become essential.

Instead of enumerating all binary resolutions, as we do in the exact algorithm, we replace $\mathfrak{T}(g)$ in line 11 of Algorithm 1 with $\widehat{\mathfrak{T}}(g)$, a subset of all possible binary resolutions. We propose several fast heuristics based on different strategies for defining this subset:

NNI: A direct alternative to enumerating all binary resolutions of a polytomy is to use tree modification techniques, such as Nearest Neighbor Interchange (NNI), to sample M binary resolutions from the entire search space, where the sample size M is selected by the user. Given an unresolved tree wherein the polytomies are the result of collapsing weak edges in the input tree, the search starts from the original binary topology. If the input tree itself was non-binary, then a starting topology is selected arbitrarily. At each iteration, NNI is applied to a randomly selected internal branch in the current tree. If the reconciliation cost of the resulting tree is equal to the best cost seen so far, the tree is added to the list of candidate resolutions. If it is lower, the current tree replaces the list of candidate resolutions. The worst case complexity of this method is $O(|V_S|^2 k_g M)$, for a single polytomy g.

Local DL: In this heuristic, we resolve each polytomy by replacing it with a single binary resolution that minimizes the DL cost, rather than the DTL cost. The quality of this heuristic will depend on how well the binary resolution that minimizes the DL cost approximates the binary resolution that minimizes the DTL cost.

Implementing the DL-resolution algorithm [14] in this context requires some modification. Recall that with DL-reconciliation, $\mathcal{M}[g]$ is calculated directly from $\mathcal{M}[g_i] \forall g_i \in C(g)$, the mappings at the children. However, under the DTL-event model, the mappings at the children of g are not finalized until pass two; instead, each child is associated with tables for *possible* species assignments. Rather than attempt all combinations of mappings at the children, we assign $\mathcal{M}[g_i] = \mathrm{argmin}_s \mathcal{K}_{g_i}[s]$, the species that minimizes the cost at g_i. The DL-resolution algorithm is then applied to find a binary resolution, T_g^{DL} of g. If there is more than one minimum cost resolution, a single resolution is selected arbitrarily.

The polytomy at g is then replaced with T_g^{DL}, a binary resolution that minimizes the DL score with these labels, resulting in $2k_g - 1$ new binary nodes. The post-order traversal then continues up the tree, calculating the entries in tables $\mathcal{K}$ and $\mathcal{H}$ at the internal nodes in T_g^{DL} using the DTL-event model. Note that since this approach yields a single binary resolution, calculating $\mathcal{H}_g$ does not require the call to the Merge function (line 13 in Algorithm 1). Each entry in $\mathcal{H}_g$ is associated with that single binary resolution, although there may still be degeneracy if the binary resolution has more than one minimum cost event history. Under the DL-event model, the time required to find an optimal resolution of a polytomy g is $O(|V_S| + k_g)$ [27]. The calculation of the cost tables for the nodes in the new binary resolution has complexity $O(|V_S|^2 k_g)$. Since this step is more computationally demanding then obtaining the DL-resolution, the worst case complexity for a single polytomy is $O(|V_S|^2 k_g)$.

Hybrid DL-NNI: The quality of the results of the *NNI* strategy, described above, are likely sensitive to the tree chosen to start the sampling process. Instead of starting with the original binary branching order, or an arbitrarily chosen one, we posited that a tree that minimizes the DL-cost might be closer, in tree space, to the optimal DTL-resolution. In the *Hybrid* heuristic, we apply the *Local DL* heuristic to generate a candidate starting tree and then sample candidate resolutions in tree space by applying *NNI* for M iterations. Since the complexity of *Local DL* is low compared to DTL-reconciliation, adding this extra step does not have an adverse effect on the running time. The worst case complexity is $O(|V_S|^2 k_g M)$ for a single polytomy.

Global DL: As a comparison to the DTL-Resolution heuristics, we also calculated the optimal DL-resolution [14, 30] of the entire input gene tree. Once DL-resolution is complete, the DTL-reconciliation cost of the resulting binary tree is calculated by reconciling it with the DTL-event model. This is in contrast to the *Local DL* strategy, described above, which uses the DL model to generate a binary resolution of each polytomy independently, but uses cost tables, obtained with the DTL model, as input on the children of each polytomy and constructs cost tables on each node of the binary resolution based on the DTL model.

2.3   Implementation

We implemented DTL-Resolution (the *Exact* method, Algorithm 1) and the heuristic strategies described above in the NOTUNG reconciliation software package. For the *Exact* method, NOTUNG finds all optimal DTL-resolutions and calculates all DTL-reconciliations for each one. All temporally feasible solutions are included in the output, up to a user-specified limit. Temporally infeasible reconciliations are discarded. No binary resolutions are generated if all solutions are temporally infeasible.

For all heuristics (except *Global DL*), for polytomies of size $k \leq 6$, which have at most 945 binary resolutions, the *Exact* method was applied and all binary resolutions of the polytomy are enumerated. The chosen heuristic method is applied for polytomies of size $k \geq 7$.

To speed up the running time in practice, several memoization techniques are implemented. When enumerating all possible binary resolutions for a polytomy, the order of the topologies is sorted such that the left subtree is kept stable if there are more topologies to check in the right subtree. This effectively maximizes memoization because we can reuse the left cost tables when searching right topologies. Although this does not improve the complexity, this significantly reduces runtime. In addition, because each NNI operation inserts a single new branch in T_G, it is therefore sufficient to recalculate only the cost tables, $\mathcal{K}$ and $\mathcal{H}$, associated with nodes on that branch and its ancestors, instead of recalculating tables for the entire binary resolution.

3   Results

To investigate the behavior of these methods, these strategies were evaluated empirically on two previously published data sets:

ALE **trees:** In order to obtain a benchmark for which events are known, we used a data set [47,48], wherein a high confidence set of events were inferred from multiple alignments of simulated sequences using a probabilistic model that integrates sequence evolution and gene events. This dataset is based on 1099 HOGENOM gene families [39] sampled from 36 cyanobacterial genomes [46]. For each gene family, we obtained the gene events and the multiple alignments (MSAs) of amino acid sequences from which they were inferred [47]. We constructed "species tree blind" gene trees from these MSAs using PhyML [17] with the LG+G+I model with four Γ-distributed rate categories. Branch support scores were obtained by bootstrapping with 100 replicates.

Barker **trees:** Gene trees for 13,854 OrthoMCL families, sampled from 49 cyanobacteria and 16 proteobacteria, were downloaded from [4]. The gene trees, and the associated species tree, were constructed as described in [10,31]. Briefly, maximum likelihood trees were constructed from amino acid sequence alignments using PhyML, following model selection with ModelGenerator [24]. Branch support was assessed using 200 bootstrap replicates.

Trees were processed using NOTUNG. All NOTUNG analyses were carried out with default event costs ($C_T = 3$, $C_D = 1.5$, $C_L = 1$). These transfer and loss costs are identical to those used with the same data sets in previous studies [10,11,23]. We chose a slightly lower duplication cost to reduce the number of degenerate solutions. These costs are also consistent with costs used in recent phylogenomic analyses (e.g., [11,44]), which were selected to minimize the total change in genome size, averaged over the species tree [11].

Trees from both datasets were rooted using DTL-event parsimony with NOTUNG. Unresolved trees were then created by collapsing branches with bootstrap values below 70%. To assess the impact of threshold choice, the Barker dataset analyses were repeated with thresholds of 80% and 90% (Table 1). Gene trees lacking weak edges for a given threshold were eliminated from the analysis at that threshold.

The resulting rooted, non-binary trees were resolved using the *Exact*, *NNI*, *Local DL*, and *Hybrid* strategies. The trees were also resolved with the *Global DL* model for comparison. This process resulted in six trees for each gene family: five rearranged trees and the original tree (i.e., the rooted binary tree prior to collapsing weak bootstrap values). All six trees were reconciled using the DTL model in NOTUNG to obtain DTL event costs (Eq. 1), which were used to assess the performance of the different binary resolution strategies. Inferred event histories were tested for temporal feasibility; only gene families for which all six trees had at least one temporally feasible reconciliation were retained for further analysis (Table 1).

Table 1. The number of unresolved gene trees as a function of branch support threshold and maximum polytomy size (k_m). F indicates the number of trees remaining, after trees that lacked a temporally feasible solution for one or more methods were removed from consideration. N_P: number of polytomies of any size in all trees, combined.

Threshold	$k_m > 2$	$2 < k_m < 7$	$k_m = 7$	$k_m > 7$	N_P
Barker trees					
70%	3672	2678	288	706	7370
70% F	3604	2650	281	673	–
80%	4280	2983	337	960	9196
80% F	4204	2955	329	920	–
90%	4882	3272	406	1204	10307
90% F	4770	3241	394	1135	–
ALE trees					
70%	1080	383	116	581	5210
70% F	813	346	90	444	–

The accuracy of the DTL-Resolution heuristics was evaluated using three different measures. First, for each gene family in the ALE, the event costs obtained with all methods were compared with the weighted sum of the high-confidence events associated with that family, calculated using Eq. 1. Second, for families in both datasets, the event costs of binary resolutions obtained with each heuristic were evaluated relative to the event costs obtained with the *Exact* method. Third, the binary tree obtained with each heuristic was compared with the binary tree obtained with the *Exact* method, using the normalized *Robinson-Foulds (RF) distance*, a tree comparison method that is frequently used to assess tree accuracy. To determine how well the heuristic strategies approximate the exact approach, we focused on gene trees with maximum polytomy size of $k_m = 7$. The space of trees with seven leaves is small enough to admit exhaustive search, so that the reduction in event cost obtained with the various heuristics can be compared with the optimal improvement in cost obtained with the *Exact* method.

For the cost-based comparisons, the per-family accuracy was quantified using the error,

$$\Delta(\widehat{\kappa}_H, \kappa^*) = \frac{\widehat{\kappa}_H - \kappa^*}{\kappa^*}, \tag{2}$$

where $\widehat{\kappa}_H$ represents the lowest DTL-reconciliation cost obtained with heuristic H and κ^* is the overall best score for the family. For the ALE trees, the best score is defined to be κ_A^*, the weighted sum (Eq. 1) of the high-confidence events inferred for that family. For comparison with the *Exact* method, for both sets of trees, the best score is κ_E^*, the optimal DTL-cost. Because *Exact* is guaranteed to find the minimum cost resolution under the Maximum Parsimony criterion, $\Delta(\widehat{\kappa}_H, \kappa_E^*)$ is always positive.

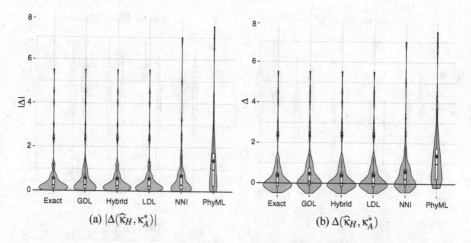

Fig. 2. Error relative to the weighted sum of high-confidence events, calculated for ALE families with a branch threshold of 70%. The *NNI* and *Hybrid* strategies were executed with $M = 1000$ samples.

Comparison with High-Confidence Events: First, to set these methods in context, we evaluated both the exact and the heuristic methods relative to the high-confidence inferred events for each ALE family (Fig. 2). All strategies reduced the magnitude of the error by a factor of two, on average, relative to the uncorrected tree (Fig. 2a). While a small number of trees had very large errors, for most trees $|\Delta(\widehat{\kappa}_H, \kappa_A^*)| < 1.0$. The greatest reductions in error were obtained with the *Exact*, *Hybrid*, and *Local DL* strategies. All strategies underestimate the number of events in some cases (Fig. 2b). The strategies for which the magnitude of the error is lowest, *Exact*, *Hybrid*, and *Local DL*, underestimate the number of events roughly half the time. *Global DL* and *NNI* have a greater tendency to overestimate than to underestimate the events.

Comparison of Heuristic Methods with the Exact Method: Next, we assessed how well various heuristics approximate the *Exact* algorithm for both the ALE trees and the Barker trees. Both datasets display similar trends (Fig. 3a), and similar behavior was observed with thresholds of 80% and 90% (Fig. 4a).

The mean, median and maximum costs obtained with all DTL-Resolution strategies are markedly lower than the corresponding costs obtained with the uncorrected PhyML trees. When no correction is applied, $\Delta(\widehat{\kappa}_H, \kappa_E^*) = 0$ for only 7% of Barker trees and 2% of ALE trees. Thus, in almost all cases, when the branch support was less than 70%, it was possible to find an alternate binary topology that agrees with all strongly supported branches and reduces the number of gene events required to explain the data.

Substantially better results are obtained with *Local DL* and *Hybrid* than with the other heuristic strategies. In fact, the *Hybrid* strategy performed as well as the *Exact* method for 84% of ALE trees and 74% of Barker trees. Heuristics based on the DL-event model performed surprisingly well, given that transfers are likely

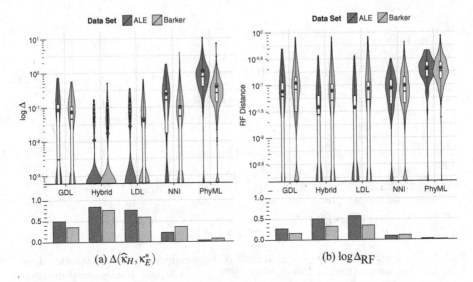

Fig. 3. Comparison of heuristic strategies with the *Exact* method for ALE and Barker trees, with a branch support threshold of 70%. The *NNI* and *Hybrid* strategies were executed with $M = 1000$ samples. Red points indicate the mean. *(a)* Top: $\log \Delta(\widehat{\kappa}_H, \kappa_E^*)$. Zero values are represented as 0.01 times the minimum non-zero error. Bottom: the proportion of trees where $\Delta(\widehat{\kappa}_H, \kappa_E^*) = 0$. *(b)* Top: Log normalized RF distance ($\log \Delta_{RF}$). Bottom: the proportion of trees where $\Delta_{RF} = 0$. (Color figure online)

common in bacterial gene family evolution. For almost half of ALE trees, the *Global DL* strategy found a binary resolution with a DTL-reconciliation cost as low as the optimal, *Exact* cost. The *Local DL* heuristic performed well for both data sets; 75% of ALE trees and 58% of Barker trees obtained optimal DTL-reconciliation costs.

The *NNI* and *Hybrid* strategies use tree modification techniques to search for a low cost resolution. Two factors that may influence the effectiveness of this strategy are the tree used as the starting point in the search and the number candidate trees that are sampled. The *Hybrid* strategy substantially outperforms *NNI* with $M = 1000$ samples (Figs. 2 and 3a), demonstrating that the starting state is, indeed, an important factor in the quality of the outcome.

To assess how the accuracy depends on the number of topologies sampled, we repeated the analysis of the Barker trees with the *NNI* and *Hybrid* strategies with sample sizes ranging from $M = 125$ to $M = 4,000$. Predictably, the error decreases as the number of samples increases for both strategies (Fig. 4b). The *Hybrid* strategy performs as well as the *Exact* strategy for three out of four trees, when $M = 1000$. The marginal gain in accuracy obtained with sample sizes greater than 1000 is limited. In contrast, for *NNI*, the mean and median error continue to improve over the entire range. Moreover, even when $M = 4000$, more than 37% of trees still have non-zero error. A sample size of 4000 corresponds to almost 40% of binary topologies with $k = 7$ leaves. With this sample size,

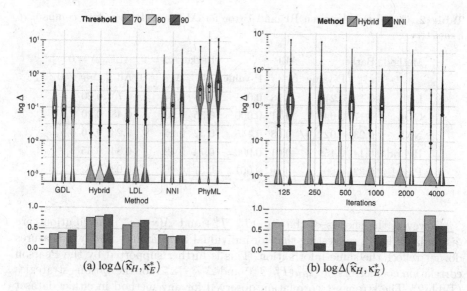

Fig. 4. Comparison of heuristic strategies with the *Exact* method for the Barker trees only. Red points indicate the mean. *(a)* Performance with branch support thresholds of 70%, 80%, and 90%. The *NNI* and *Hybrid* strategies were executed with $M = 1000$ samples. Top: $\log \Delta(\widehat{\kappa}_H, \kappa_E^*)$. Zero values are represented as 0.01 times the minimum non-zero error. Bottom: the proportion of trees where $\Delta(\widehat{\kappa}_H, \kappa_E^*) = 0$. *(b)* Performance of the *NNI* and *Hybrid* strategies with sample sizes from $M = 125$ to $M = 4000$; branch support threshold of 70%. Top: $\log \Delta(\widehat{\kappa}_H, \kappa_E^*)$. Zero values are represented as 0.01 times the minimum non-zero error. Bottom: the proportion of trees where $\Delta(\widehat{\kappa}_H, \kappa_E^*) = 0$. (Color figure online)

NNI does not represent a dramatic reduction in computational load relative to exhaustive search, yet almost half the trees could be further improved. With larger polytomy sizes, the fraction of tree space that can be sampled in a realistic time frame will dwindle. For example, for a polytomy of size eight, 4000 NNI modifications will sample less than 3% of possible topologies.

Robinson-Foulds Distance: We also compared the resolved binary trees obtained with the various strategies using normalized RF distance (Δ_{RF}), a measure of the difference in the topologies of two binary trees with the same leaf labels. Δ_{RF} is the number of bi-partitions that are present in one tree, but not the other, normalized by the total number of bi-partitions. For each family in each data set, we calculated $\Delta_{RF}(T_H^b, T_E^b)$, the normalized RF distance between the binary resolution obtained with heuristic H and the binary resolution obtained with the *Exact* method (Fig. 3b).

The overall trends observed with Δ_{RF} are similar to those observed with $\Delta(\widehat{\kappa}_H, \kappa_E^*)$. Most uncorrected PhyML trees display some dissimilarity with the *Exact* binary resolution. For corrected trees, the binary resolutions obtained with the *Local DL* and *Hybrid* strategies are most similar to the *Exact* binary resolution.

Table 2. Correlation between RF and Error for ALE and Barker data compare to Exact tree

Method	Barker		ALE		Barker, $\Delta = 0$		ALE, $\Delta = 0$	
	r	P-value	r	P-value	All	$\Delta_{RF} \neq 0$	All	$\Delta_{RF} \neq 0$
LDL	0.13	0.214	0.18	0.001	0.58	0.24	0.75	0.20
GDL	0.17	0.091	0.21	0.000	0.35	0.20	0.48	0.20
NNI	0.23	0.022	0.08	0.145	0.32	0.25	0.23	0.16
Hybrid	0.14	0.159	−0.01	0.906	0.74	0.42	0.84	0.35
PhyML	0.32	0.001	0.19	0.001	0.07	0.06	0.03	0.02

However, the shapes of the $\Delta_{RF}(T_H^b, T_E^b)$ and $\Delta(\widehat{\kappa}_H, \kappa_E^*)$ distributions are dissimilar (Fig. 3), suggesting that for individual gene families, the two measures do not reflect the same information. This is further supported by the Pearson correlation coefficients of $\Delta_{RF}(T_H^b, T_E^b)$ and $\Delta(\widehat{\kappa}_H, \kappa_E^*)$ for the various strategies (Table 2). The strongest correlation observed for any method in either dataset is 0.32. The other correlations are even weaker and only half are significant at the 0.005 level.

Several factors may be contributing to the lack of correlation between $\Delta(\widehat{\kappa}_H, \kappa_E^*)$ and Δ_{RF}. First, degeneracy may inflate the reported RF distance. If there are multiple optimal binary resolutions, but only one, chosen arbitrarily, is reported by each method, then a non-zero RF distance can result, even if both methods find the same set of binary resolutions. Reconciliation cost is not sensitive to degeneracy in this way. If a heuristic method finds at least one optimal resolution, then $\Delta(\widehat{\kappa}_H, \kappa_E^*)$ will be zero, since it depends on the event cost, not the tree topology.

Recall that the heuristic strategies are not guaranteed to return all optimal binary resolutions. The *Global DL* and *Local DL* strategies will find one binary resolution, which may, or may not, be optimal. The number of optimal binary resolutions found by the *NNI* and *Hybrid* strategies will depend on the sampling process. In fact, Table 1 shows that for roughly 20% to 40% of trees, $\Delta(\widehat{\kappa}_H, \kappa_E^*) = 0$, but $\Delta_{RF}(T_H^b, T_E^b) \neq 0$. This suggests that the fact that the heuristics only find a subset of the binary resolutions is contributing to the difference between error and RF distance.

Even when all resolutions found by the heuristics are compared against all optimal resolutions found by the *Exact* method, event-cost error and RF distance may be poorly correlated because they capture different information. A pair of topologies with similar event histories may have few bi-partitions in common and vice versa. For example, the trees $(A, (B, (C, (D, E))))$ and $(B, (C, (D, (A, E))))$ differ by a single transfer, but have no bi-paritions in common. Zheng and Zhang [53] recently showed that RF distance is not equivalent to event-based distances in a DL model. This is likely also true for event distances based on transfers.

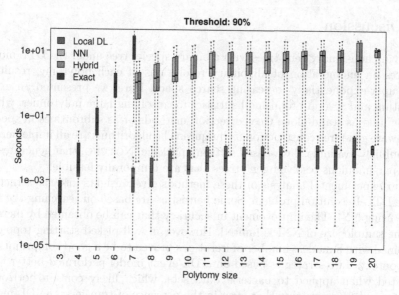

Fig. 5. Run-time: time required (in seconds) to process each polytomy in the Barker trees, with branch support threshold of 90%, using the *Exact*, *Local DL*, *NNI*, and *Hybrid* strategies. The *NNI* and *Hybrid* strategies were executed with $M = 1000$ samples.

Run-Time: We recorded the time required to process each polytomy of size $k < 8$ for the *Exact* method, and $k > 6$ for other heuristics. As expected, the run-time for the *Exact* method increases dramatically with polytomy size. The *NNI* and *Hybrid* methods, with $M = 1,000$ iterations, show very similar trends; indeed, the run for these methods-time is a factor 1000 greater than that of the *Local DL* method. For $k = 6$, the *Exact* method considers 945 binary resolutions, which is comparable to the 1000 resolutions used in *NNI* and *Hybrid*. This is reflected in the similar run-times for the three methods, suggesting that all three methods take roughly the same amount of time to recalculate the cost tables for each binary resolution. Thus, the major factor for run-time is the number of binary resolutions sampled.

The worst case complexity for resolving a polytomy of size k with *Hybrid* and *NNI* is $O(|V_S|^2 kM)$. However, in Fig. 5, the median run-times of *Hybrid* and *NNI* remain almost flat as polytomy size increases. This could be caused by memoization techniques, which significantly reduce the number of new nodes for which cost tables must be recalculated. It is also possible that run-times for these experiments are not in the asymptotic regime of the big-O complexity.

The time required to process polytomies of the same size is extremely variable for all four methods. This is because calculating the cost table for a gene node may not always reflect the worst case complexity (i.e., $O(|V_S|^2)$). Entries in $\mathcal{H}_g$ and $\mathcal{K}_g$ are calculated by considering all possible combinations of species for the mappings at the children of g. When all species are possible, there are $|V_S|^2$ combinations to consider; when only a few species are possible, this value is much smaller.

4   Discussion

Resolving non-binary gene trees with a binary species tree under the DTL model requires an enumeration of all binary resolutions of each polytomy, resulting in a super-exponentially increasing search space. Here, we presented an exact algorithm and several promising heuristics for resolving large polytomies, which can be used as stand-alone tree correction methods or as subroutines in species tree-aware gene tree construction techniques. The algorithms are all implemented in a publicly available, Java-based software package, NOTUNG, that generates all most parsimonious resolved gene trees that are temporally feasible.

Our experimental analysis of these methods gives insights into the structure of the DTL-Resolution problem. Some heuristics are based on searching the tree space using NNI. The improvement in accuracy that can be obtained by increasing the sample size of NNI is limited. However, a well picked starting topology, combined with a modest number of iterations proved to be a good compromise in this context. Two strategies based on a DL-event model performed better than expected when applied to bacteria data sets, which likely contain horizontal transfer events, suggesting that even in the presence of transfers, the DL-model has some utility as a heuristic.

These preliminary results introduce interesting questions that could be investigated further, such as a better way to sample the search space. Using NNI, the newly proposed gene tree is nearly identical to the previous tree. As a result, NNI may be too fine grained when searching for resolutions where horizontal gene transfer can introduce more dramatic changes to the topology. Subtree Prune and Regrafting (SPR) moves propose larger jumps between topologies in tree space, which may mimic transfer moves more realistically [19]. Thus, heuristics based on SPR, and possibly other tree modification procedures, are good directions for future research. Also, instead of randomly sampling the tree space, a hill-climbing approach guided by a figure-of-merit could be employed when considering proposed topology modifications.

We assessed the performance by comparing DTL-event costs obtained using each heuristic with the minimum DTL-event cost obtained using the exact method. We also calculated the Robinson-Foulds (RF) distance between the resolved trees obtained with the heuristic and exact methods and observed poor correlation between the two measures. Prior studies have shown that several metrics based on the DL model are not equivalent to RF distance [53]. Our empirical observations suggest that a theoretical comparison of measures based on DTL-events and RF distances may be a similarly fruitful direction.

Finally, this class of "species-tree" aware gene tree correction methods assumes that incongruence is due to either uncertainty or gene events. Yet other processes, such as gene conversion, incomplete lineage sorting, migration, hybridization, and introgression, can also result in gene tree incongruence. Thus, if branches in the gene tree are weak, but reflect the true evolutionary history, tree correction methods could introduce error, rather than decrease it. Methods that account for these processes are an important direction for future work.

Acknowledgments. We thank Annette McLeod for help with figures.

Funding. This material is based upon work supported by the National Science Foundation under Grant No. DBI-1262593. Any opinions, findings, and conclusions or recommendations expressed in this material are those of the author(s) and do not necessarily reflect the views of the National Science Foundation.

References

1. Anisimova, M., Gil, M., Dufayard, J.F., Dessimoz, C., Gascuel, O.: Survey of branch support methods demonstrates accuracy, power, and robustness of fast likelihood-based approximation schemes. Syst. Biol. **60**(5), 685–699 (2011)
2. Bansal, M.S., Alm, E.J., Kellis, M.: Efficient algorithms for the reconciliation problem with gene duplication, horizontal transfer and loss. Bioinformatics **28**, i283–i291 (2012)
3. Bansal, M.S., Wu, Y.C., Alm, E.J., Kellis, M.: Improved gene tree error correction in the presence of horizontal gene transfer. Bioinformatics **31**, 1211–1218 (2015)
4. Barker, D.: Gene trees for orthologous groups from: the evolution of nitrogen fixation in cyanobacteria (2012). Edinburgh DataShare. doi:10.5061/dryad.pv6df
5. Boussau, B., Szöllősi, G.J., Duret, L., Gouy, M., Tannier, E., Daubin, V.: Genome-scale coestimation of species and gene trees. Genome Res. **23**, 323–330 (2013)
6. Chang, W.-C., Eulenstein, O.: Reconciling gene trees with apparent polytomies. In: Chen, D.Z., Lee, D.T. (eds.) COCOON 2006. LNCS, vol. 4112, pp. 235–244. Springer, Heidelberg (2006). doi:10.1007/11809678_26
7. Chaudhary, R., Burleigh, J.G., Eulenstein, O.: Efficient error correction algorithms for gene tree reconciliation based on duplication, duplication and loss, and deep coalescence. BMC Bioinformatics **13**(Suppl 10), S11 (2012)
8. Chauve, C., El-Mabrouk, N., Guéguen, L., Semeria, M., Tannier, E.: Duplication rearrangement and reconciliation: a follow-up 13 years later. In: Chauve, C., El-Mabrouk, N., Tannier, E. (eds.) Models and Algorithms for Genome Evolution, pp. 47–62. Springer, London (2013). doi:10.1007/978-1-4471-5298-9_4
9. Chen, K., Durand, D., Farach-Colton, M.: Notung: a program for dating gene duplications and optimizing gene family trees. J. Comput. Biol. **7**(3/4), 429–447 (2000)
10. Darby, C.A., Stolzer, M., Ropp, P.J., Barker, D., Durand, D.: Xenolog classification. Bioinformatics **33**(5), 640–649 (2017)
11. David, L.A., Alm, E.J.: Rapid evolutionary innovation during an Archaean genetic expansion. Nature **469**, 93–96 (2011)
12. Donati, B., Baudet, C., Sinaimeri, B., Crescenzi, P., Sagot, M.F.: EUCALYPT: efficient tree reconciliation enumerator. Algorithms Mol. Biol. **10**(1), 3 (2015)
13. Doyon, J.-P., Scornavacca, C., Gorbunov, K.Y., Szöllősi, G.J., Ranwez, V., Berry, V.: An efficient algorithm for gene/species trees parsimonious reconciliation with losses, duplications and transfers. In: Tannier, E. (ed.) RECOMB-CG 2010. LNCS, vol. 6398, pp. 93–108. Springer, Heidelberg (2010). doi:10.1007/978-3-642-16181-0_9
14. Durand, D., Halldorsson, B., Vernot, B.: A hybridmicro-macroevolutionary approach to gene tree reconstruction. J. Comput. Biol. **13**(2), 320–335 (2006). A preliminary version appeared in RECOMB 2005, 250–264

318   H. Lai et al.

15. El-Mabrouk, N., Ouangraoua, A.: A general framework for gene tree correction based on duplication-loss reconciliation. In: Proceedings of the Workshop on Algorithmics in Bioinformatics (WABI). (2017, in press)
16. Górecki, P., Eulenstein, O.: Algorithms: simultaneous error-correction and rooting for gene tree reconciliation and the gene duplication problem. BMC Bioinform. 13(Suppl 10), S14 (2012)
17. Guindon, S., Dufayard, J.F., Lefort, V., Anisimova, M., Hordijk, W., Gascuel, O.: New algorithms and methods to estimate maximum-likelihood phylogenies: assessing the performance of PhyML 3.0. Syst. Biol. 59, 307–321 (2010)
18. Hallett, M., Lagergren, J., Tofigh, A.: Simultaneous identification of duplications and lateral transfers. In: Proceedings of the 8th International Conference on Research in Computational Biology, RECOMB 2004, pp. 347–356. ACM Press, New York (2004)
19. Hill, T., Nordström, K.J.V., Thollesson, M., Säfström, T.M., Vernersson, A.K.E., Fredriksson, R., Schiöth, H.B.: Sprit: Identifying horizontal gene transfer in rooted phylogenetic trees. BMC Evol. Biol. 10, 42 (2010)
20. Huson, D., Rupp, R., Scornavacca, C.: Phylogenetic Networks: Concepts, Algorithms and Applications. Cambridge University Press, Cambridge (2011)
21. Huson, D.H., Scornavacca, C.: A survey of combinatorial methods for phylogenetic networks. Genome Biol. Evol. 3, 23–35 (2011)
22. Jacox, E., Chauve, C., Szöllősi, G.J., Ponty, Y., Scornavacca, C.: ecceTERA: comprehensive gene tree-species tree reconciliation using parsimony. Bioinformatics 32, 2056–2058 (2016)
23. Jacox, E., Weller, M., Tannier, E., Scornavacca, C.: Resolution and reconciliation of non-binary gene trees with transfers, duplications and losses. Bioinformatics 33, 980–987 (2017)
24. Keane, T.M., Creevey, C.J., Pentony, M.M., Naughton, T.J., McInerney, J.O.: Assessment of methods for amino acid matrix selection and their use on empirical data shows that ad hoc assumptions for choice of matrix are not justified. BMC Evol. Biol. 6, 29 (2006)
25. Kordi, M., Bansal, M.S.: Exact algorithms for duplication-transfer-loss reconciliation with non-binary gene trees. In: ACM International Conference on Bioinformatics, Computational Biology, and Health Informatics, pp. 297–306 (2016)
26. Kordi, M., Bansal, S.: On the complexity of duplication-transfer-loss reconciliation with non-binary gene trees. IEEE/ACM Trans. Comput. Biol. Bioinform. 14(3), 587–599 (2017)
27. Lafond, M., Chauve, C., Dondi, R., El-Mabrouk, N.: Polytomy refinement for the correction of dubious duplications in gene trees. Bioinformatics 30, i519–i526 (2014)
28. Lafond, M., Noutahi, E., El-Mabrouk, N.: Efficient non-binary gene tree resolution with weighted reconciliation cost. In: Grossi, R., Lewenstein, M. (eds.) 27th Annual Symposium on Combinatorial Pattern Matching (CPM 2016), Leibniz International Proceedings in Informatics (LIPIcs), vol. 54, pp. 14:1–14:12. Schloss Dagstuhl-Leibniz-Zentrum fuer Informatik, Dagstuhl, Germany (2016)
29. Lafond, M., Semeria, M., Swenson, K.M., Tannier, E., El -Mabrouk, N.: Gene tree correction guided by orthology. BMC Bioinform. 14(Suppl 15), S5 (2013)
30. Lafond, M., Swenson, K.M., El-Mabrouk, N.: An optimal reconciliation algorithm for gene trees with polytomies. In: Raphael, B., Tang, J. (eds.) WABI 2012. LNCS, vol. 7534, pp. 106–122. Springer, Heidelberg (2012). doi:10.1007/978-3-642-33122-0_9

31. Latysheva, N., Junker, V.L., Palmer, W.J., Codd, G.A., Barker, D.: The evolution of nitrogen fixation in cyanobacteria. Bioinformatics **28**(5), 603–606 (2012)
32. Ma, W., Smirnov, D., Forman, J., Schweickart, A., Slocum, C., Srinivasan, S., Libeskind-Hadas, R.: DTL-RnB: algorithms and tools for summarizing the space of DTL reconciliations. IEEE/ACM Trans. Comput. Biol. Bioinform. (2016, in press)
33. Nakhleh, L.: Evolutionary phylogenetic networks: models and issues. In: Heath, L., Ramakrishnan, N. (eds.) The Problem Solving Handbook for Computational, pp. 125–158. Springer, Heidelberg (2010). doi:10.1007/978-0-387-09760-2_7
34. Nakhleh, L.: Computational approaches to species phylogeny inference and gene tree reconciliation. Trends Ecol. Evol. **28**, 719–728 (2013)
35. Nakhleh, L., Ruths, D.: Gene trees, species trees, and species networks. In: Guerra, R., Goldstein, D. (eds.) Meta-Analysis and Combining Information in Genetics and Genomics, pp. 275–293. CRC Press, Boca Raton (2009)
36. Nguyen, T.H., Ranwez, V., Pointet, S., Chifolleau, A.M.A., Doyon, J.P., Berry, V.: Reconciliation and local gene tree rearrangement can be of mutual profit. Algorithms Mol. Biol. **8**(1), 12 (2013)
37. Noutahi, E., Semeria, M., Lafond, M., Seguin, J., Boussau, B., Guéguen, L., El -Mabrouk, N., Tannier, E.: Efficient gene tree correction guided by genome evolution. PLoS ONE **11**, e0159559 (2016)
38. Ovadia, Y., Fielder, D., Conow, C., Libeskind-Hadas, R.: The cophylogeny reconstruction problem is NP-complete. J. Comput. Biol. **18**, 59–65 (2011)
39. Penel, S., Arigon, A.M., Dufayard, J.F., Sertier, A.S., Daubin, V., Duret, L., Gouy, M., Perrière, G.: Databases of homologous gene families for comparative genomics. BMC Bioinform. **10**(Suppl 6), S3 (2009)
40. Rasmussen, M.D., Kellis, M.: A Bayesian approach for fast and accurate gene tree reconstruction. Mol. Biol. Evol. **28**, 273–290 (2011)
41. Scornavacca, C., Jacox, E., Szöllősi, G.J.: Joint amalgamation of most parsimonious reconciled gene trees. Bioinformatics **31**, 841–848 (2015)
42. Sjöstrand, J., Sennblad, B., Arvestad, L., Lagergren, J.: DLRS: gene tree evolution in light of a species tree. Bioinformatics **28**, 2994–2995 (2012)
43. Sjöstrand, J., Tofigh, A., Daubin, V., Arvestad, L., Sennblad, B., Lagergren, J.: A Bayesian method for analyzing lateral gene transfer. Syst. Biol. **63**(3), 409 (2014)
44. Stolzer, M., Lai, H., Xu, M., Sathaye, D., Vernot, B., Durand, D.: Inferring duplications, losses, transfers, and incomplete lineage sorting with non-binary species trees. Bioinformatics **28**, i409–i415 (2012)
45. Swenson, K.M., Doroftei, A., El-Mabrouk, N.: Gene tree correction for reconciliation and species tree inference. Algorithms Mol. Biol. **7**, 31 (2012)
46. Szöllősi, G.J., Boussau, B., Abby, S.S., Tannier, E., Daubin, V.: Phylogenetic modeling of lateral gene transfer reconstructs the pattern and relative timing of speciations. Proc. Natl. Acad. Sci. U.S.A. **109**, 17513–17518 (2012)
47. Szöllősi, G.J., Rosikiewicz, W., Boussau, B., Tannier, E., Daubin, V.: Data from: efficient exploration of the space of reconciled gene trees (2013). Dryad Digital Repository. doi:10.5061/dryad.pv6df
48. Szöllősi, G.J., Rosikiewicz, W., Boussau, B., Tannier, E., Daubin, V.: Efficient exploration of the space of reconciled gene trees. Syst. Biol. **62**, 901–912 (2013)
49. Thomas, P.D.: GIGA: a simple, efficient algorithm for gene tree inference in the genomic age. BMC Bioinform. **11**, 312 (2010)
50. Tofigh, A., Hallett, M., Lagergren, J.: Simultaneous identification of duplications and lateral gene transfers. IEEE/ACM Trans. Comput. Biol. Bioinf. **8**, 517–535 (2011)

51. Vilella, A.J., Severin, J., Ureta-Vidal, A., Heng, L., Durbin, R., Birney, E.: Ensemblcompara genetrees: complete, duplication-aware phylogenetic trees in vertebrates. Genome Res. **19**, 327–335 (2009)

52. Wapinski, I., Pfeffer, A., Friedman, N., Regev, A.: Automatic genome-wide reconstruction of phylogenetic gene trees. Bioinformatics **23**, i549–i558 (2007)

53. Zheng, Y., Zhang, L.: Are the duplication cost and robinson-foulds distance equivalent? J. Comput. Biol. **21**, 578–590 (2014)

54. Zheng, Y., Zhang, L.: Reconciliation with non-binary gene trees revisited. In: Sharan, R. (ed.) RECOMB 2014. LNCS, vol. 8394, pp. 418–432. Springer, Cham (2014). doi:10.1007/978-3-319-05269-4_33

Author Index

Aganezov, Sergey 179
Alekseyev, Max A. 156, 179
Alexeev, Nikita 156
Altman, Tahel 31
Atar, Shimshi 31
Avdeyev, Pavel 156

Benaroya, Rony Oren 31
Braga, Marília D.V. 76

Chauve, Cedric 14
Chindelevitch, Leonid 256
Chou, Jed 277

Doerr, Daniel 197
Durand, Dannie 298

Elworth, Ryan A. Leo 213

Feijão, Pedro 14

Górecki, Paweł 101
Goz, Eli 31

Hoshino, Edna A. 76

Julander, Justin 31

Lai, Han 298

Mai, Uyen 116
Mane, Aniket 14

Martinez, Fábio V. 76
Medeiros, Gabriel L. 76
Meidanis, João 256
Mirarab, Siavash 53, 116

Nakhleh, Luay 213
Nute, Michael 277

Paszek, Jarosław 101
Pulicani, Sylvain 141

Rivals, Eric 141
Rong, Yongwu 156
Rubert, Diego P. 76

Sankoff, David 1
Sayyari, Erfan 53
Schulz, Tizian 197
Simonaitis, Pijus 141
Stolzer, Maureen 298
Stoye, Jens 76, 197
Swenson, Krister M. 141

Tsalenchuck, Yael 31
Tuller, Tamir 31

Vachaspati, Pranjal 232

Warnow, Tandy 232

Zhang, Chao 53
Zhang, Yue 1